Statistics for Making Decisions

Statistics for Making Decisions

Nicholas T. Longford

CRC Press
Taylor & Francis Group
Boca Raton London New York

CRC Press is an imprint of the
Taylor & Francis Group, an **informa** business
A CHAPMAN & HALL BOOK

First edition published 2021
by CRC Press
6000 Broken Sound Parkway NW, Suite 300, Boca Raton, FL 33487-2742

and by CRC Press
2 Park Square, Milton Park, Abingdon, Oxon, OX14 4RN

Library of Congress Control Number: 2020950454

ISBN: 9780367342678 (hbk)
ISBN: 9780429324765 (ebk)

Typeset in Computer Modern font
by KnowledgeWorks Global Ltd.

We just ask them for the best estimates,
and then we make our decisions.
(Anon.)

Contents

Preface

As an aspiring academic statistician, I spent all my early efforts on modelling, firmly believing that discovering all the truths was only a matter of identifying suitable models and implementing methods for fitting them. The latter task was particularly challenging in the pre-`Splus` era of dinosaur-like mainframes and software with limited scope and capacity. My senior colleagues, at the Educational Testing Service in particular, planted in me the seed of the conviction that data cannot be taken for granted, and obtaining high-quality data tailored to what we want to know, in a timely and affordable manner, is the ultimate goal of statistics. In brief, design has a supremacy over, and far greater potential than modelling and data analysis. This led me to a renewed appreciation of survey design. And that is just an analytical version of a bus ride from methods for dealing with missing data, multiple imputation as a near-universal remedy and the EM algorithm as a powerful framework for thinking about problems for which we do not have an off-the-shelf solution. Soon after establishing myself as a statistical consultant I discovered the clients as a factor in how I should think, argue and operate, and how to respond to the profound mistrust and dissatisfaction with statistics interpreted narrowly and practiced as application of the 'correct' hypothesis test. That is where I am now, as narrow-minded as ever, thinking that every problem in statistics is about making a decision, following decades of conviction that every problem was about finding the right (multilevel) model, that every problem was a missing-data problem, that everything is possible with a powerful computer (and `R`), and so on.

Like countries have constitutions, and as do some professions, the medical and legal ones in particular, I put forward one for statistics, that it is

> the science and profession of making purposeful decisions in the presence of uncertainty and with limited resources.

In this proposal, I want to emphasise that decision is the central intellectual activity in our everyday lives, both private and public, in business, public service and research. Assisting in this activity is at the core of the statistics profession when some uncertainty prevails and the information in our possession is incomplete. The 'limited resources' in the statement is a reference to the importance of design, reflecting the expense and difficulty of collecting (primary or supplementary) information. Here, the resources should be interpreted broadly; they comprise not only funding, but also time, expertise, physical and manpower limitations and, not in the least, goodwill of the recruiting agent and the respondent.

Choice, in the context of a decision, is studied in several subject areas, foremost in economics. I want to distance myself from one general concern in these studies, namely the study of how human subjects choose from a set of available courses of action, and whether these choices are rational. My concern is solely with proposing or prescribing to subjects, called clients, how to choose, based on rules derived from and for the specific client's perspective, value judgements, priorities, or remit. That is, in its ideal form, the analysis starts by encoding the perspective, in the form of the harm, loss or damage done by selecting one course of action when another should have been selected. It then follows by evaluating the expected magnitude of this loss and concludes by electing the course of action that is least harmful, or most advantageous, in expectation.

Decision theory is the subject of several monographs and textbook chapters. DeGroot (1970), Berger (1985) and Lindley (1985) appeal to its general application, the latter with simple examples and with minimal mathematical equipment. More recent treatment of the subject is by Liese and Miescke (2008), Parmigiani and Inouye (2009), Longford (2013) and Peterson (2017). This book focuses on distinctly ordinary problems that have the following formulation. The client, who sponsors the analysis, contemplates a shortlist of courses of action (options), and the purpose of the analysis is to propose one of them that best fits the client's criterion of greatest benefit or least loss.

When there are only two options, hypothesis testing would seem to be appropriate. This we dismiss outright because the test has no means of incorporating the perspectives, value judgements and remits of the client. (The client pays—we work for the client.) The methods dealt with in this book drop a lot, but not all, of the theoretical background and reduce the attention to settings encountered in statistical practice most frequently. Some assumptions may appear as too restrictive for the theoretician, but are constructive for the practically oriented analyst. We highlight the role of the client in the analysis and emphasise the subjective nature of the analysis.

Historically, decision theory and its applications have been firmly within the domain of the Bayesian paradigm. The development in this book is equally well suited for the frequentist and the Bayesian. It is intended not for *some* problems but for all of them that are of some consequence, with no exceptions.

Chapter 1 sets the scene and Chapter 2 introduces the principal terms and definitions. Chapters 3–5 present the theory, first for normally distributed estimators, then for estimators and statistics with other familiar distributions (Student t, chi-squared, F, binomial and Poisson) and for problems with more than two options. Chapter 6 addresses the issue of study design, reduced to setting the sample size of a planned study but packaged in an open-ended sensitivity analysis. Chapters 7–10 deal with distinct areas of application in which, I believe, not much persuasion is necessary that their problems are about electing one of the available options, and that the gains and losses have to be integrated in the analysis. Chapter 11 treats the problem of model selection from the decision-making perspective and develops an alternative that uses no model fit comparisons and aims at estimation with small mean squared error more directly.

The book is intended for a postgraduate semester course with lectures supplemented by practical classes and assignments of the exercises that accompany each chapter. There is ample scope for digression to computational and programming issues, as well as further theory. The first six chapters are essential and have to be read in a sequence. The next four chapters are largely independent and one or two of them can be dropped if they do not fit the reader's interest. Chapter 11, proposing a radical alternative to model selection, is of general interest.

A foundation course in calculus and linear algebra, with some experience of differentiation and integration, as well as matrix operations associated with ordinary least squares, are essential prerequisites. The text makes frequent references to computing in R and some exercises require computing and graphics, which could in principle be accomplished in other software, but I have thought about them in R only. Computing is also invaluable for exploring and confirming the properties and conclusions derived analytically. I have a suite of functions and their applications that implement all the computing and graphics described and reproduced in the text, and would be happy to share it with the reader, but I believe that greater benefit will be derived by the reader developing a personal (and personalised) library. The book is not for a beginner in statistics. The student or reader has to be familiar with ordinary regression, the most common distributions in statistics, maximum likelihood estimation and analysis of variance, hypothesis testing and the frequentist and Bayesian paradigms, although it is necessary to be conversant only in one of them.

I regard hypothesis testing as a distraction from and a barrier to good statistical practice. Its ritualised application should be resisted from the position of strength, by being well acquainted with all its theoretical and practical aspects. I very much hope that after a few chapters the reader will be converted to my view that the rightful place for hypothesis testing is in a museum, right next to the steam engine.

Author

Nicholas T. Longford is a senior statistician in the Neonatal Data Analysis Unit at Imperial College London, England. His career includes appointments at the Educational Testing Service, Princeton, NJ, USA, De Montfort University, Leicester, England, and a spell of self-employment as the Director of SNTL Statistics Research and Consulting (UK and Spain). He is the author of six other monographs in statistics and the sole author or coauthor of over one hundred articles in peer-reviewed statistics journals. He is an editor of *StatsRef* (Wiley), reviewer for mathematical reviews and an associate editor of *Statistical Methods in Medical Research* and *Journal of Educational and Behavioral Statistics*.

1

First steps

1.1 What shall we do?

Making decisions is a common mental activity in our everyday lives. In any one instance, we contemplate the courses of action, or options, that are available to us. We inquire and deliberate about the likely profit or advantage that we would gain, or the harm, damage or injury that we might incur, or what we would lose, by choosing either of these options. The gains and losses usually depend on the actions taken by other people and institutions, as well as on the current, recent past and future states of the nature and the environment in which we live and operate. Our information about these states and actions is incomplete—we are not privy to all the intentions of others and circumstances and other details related to the options available to us. In some settings, we have some limited means of learning more about the circumstances, and thus informing the decision—help us to choose the option that is likely to fit best our desires, goals or remits and cause least harm, damage or loss.

This general description covers a wide range of scenarios, from the mundane to the momentous. When crossing a busy road we pay attention to the traffic, observe the direction in which the vehicles are moving and assess their speed. We set off to cross the road when we judge it safe to do so, even allowing for something untoward to happen, such as being slowed down by a stumble or impeded by an obstacle, and allow ample time and distance for the drivers of approaching vehicles to react if necessary.

As another example, a proposal for marriage follows (or should follow) a careful deliberation of the match with one's acquaintance, how he or she is likely to develop over the years of cohabitation, how well you would support one another and how you would cope with the vicissitudes, illnesses, hardships, disappointments and discord you are likely to encounter in the future that is full of optimistic plans and hopes, but also entails a lot of uncertainty. And, until a divorce or death, it precludes marrying someone else.

Some decisions are difficult to make because they involve comparisons of apparently incomparable values. One such comparison is 'life or money'—how much are you willing to pay to (possibly) extend or improve your life, or the life of a patient. It entails many imponderables. The claim that a particular course of action, such as a costly or inconvenient treatment, or a complex surgery, would extend the life may be in balance even after carefully weighing the potential advantages and drawbacks. For example, a surgery on a patient's vital organ is highly likely to improve the functioning of the organ but may adversely affect other organs, the heart and brain in particular. We are rarely in full control of our immediate environment or of events affecting us (our fate). Our understanding is usually limited, and so our deliberation is full of 'maybes, ifs and buts'.

While deciding on one matter, other matters are not waiting patiently for their turn in an orderly queue, nor are newly arising matters obediently joining the end of the queue. They have to be dealt with at the time of their calling, placing constraints on our resources: time, money, mental capacity, distracting us from activities we are currently pursuing or would prefer to pursue. The choices we have to make are structured: some are in a time sequence and depend on the choices made earlier, some fall into natural groups (how often do you cross the road, at a variety of locations?) or packages (driving a car and conducting a job interview). For some it is essential to define a set of rules or a policy, or to adopt a policy formulated by others (the Driving Code and instructions and advice of a coach given to an athlete), because decisions have to be made instantly, permitting no deliberation or consultation. In brief, we have a limited capacity to handle the many decision-related tasks that are constantly pressing on us. We might be well served by a strategy for distributing our resources so that we get the best deal from a package of decisions, even if some problems could have been dealt with better, at the expense of poor responses to the others, due to not paying full attention to them. Part of the strategy may be investment in our decision-making capacity (such as by education and training), possibly a digression in the short term that may be amply rewarded later. Our capacity to deal with these tasks is not static; it is evolving gradually as we accumulate experience, information and insight. Our priorities and value judgements are also evolving and they influence our decisions by our appraisal of the consequences of the options available to us.

This volume carves out a small segment of this general class of problems and treats them with statistical methods that are not novel but have fallen into disuse in the mainstream statistical practice because they can be meaningfully applied only after thorough introspection and assessment, combined with careful calculation and exploration. One may be unwilling to do or take part in this, finding it unpleasant, too demanding, vexing or contentious, and disclosure of the results may cause some discomfort. The approach proposed runs counter to the prevailing way of conducting statistical evaluations, in which a dataset is analysed objectively, in detachment from and without any interference by the party that has a stake in the outcome of the analysis. In contrast, we want to integrate in the analysis this party's perspective, which includes their value judgements, priorities and remit.

We assume that every analysis is associated with such a party, called the *client*, who sponsors or provides some other form of incentive for the analysis and supplies the purpose—to assist in the process of promoting or achieving a specified goal (agenda) in production, provision of services, pursuit of peace, comfort and happiness, academic research and exploration, or some other gainful activity. A relevant analysis responds to this client-specific purpose, and we do not consider conducting any analysis that does not do so, or is not linked to a well-identified client. An objective analysis deals with the purpose, but ignores the elements of the client's perspective that are not incorporated in the purpose.

We want to promote a transparently subjective analysis or, more generally, statistical practice, in which the client's perspective, value judgements, priorities and remits are treated on par with the data that would suffice for a more conventional objective analysis. The motivation for this is the professional credo of a statistical consultant

> to serve the best interests of the client as regards information and the associated uncertainty.

It is an adaptation of similar standards in the medical and legal professions that have a firm focus on the patient and client, respectively.

Example 1

To flesh out this viewpoint, consider a simple question: 'Is the performance of a production unit satisfactory?' Suppose everybody involved (all the stakeholders) agree on the meaning of the term 'satisfactory performance', for instance, by reference to an established industrial standard. Suppose the unit produces items that are intended to be identical in all relevant aspects, and the standard for them is that the probability p that a production item has satisfactory quality and functionality exceeds a specified value $p_0 = 0.96$; that is, $p > p_0$. The probability p is not known but is estimated, as well as it can be, by $\hat{p}$, by inspecting the items produced in a given period of time; $\hat{p} = n_s/n$, where $n = 1500$ is the number of items inspected and n_s is the number of them that are found satisfactory by an (infallible) inspection. We leave for later the issues of how n should be set and how n items should be selected from the production output.

With the counts n and n_s established, this would seem to be a straightforward problem, resolved by testing the hypothesis that $p \leq p_0$ against the alternative that $p > p_0$. If we reject this hypothesis, then we have evidence in support of the alternative. This would be the case, for instance, for $n_s = 1490$. The t statistic is equal to $(1490 - 0.96 \times 1500)/\sqrt{0.04 \times 0.96 \times 1500} = 6.59$, which is in the distant tail of the standard normal distribution (p value less than 10^{-6}).

If we fail to reject the hypothesis, then we have to improvise. By way of an example, suppose $n_s = 1450$, so that the t statistic is 1.32. Now $\hat{p} > p_0$, but by a margin that is not sufficient for supporting the alternative. The logically correct interpretation of such an outcome is that we are in a state of ignorance because we have established evidence of neither the hypothesis nor the alternative. In practice, such a conclusion is not tenable—it is hard to accept a statement that, despite an uncontroversial analysis of a relatively simple problem, we have come up empty-handed, without a firm and credible verdict as to whether the production process is satisfactory or not. The logical error of concluding (or implying) that the hypothesis is well supported, and proposing to carry on as if it were valid, is often regarded as a suitable compromise. We find this practice objectionable. A reader not familiar with hypothesis testing should not be discouraged at this point and can return to it when hypothesis testing is discussed in greater detail in Section 2.7.   □

Suppose the analysis is for a client who has a particular agenda, described by a list of contemplated courses of action (available options). The consequences of choosing each of them depend on the reality known to us only partially. These consequences are characterised in Example 1 by the relation of p to p_0: whether $p > p_0$, and the analysis concludes this or not, and whether $p \leq p_0$, and the analysis concludes this or not. Thus, there are four scenarios, formed by the two *states* (satisfactory and unsatisfactory quality of production) combined (or, crossed) with two possible conclusions of the analysis. These conclusions, called *verdicts*, are linked to the available options: to make no changes, which is appropriate if $p > p_0$, and to overhaul the production process otherwise.

No method of selecting one of the two courses of action is foolproof because the value of p cannot be established with absolute precision. We have to admit that the verdict of the analysis may be erroneous. This is not an error caused by the analyst's negligence, incompetence or poor expertise, but a consequence of the limited capacity of the production unit, and limited time available for inspection. These are limits on the number of items that can be inspected, and therefore also on the precision with which p can be estimated. When the inspection is not seamless, instant and cheap, the resources available for conducting it have to be considered. These resources are not only funds, but also expertise (manpower), time, availability of materials, lost production (and related income) and goodwill of all the parties whose cooperation is essential in the enterprise of collecting the relevant information (data).

If we conclude that $p > p_0$, we may fail to discover that the production unit's performance is unsatisfactory. We refer to this type of error as *failure to discover*. Conversely, if we conclude that $p \leq p_0$, we will in effect advise the client to overhaul the production line. If this advice is inappropriate, then the overhaul is unnecessary. This type of error is called *false discovery*. If the contemplated overhaul is inexpensive and disrupts the production for only a short time, the client is more averse to failure to discover. On the other hand, if the overhaul is costly and the failure to satisfy the standard has no serious consequences, then the client is more averse to false discovery.

These two contrasting scenarios expose a structural deficiency of the hypothesis test as a general and objective method. It yields the same answer, with no regard for the client's perspective and circumstances. In brief, it is impervious to the consequences (ramifications) of the two kinds of error: the cost of the overhaul (together with the disruption it entails) and the income and reputation lost by substandard production. No right-minded manager would (nor should) take seriously any advice that ignores these two factors.

The hypothesis test may be a good fit for a particular perspective, or set of consequences, but it is a hit-or-miss for any given client. A profound short-coming of a lot of statistical practice is that this 'objective' perspective is not elucidated. It is adopted as the default to spare us delving into details of the client's context and purpose. Some clients are content that this perspec-tive is not dwelt on, that the analyst gets on with the task without raking through any discomforting intricacies of the client's business as it pertains to the problem. Some of its aspects may be confidential or not fully elabo-rated, reinforcing the incentive to conduct an objective (detached) analysis. We regard this incentive as false, an obstacle to a more relevant analysis.

This monograph deals with decisions in statistics. Its aim is to respond to a client's problem formulated generally as:

> What should I, the client, do next, given the resources, limitations and options I have?

The term 'resources' is added as a condition in this question to emphasise that we are concerned not only with analysis, interpreted narrowly as the processing of a dataset already collected, with a specified purpose. We are also interested in designing studies, that is, prescribing how data should be collected (or generated) for such an analysis. By 'resources' we mean not only monetary funds, but also time, space, equipment, manpower with the relevant expertise, including their training, hindrance to the conduct of normal business and the related inconvenience, and goodwill of all the parties involved, the respondents and their representatives in particular. Limitations can be interpreted as the negative of resources or an upper bound on the resources that can be released for the study. They include barriers imposed by law, ethical standards and conventions, and the laws of nature.

Throughout we assume that the options are specified. We also assume that in the imagined setting of no uncertainty about the reality the choice of the optimal option would be straightforward. For example, if it were known that $p \leq p_0$, then overhaul of the manufacturing process would be unquestionably the right course of action. We want to devise a method of analysis that con-cludes with a decision, referred to as the verdict and defined as the choice of one of the specified options. This is in contrast with an analysis that concludes with an inferential statement, such as an estimate and the associated standard error, or the verdict of a hypothesis test, discussed in Section 2.7, which would require further processing, commonly referred to as *interpretation*.

Most of the developments in this volume address the problem of making a single decision. In practice, we have to make many decisions. They are interrelated, arising in sequences or simultaneously, one affecting the list of options to choose from in the next decision, altering the consequences of the inappropriate decisions, and sometimes enabling us to gain information (to learn) that helps us make better decisions in the future. However, this problem is common to all approaches in which analyses are conducted in a sequence, one at a time. We operate in a chaotic jumble of problems, but can deal only with one or a few of them at any one time.

Another limiting assumption we make is that of stability. We assume that the consequences of the errors and other factors do not change abruptly. Otherwise, our conclusions or recommendations (advice), to be implemented soon and be applicable for some time in the future, are based on assumptions that have become or will soon become obsolete. A compromise can be struck by anticipating how these factors may change in the (near) future, and conduct the analysis for a range of such plausible scenarios. If this range is wide, we are likely to conclude that 'it depends'; for some scenarios one conclusion, and for others a different one is arrived at. In brief, we assume that the setting of the analysis and the client's perspective are not swinging around wildly out of control.

1.2   The setting

We are in the role of a (statistical) analyst working for a client. A course of action contemplated by the client is a well-defined activity, such as the purchase of a specified item from a given supplier, entry into a new business venture, application of a process, and the like. Inaction, doing nothing, may also be a course of action. We assume that none of these courses of action have more than one version. An example of versions of an action is the choice of alternative health-care providers for a surgery and the choice of when to undergo the surgery: immediately, to suit family or workplace commitments, or after the appearance or certain symptoms.

We consider a finite list, or set, of courses of action $\mathbf{A} = (A_1, \ldots, A_K)$, also referred to as options, which have the following four properties. They are:

- exclusive – no pair of these actions can be taken at the same time;

- complementary – it is impossible to take none of these actions;

- indivisible – an option can be either taken in full or not taken at all;

- irreversible – the consequences of adopting a course of action cannot be cancelled out.

In brief, the choice of exactly one of these K options is unavoidable, partial application of an option is not possible, and once the selected option is applied its effects cannot be neutralised or undone. The options form a partition (division) of the entire space of possibilities—one option is realised and all the others are not. For example, submitting and not submitting a bid in response to a call for proposals by a research foundation are a pair of such options or actions. If there is a way of submitting an incomplete bid, it may be regarded as a third option. As an alternative, it could be classified as not submitting a bid, if it would certainly be rejected at an early screening stage, and would not be regarded as a genuine submission. If inaction is an option it is treated on par with the other options.

When there are K options, we assume that there are K possible states, $\mathbf{S} = (S_1, \ldots, S_K)$; action A_k is optimal when the state that is realised is S_k; $k = 1, \ldots, K$. For a sharper focus, we consider first the simplest nontrivial case, in which $K = 2$. Such a problem can be described as a dilemma: to take a particular action or not to take it. In the case of responding to a call for proposals, submitting a bid is the optimal course of action if our submission would be selected and working on the contracted project would turn out to be rewarding. Not applying is the better course of action if the complement is the case, if our submission would be selected but the contract would not be rewarding (for instance, we would fail to complete the project or could not engage in some more attractive activity), or if we would not be selected. Without knowing the outcome of the contemplated submission, not submitting it is advisable if it would likely be rejected because our effort to prepare a bid would be wasted and we would be dejected by the failure. The fundamental difficulty is that we cannot predict the disposition of the selectors (reviewers) towards our bid, the quality of the bids submitted by other applicants, nor the outcome of the negotiations after a conditional acceptance of the bid.

The consequences of the 'ifs and buts' are without any contention, although applicants may have different perspectives. Some may regard it a great kudos just to be selected in a preliminary round, helping them to refine the art of bidding, benefit from the feedback and prepare a better proposal at the next opportunity. Others may be more averse to submitting bids that are likely to fail. For example, preparing a bid may amount to a diversion of one's resources from more rewarding activities, in which experience may be gained that would enable to submit a stronger bid in a future call. We do not want to start an argument about which perspective is more appropriate but want to highlight the plausible differences and convince the reader that a single solution may not be best suited for all the plausible and reasonable perspectives. A solution has to be tailored to the client, and the client's perspective has to be elicited to inform the analysis.

1.2.1 Losses and gains

The first task is to place the consequences of the contemplated courses of action on an ordinal scale, akin to a currency. This currency for error shares with a monetary currency the feature that we want to be frugal with it. With $K = 2$, as in the case of bidding for a contract, it means defining the four values L_{vs}, for verdicts $v = V_1$ and V_2, and states $s = S_1$ and S_2. See Table 1.1.

TABLE 1.1
The loss matrix for two states and two verdicts.

	State	
Verdict (selected action)	S_1 (Approved)	S_2 (Rejected)
V_1 (Submit a bid, A_1)	L_{11}	L_{12}
V_2 (Do not submit a bid, A_2)	L_{21}	L_{22}

We draw a distinction between the verdicts V and courses of action A. A verdict, say V_1, is the statement, proposal or advice to take the course of action A_1, that is, the choice of the option A_1. It is arrived at by an analysis. The courses of action are declared in advance, as integral elements of the problem formulation. In general, we have no means of influencing which state occurs but we are in full control of selecting a verdict.

The values L_{vs} in Table 1.1 are referred to as losses. In practice, only the differences $D_1 = L_{12} - L_{22}$ and $D_2 = L_{21} - L_{11}$ matter; they characterise what is at stake with the respective verdicts V_1 and V_2. Thus, no generality is lost by setting $L_{11} = L_{22} = 0$. Then $L_{12} = D_1$ and $L_{21} = D_2$. This reflects the idea that we lose nothing if we select the appropriate course of action, or that we relate the losses to the optimal choice. Without this convention, L_{11} and L_{22} could be interpreted as the cost of existence, a characteristic of the state not affected by the verdict. In any meaningful setting, $D_1 > 0$ and $D_2 > 0$.

Instead of losses, we may consider gains that would result from the choice of the appropriate course of action over the inappropriate one. For this, we replace the L-notation with G-notation in the 2×2 table of verdicts and states. We lose no generality by setting $G_{21} = G_{12} = 0$; we gain nothing when we make the inappropriate choice. There is no advantage in using L over G. Our convention is to use L.

The next nontrivial task is to establish how likely the two states are. Here we have to draw on intelligence combined with relevant data. If a state is highly unlikely, then we discount the corresponding verdict, which would be optimal if this state did occur. In contrast, if a state is very likely we give more weight in our consideration to the corresponding verdict. The stakes, as described by the losses L_{12} and L_{21}, are another factor in the deliberations. In the bidder's dilemma, if we do not mind being rejected by the funding agency and would very much like to gain a contract, it is appropriate to submit bids liberally. In this case, the loss associated with rejection is small in relation to not gaining a contract because of failure to apply.

When we are averse to rejections and regard composing a proposal and an accompanying letter and completing an application as unpleasant, distracting or time consuming (resource demanding) chores, or are not in a rush to gain a research contract, then we submit bids more selectively, focusing on calls where we rate our chances higher and where there is more to be gained. Here the loss incurred by a failed application is greater than the loss due to not winning the contract that could have been won by an application. In brief, different perspectives may result in exercising, in both instances rationally, different options.

1.2.2 States, spaces and parameters

The finite list of states $\mathbf{S} = (S_1, \ldots, S_K)$, which we are concerned with, is often a coarse (grouped-up) version of a continuum of possible states, $\boldsymbol{\Theta} = (\Theta_1, \ldots, \Theta_K)$, such as a few non-overlapping intervals that cover the entire real axis, $\Theta_1 \cup \ldots \cup \Theta_K = (-\infty, +\infty)$. Statistical inference is more comfortable with continua, and Euclidean spaces in particular. In contrast, decisions involve small finite sets of options, frequently just two of them.

The switch from a continuum ($\boldsymbol{\Theta}$) to a finite set $\mathbf{S}$ is straightforward when there is no uncertainty, but not otherwise. For example, we can decide instantly whether a parameter θ is positive or negative when the value of θ is known, but basing a decision on a guess or estimate $\hat{\theta}$ is not straightforward, especially when $\hat{\theta}$ happens to be close to zero.

The problem of $K > 2$ options can be converted to a series of problems with two options each. For example, if we have three options, A, B and C, then we choose first between A and B, and then pit the selected option against C.

A partition of a space $\boldsymbol{\Theta}$ is defined as a collection of sets Θ_k, $k \in \mathcal{K}$, not necessarily finite, such that they are disjoint,

$$\Theta_k \cap \Theta_{k'} = \emptyset$$

for any pair $k \neq k'$ in the index set $\mathcal{K}$, and they cover the space,

$$\bigcup_{k \in \mathcal{K}} \Theta_k = \boldsymbol{\Theta}.$$

We will use only finite partitions, in which $\mathcal{K} = (1, 2, \ldots, K)$, and they will be of the H-dimensional Euclidean space $\boldsymbol{\Theta}$. When well motivated, we will use a different set of indices. For example, $\mathcal{K} = (A, B)$ is suitable for the choice of one of alternative (medical) treatments A and B, and $\mathcal{K} = (+, -)$ may be used when the value of a particular quantity is either positive or negative. The ambivalence about the possible value of zero will be explained in specific examples. The K options of a decision-making problem are a partition of the space of all possibilities.

In most cases we will deal with the simplest nontrivial case, $H = 1$ and $K = 2$. The *realised state* is a point θ in $\boldsymbol{\Theta}$ ($\theta \in \boldsymbol{\Theta}$), but we are interested only in the identity of the set Θ_k, that is, in the index k for which $\theta \in \Theta_k$. We refer to θ as a *parameter*. It has two forms, as a way of summarising and characterising the relevant reality, and its (numerical) value. For example, θ

may be the median of a variable, such as the annual household income in a given country in 2019, implying a way how its value might be established in some circumstances. The other form of this parameter is the value itself, such as 25 352 Euro. If the interest is merely in establishing whether θ exceeds 24 000 Euro (2000 Euro per month), we can instantly respond with an affirmative.

In most common problems, $\Theta = (-\infty, +\infty)$, or its subinterval, and the states $\Theta_1, \ldots, \Theta_K$ are intervals delimited by a set of cutpoints $T_1 < T_2 < \cdots < T_{K-1}$. It is convenient to introduce $T_0 = -\infty$ and $T_K = +\infty$, so that $\Theta_k = (T_{k-1}, T_k)$, $k = 1, \ldots, K$. Whether we assign the cutpoint T_k, $k = 1, \ldots, K - 1$, to Θ_k or Θ_{k+1} is immaterial. We will be preoccupied first with the setting of $K = 2$ and $T_1 = 0$.

An established approach to identifying the realised state k comprises two steps:

1. estimate θ (by $\hat{\theta}$);

2. identify k for which $\hat{\theta} \in \Theta_k$.

In words, having estimated the parameter we set aside any distinction between the value of the parameter and its estimate. Note that the verb 'to identify' is used in this description inconsistently. Identification (or its synonym, determination) is incompatible with uncertainty and incorrectness, yet we admit that the index k, and therefore state Θ_k, may be selected incorrectly because they are subject to uncertainty entailed in estimation. Thus, an 'approach to identifying' concludes with a fallible verdict (an informed guess), and that is not identification. Step 2 does involve identification, but it is for the state for $\hat{\theta}$, and not for the state for the unknown θ. The approach is more accurately described as an (imperfect) attempt to identify. In this attempt we want to incur as little damage, harm or loss caused by incorrect selection of k as possible.

1.2.3   Estimation. Fixed and random.

For the moment, it suffices to say that estimation (of a parameter θ) is a process of operating on data and forming a summary $\hat{\theta}$ that would be regarded as our guess of the value of θ. This guess or, more precisely, the method or formula (computer algorithm or programme) for deriving it, is called an *estimator*. A simple example is the arithmetic mean, defined for a vector of outcomes $\mathbf{y} = (y_1, y_2, \ldots, y_n)^\top$ as $\hat{\theta} = \frac{1}{n}(y_1 + y_2 + \cdots + y_n)$, where the value of n is implied by the notation. Whatever the values in $\mathbf{y}$, the task of evaluating $\hat{\theta}$ is well defined, without any contention or ambiguity.

The dataset $\mathbf{y}$ is condensed by the estimator $\hat{\theta}$ to $\hat{\theta}(\mathbf{y})$—the estimator applied to the dataset. The value obtained, after $\mathbf{y}$ has been observed, is called the *estimate*. By way of an example, suppose $\mathbf{y} = (7.2, 2.7, 4.7, 3.4)^\top$, so that $n = 4$. The mean is $\hat{\theta} = 4.5$. We have to accept that the same notation, $\hat{\theta}$, is used for both the estimator (a formula) and the estimate (a numerical value).

Moreover, we have the same notation, $\mathbf{y}$, for a dataset when its collection is contemplated or planned (using a specific design), and after the dataset is collected (is *realised*).

At the design stage, before $\mathbf{y}$ is realised, $\mathbf{y}$ and the estimator $\hat{\theta}$ are *random*—we are uncertain about which members of the population will be selected to the sample, and therefore we are uncertain about the elements of $\mathbf{y}$, and consequently also about the value of $\hat{\theta}$. As soon as a sample is selected (the study is realised), the dataset $\mathbf{y}$ is fixed, and so is the estimate $\hat{\theta}$.

1.3 Study design

The design of a study is a prescription for how data for the study should be generated—collected and recorded. Important aspects of the design are subjects that will be engaged in the study and the variables that will be recorded on these subjects. The subjects are selected or recruited from a specified population. The design may specify the number of subjects to be engaged (the sample size), or the number of subjects to be approached, with the understanding that some of them may decline participation, as is their right. The subjects may be administered specified treatments (interventions or exposures), and the treatment regime (level of exposure, its frequency and timing, or the like) is assigned by a prescribed process, such as completely at random or by each subject's choice.

The design of a study is important because the analysis of the data recorded by the study considers the variety of datasets that would be recorded in the hypothetical setting of complete re-runs (replications) of the study. For example, the sample size may be fixed (e.g., set to 28), the numbers of subjects assigned to the treatments may also be fixed, or some restrictions imposed, such as (nearly) equal representation of both sexes in either treatment group.

1.4 Exercises

1.1. By way of revision, formulate in your own words the definitions of the terms client, courses of action, state, verdict, loss and gain, and give examples. What is the role of the client in the analysis that concludes with a verdict?

1.2. Describe an instance in which you had to make a decision in the recent past. Give details of the options you had, the advantages, benefits and harms or losses that were at stake and the uncertainties involved, your preferences and the associated costs. Describe what kind of 'detective work' you did to inform your decision. Summarise the time and effort you and those whom you had in confidence spent, and the expense you incurred on making the decision. Keep to yourself what you decided to do and why.

1.3. Give the description from Exercise 1.2 to a friend or colleague of yours and ask him or her to decide on your behalf what you should have done. Having done so, now ask the colleague to imagine that he or she is in this situation, and how he or she would decide.

1.4. Compare the three decisions made in Exercises 1.2 and 1.3. Discuss why they are not identical, and if they are identical, then why they might not have been if you involved someone else. Relate the conclusions to the terms introduced in this chapter (see Exercise 1.1).

1.5. Suppose you are on holiday in a city and you have spotted a landmark that you want to see as the next thing. How would you deliberate about getting there on foot? How would your choices (route) be affected if you were with a partner or with your family with small children? Describe carefully your perspective and value judgements and other factors that would or might influence your choices. If you find this task too abstract, consider a particular setting, such as the Coliseum in Rome or Charles Bridge in Prague.

1.6. Read the letter in Longford (2007) and comment on what you would do if you were in the role of the various actors in the story: the athlete, his coach, the sponsoring bank's CEO and the bank's statistician. How would you advise the athlete if you had an opportunity (and the remit) to do so? (Look up the statistical technical terms used in the article in the literature of your choice.)

1.7. Discuss why a medical consultant might present a patient a few alternative courses of treatment and not decide him- or herself as to which of them is the best.

1.8. Suppose you or a person whom you are advising has to choose between buying a car, hiring a car on occasions, and relying entirely on public transport (including taxis). Compile a list of factors that are relevant to the choice and order them according to their importance. If you do this exercise in a (small) group, record all the disagreements you have as well as instances when somebody changes their mind.

1.9. An exam comprises many questions and they are scored one point each for the correct answer, and no point for an incorrect answer. To pass it, you have to answer correctly at least 80% of the questions. Relate the issue of revision in preparation for the exam to the problem of establishing whether the production line satisfies a given standard (Example 1). Describe the losses with the two kinds of incorrect verdicts.

1.10. Consider the problem in Exercise 1.9, but with a sequence of exams or tests that you have to take within a short period, say, of two weeks.

1.11. Describe some instances when you found yourself in the position of a decision overload—when you had to pick which problems to attend to first and set others aside. Consider how you made the choices then and whether you would make the same choices today.

1.12. Imagine that your country is at war. The Ministry of Defence supports a recruitment campaign for the Armed Forces. Set out the conditions that would make it an attractive proposition for you to enlist. Construct a perspective (of another person) in which enlistment would not be attractive. Construct the verdict-by-state table (as in Table 1.1) for this dilemma and describe the difficulties in setting the values of the losses L_{vs}.

1.13. Check that the division of the problem with $K > 2$ options to a set of $K - 1$ problems, each with two options, does not affect the result when the choices involve no uncertainty. That is, among the options A, B and C, the same option will be identified as the best whether we start by comparing A with B, and then pit the 'winner' against C, or start by comparing B with C, and then compare the 'winner' with A. Consider what happens when each comparison may be subject to error.

1.14. Suppose in the example of the median annual household income, the value of θ is not established, but is estimated by 24 257 Euro. Why would somebody (of perfectly sound mind and acting rationally) conclude that the median is not greater than 24 000 Euro? Feel free to criticise the statement of this exercise as incomplete, and propose a completion.

1.15. Check that the courses of action in any of the previous exercises satisfy the four properties marked by bullets in Section 1.2. Discuss how the set-up would break down if one of these conditions were not met.

1.16. Discuss the difficulties you have experienced or one might experience in the process of looking for a flat (apartment) or a room to rent. Discuss them in the context of making decisions, under time pressure and other constraints. Describe the uncertainties involved and the scope for making a choice that you might later regret.

1.17. Could you specify the losses associated with the wrong choice(s) in Exercise 1.16? Would you be more comfortable with giving ball-park figures instead? Convert these ball-park figures to plausible ranges. Do you think that somebody who understood your predicament would be satisfied that you have thought hard enough in coming up with these plausible values? Compare the circumstances of this textbook exercise, posed long after the event, with dealing with the same task without any benefit of hindsight and under time pressure.

1.18. Consider the setting of two courses of action (options), A and B. The state for which A is preferred, S_A, has a far greater probability than the state for B, S_B; $P(S_A) \gg P(S_B)$. Can you envisage any circumstances in which it would be appropriate to select B?

1.19. Set the problem of deciding which mean, μ_1 or μ_2, is greater in the context of the terminology introduced in this chapter. In particular, what are the states and their cutpoint(s)? Would the problem be dealt with differently if the variables involved were continuous or binary?

2

Statistical paradigms

When there is uncertainty, or our knowledge is imperfect, errors are inevitable. The science of statistics entails two kinds of effort: reducing the error as much as possible and ameliorating its effects, that is, managing the error. This chapter gives a condensed background to error reduction and discusses estimation and hypothesis testing. Its purpose is to prepare the ground for error management, our main concern in the following three chapters.

The estimation error is defined as the deviation of the estimate or estimator from the parameter $\Delta\theta = \hat{\theta} - \theta$. Usually, we consider it as a random variable. It is fixed only after its value is established, when $\hat{\theta}$ has been evaluated. If the value of θ is also disclosed, then the value of $\Delta\theta$ becomes known. At that point, estimation of θ loses all its purpose except for assessing how well we would have done in the hypothetical setting of θ being unknown.

The value of θ may be unknown at the time of its estimation, t_E, and then established at a later time point, t_F. In such a case, estimation is useful in the time period between t_E and t_F. Nevertheless, an appraisal of the method employed for estimating θ (applying $\hat{\theta}$) would be unfair if some information relevant for estimating θ became available in the period (t_E, t_F), and it could have contributed to improved estimation of θ. Another element of unfairness is that the properties of $\Delta\theta$, as a random variable, are assessed by its single realisation. But no other realisation is available.

In place of the difference $\hat{\theta} - \theta$, another contrast can be used. Some of these contrasts are the differences after a monotone transformation. For example, when both $\hat{\theta}$ and θ are positive, their log ratio is the error in estimating $\log(\theta)$ by $\log(\hat{\theta})$:

$$\log\left(\frac{\hat{\theta}}{\theta}\right) = \log\left(\hat{\theta}\right) - \log\left(\theta\right) .$$

The tasks of estimating θ and $\log(\theta)$ are in general different. For instance, we may adopt the difference $\hat{\theta} - \theta$ as the estimation error in one task and the difference $\widehat{\log(\theta)} - \log(\theta)$ in the other. Here we place the wide circumflex $\widehat{}$ over $\log(\theta)$, and not solely over θ, because it is not appropriate to exchange the order of the operations of estimation and (nonlinear) transformation, unless $\hat{\theta} = \theta$.

A common feature of the various estimation errors is that $\Delta\theta = 0$ only when $\hat{\theta} = \theta$, when the estimate is right on the mark. In general, greater absolute value $|\Delta\theta|$ is regarded as a greater (and more serious) error, although

15

errors of the same magnitude, $\Delta\theta$ and $-\Delta\theta$, have profoundly different conse-quences in some settings.

The quality of an estimator is defined by a characterisation of the error $\Delta\theta$, regarded as a random variable. This characterisation is based on an analytical perspective, called the paradigm. There are two firmly established paradigms in statistics, frequentist and Bayesian. Their essential difference is in how they treat $\hat{\theta}$ and θ.

2.1   Frequentist paradigm

In the frequentist paradigm, θ is regarded as an unknown constant. An infinite sequence of hypothetical replications of $\mathbf{y}$ is considered. Each replication is an application of the same design; it yields a replicate value of $\hat{\theta}$. The limit of the averages of $f(\Delta\theta)$ is evaluated. We define next the terms 'replication' and 'design', expanding on their introduction in Section 1.3, and discuss the choice of the function f.

A replication is defined as an independent repeat of the processes of data generation and evaluation of an estimator. Independence refers to no influence of (nor learning from) any previous repeats, and it implies no prescience of any future repeats. The qualifier 'hypothetical' indicates that we do not intend to realise any replications, but engage in a theoretical exercise of evaluating (deriving analytically), approximating, or merely speculating about the values of $\Delta\theta = \hat{\theta} - \theta$, how frequent each plausible value would be and, in particular, how close to zero they, or $f(\Delta\theta)$, would be on average.

A design is a prescription for conducting a study or its part, such as selec-tion of the units for the study (sampling design), procedures to which these units are subjected (treatment, intervention or exposure) and data collection, including measurement instruments and scales to be used. A design is de-scribed in its protocol. The ideal protocol leaves no uncertainty about how the study should be conducted and can resolve any conceivable query about the minutiae of the conduct. Often one party compiles the protocol and an-other party implements it—realises the study, so the protocol is the principal document or guide for the conduct of the study.

Had a different party implemented the study, the data collected would be different, but the differences would arise only in its elements that the protocol prescribes to be random or deliberately leaves to chance. For example, the protocol may prescribe the number of subjects (respondents) to be recruited for the study as well as how they are to be recruited and from which pop-ulation, but does not prescribe the identities of the recruits. In a different design, the number of attempts to recruit may be prescribed. In that case, if not all attempts are successful, the number of recruits (respondents) is left to chance.

The most common choice of the function f for summarising the errors $\Delta\theta$ is the square $f(\Delta\theta) = (\Delta\theta)^2$. The limit of averages, as the number of replications grows above all bounds, is called the expectation and is denoted by the symbol E. So, a common characterisation of an estimator is the mean squared error (MSE), written as

$$\mathrm{E}\left(\Delta\theta^2\right) = \mathrm{E}\left\{\left(\hat{\theta} - \theta\right)^2\right\};$$

the argument $\Delta\theta^2$ is to be interpreted as $(\Delta\theta)^2$, not as $\Delta(\theta^2)$. We also use the notation $\mathrm{MSE}(\hat{\theta}; \theta)$ and refer to θ as the estimand or the *target* (of estimation). The target is included as an argument of MSE because, at least in principle, an estimator $\hat{\theta}$ could be considered for several distinct targets θ. The corresponding MSEs usually differ. In any case, it is useful to indicate, together with the estimator, the parameter that is estimated.

The expectation is not defined for some variables. However, estimators for which the expectation or the MSE are not defined are of no practical importance.

2.1.1 Bias and variance

The bias of an estimator $\hat{\theta}$ of the target θ is defined as the difference $\mathrm{B}(\hat{\theta}; \theta) = \mathrm{E}(\hat{\theta}) - \theta$. Estimator $\hat{\theta}$ is said to be unbiased for θ if $\mathrm{B}(\hat{\theta}; \theta) = 0$, and biased otherwise. That is, the estimation errors of an unbiased estimator are balanced around zero. It does not imply that they are distributed symmetrically around zero. For example, suppose an estimator has estimation error $\Delta\theta = 2$ in 25% of replications, no error in 25%, and error of -1 in 50% of replications. This estimator is unbiased.

The sampling variance of an estimator is defined as

$$\mathrm{var}\left(\hat{\theta}\right) = \mathrm{E}\left[\left\{\hat{\theta} - \mathrm{E}\left(\hat{\theta}\right)\right\}^2\right].$$

For an unbiased estimator, the sampling variance and MSE coincide. In general,

$$\mathrm{MSE}\left(\hat{\theta}; \theta\right) = \mathrm{var}\left(\hat{\theta}\right) + \left\{\mathrm{B}\left(\hat{\theta}; \theta\right)\right\}^2.$$

For a given dataset $\mathbf{y}$, regarded as a random variable, an estimator $\hat{\theta}_1$ is said to be *more efficient* than estimator $\hat{\theta}_2$ for a target θ, if $\mathrm{MSE}(\hat{\theta}_1; \theta) < \mathrm{MSE}(\hat{\theta}_2; \theta)$. An estimator $\hat{\theta}$ is said to be *efficient* for θ if there is no estimator that is more efficient for θ than $\hat{\theta}$. An estimator efficient in a given class of estimators is defined similarly. An example of such an estimator is the minimum-variance unbiased estimator (of a particular target). It is efficient in the class of unbiased estimators (of the target).

We prefer estimators with smaller MSE, and efficient estimators are generally regarded as superior to others. Unbiased estimators also hold some appeal, although we believe that it is somewhat overrated. The contrary viewpoint suggests the following two-stage strategy. First we reduce our attention to unbiased estimators. Then we search among them for the estimator with the smallest variance. Such a minimum-variance unbiased estimator is not always efficient because in the search we have ignored some biased estimators that have small variances.

The bias of an estimator $\hat{\theta}$ could be adjusted for by subtracting the bias, that is, by applying $\hat{\theta} - B(\hat{\theta}; \theta)$. However, the bias is usually not known and has to be estimated by $\hat{B}$. As we adjust the original estimator $\hat{\theta}$ by the (estimated) bias, to $\hat{\theta}' = \hat{\theta} - \hat{B}$, we also alter the variance $\mathrm{var}(\hat{\theta}') \neq \mathrm{var}(\hat{\theta})$. The bias-corrected estimator $\hat{\theta}'$ may have greater MSE than the original biased estimator $\hat{\theta}$. The correction for bias is in that case counterproductive in the effort to reduce the MSE.

We emphasise that assessing the quality of an estimator by its MSE is a convention that is often reasonable and well suited, but there are exceptions. An undisputed advantage of the MSE is its analytical tractability, but that may be irrelevant to (the perspective of) a particular client. Nevertheless, a lot of flexibility is enabled by transformations, considering $\mathrm{MSE}\{g(\hat{\theta}); g(\theta)\}$ for a monotone function g. Of interest in this context are only nonlinear functions g. Note, however, that the properties of no bias and efficiency are not preserved by such transformations.

2.1.2 Distributions

An estimator, linked to a particular design, is described by its bias and variance, which together determine the MSE. A more detailed description of an estimator is provided by its sampling distribution. We define distribution through the distribution function. The distribution function of a random variable X is defined as the function of the probabilities

$$F(x) = P(X \leq x)$$

for $x \in (-\infty, +\infty)$. It is a nondecreasing function, with limits of zero and unity at $-\infty$ and $+\infty$, respectively. For categorical (discrete) variables, which have positive probabilities at isolated values, such as some or all the integers, this function is piecewise constant, with jumps at the (isolated) points of positive probability. (The qualifier 'isolated' is defined at the end of this section.) In general, a distribution function can have only countably many points at which it is discontinuous.

A distribution is said to be absolutely continuous if the corresponding distribution function is differentiable. A distribution with a continuous distribution function F that is not necessarily differentiable throughout is called continuous. Distributions that are continuous but not absolutely continuous can be constructed but they are of no practical importance. As a result, no confusion arises by dropping the qualifier 'absolutely'.

The differential of an absolutely continuous distribution function is called the density:

$$f(x) = \frac{\partial F(x)}{\partial x}.$$

Most densities encountered in practice are themselves differentiable. Some have analytical expressions that can be studied by elementary tools and can be grouped into classes or families according to their forms. For example, the normal density is defined by the expression

$$\varphi(x; \mu, \sigma) = \frac{1}{\sqrt{2\pi\sigma^2}} \exp\left\{-\frac{(x-\mu)^2}{2\sigma^2}\right\},$$

where μ and σ^2 are parameters; μ has a real value and σ^2 a positive value. A distribution is said to be normal if it has a density, and this density is normal. Examples of other classes of distributions are chi-squared, gamma and beta. A class of distributions is said to be a subclass of another class of distributions if it is entirely subsumed in it. For example, exponential and chi-squared distributions are also gamma distributions and the standard uniform distribution on the interval $(0, 1)$ is a beta distribution. The distribution of an estimator is referred to as its sampling distribution.

The support of a continuous distribution is defined as the set of all points where its density is positive. The support of a univariate distribution is usually an interval, finite, one-side infinite, such as $(0, +\infty)$, or the entire real axis, $(-\infty, +\infty)$.

A real value x is said to be isolated in a set Θ if $x \in \Theta$ and there is an open set that contains x but no other element of Θ. (Instead of open sets the condition in this definition can be reduced to open intervals.) For example, every integer is isolated in the set of all real numbers. A distribution is said to be discrete if the set of all points that have positive probabilities comprises isolated points, and these probabilities add up to unity. Examples of discrete distributions with support on integers are binomial, geometric, uniform and Poisson.

2.1.3 Sampling from finite populations

A population is defined as a collection of units called members. Its formal definition is a rule that arbitrates about any item or entity whether it is or is not a member of this population. A design for sampling from a population (a sampling design) is defined by a function that assigns to any subset of the population a probability that it would form the sample. These probabilities add up to unity. Some, or even a vast majority of these subsets may have zero probability. For example, a sampling design may have a fixed size, say, n. In such a design, every subset of size different from n has zero probability.

A simple random sample of size n from a finite population of size N is defined by the probabilities $1/\binom{N}{n}$ for every subset of size n; all other subsets are ruled out from forming the sample. In some settings, a member can be included in a sample more than once; such a sampling design is said to be 'with replacement'. We work mostly with sampling designs that permit no replacement — a member can be included in a sample at most once. In most cases, we consider only simple random sampling; the qualifier 'simple' is usually dropped.

Many populations encountered in practice are finite but very large. It is practical to regard them as infinite. An important rationale for this is that a variable defined in a finite population (of size N) is discrete because it can have at most N distinct values. It is convenient to relate such a discrete variable to a continuous variable. Formally, this can be justified by reference to a *superpopulation*, a population of infinite size, from which the population we consider is assumed to have been generated as a representative sample. By 'representative' we mean that the distribution of the values of the variable considered (or of any other variable) is very similar to its counterpart in the superpopulation.

Rounding converts a continuous variable to a discrete one. Some variables are discrete, but have a lot of distinct values, without any pronounced gaps between them. Variables expressed in monetary units (e.g., annual income) are a case in point. It is practical to regard such variables as continuous, and if a formal approach is required, then the actual values can be regarded as rounded values of a genuinely continuous variable. For instance, when quoting a price (in the United Kingdom), fractions of a penny are ignored. Any method in which the results would be materially affected by how such fractions are dealt with is highly problematic unless there are so many observations that the fractions they involve add up to a nontrivial amount.

2.2   Bayesian paradigm

In the Bayesian paradigm, the data $\mathbf{y}$ is regarded in the analysis as given (fixed), having been observed, and the target θ or, more generally, a vector or set of parameters, is regarded as random. At the design stage, the distribution of the data is specified conditionally on the value of θ. This set of distributions, parametrised by θ, is called the *model*. We use the notation $(\mathbf{y} \mid \theta)$ for this set, interpreted as conditional distributions given the value of θ, for θ attaining all its possible values. After $\mathbf{y}$ is realised, the conditional distribution of $(\theta \mid \mathbf{y})$, with the roles of θ and $\mathbf{y}$ interchanged, is derived by means of the Bayes theorem. This distribution is called *posterior*. The two conditional distributions are linked by the *prior* distribution for θ,

$$f_{\text{post}}(\theta \mid \mathbf{y}) = \frac{f_{\text{mod}}(\mathbf{y} \mid \theta)\, f_{\text{pri}}(\theta)}{\displaystyle\int f_{\text{mod}}(\mathbf{y} \mid \theta)\, f_{\text{pri}}(\theta)\, \mathrm{d}\theta} , \qquad (2.1)$$

where f is the generic notation for a density, and its subscript indicates that the density is prior, for the model, or posterior. Evaluations are greatly simplified when f_{mod} depends on the data $\mathbf{y}$ through a univariate statistic, say t, so that $f_{\mathrm{mod}}(\mathbf{y}\,|\,\theta) = f'_{\mathrm{mod}}\{t(\mathbf{y})\,|\,\theta\}$. We say that t is a sufficient statistic. Sufficient statistics are not unique; if t is sufficient, then so is any of its one-to-one transformations, and nonconstant linear transformations in particular. Not all models, or densities f_{mod}, have univariate sufficient statistics. Some have a small number of statistics, a vector $\mathbf{t}$, that are sufficient, so that $f_{\mathrm{mod}}(\mathbf{y}\,|\,\theta) = f'_{\mathrm{mod}}\{\mathbf{t}(\mathbf{y})\,|\,\theta\}$. By this definition, disregarding the vague qualifier 'small', even $\mathbf{y}$ is sufficient. But advantages accrue only when the number of sufficient statistics is small and does not depend on the sample size n.

All inferences about θ are based on the posterior distribution. Its expectation can be regarded as an estimate and its variance as a measure of uncertainty, closely related to the sampling variance $\mathrm{var}(\hat{\theta})$ in the frequentist paradigm. A function or transformation g intended originally for θ might now be applied to this estimate, $\hat{\theta} = \mathrm{E}_{\mathrm{post}}(\theta\,|\,\mathbf{y})$, evaluating $g(\hat{\theta})$, but it is far better to apply it to the posterior distribution, and then summarise the resulting distribution, for instance, by its expectation, $\mathrm{E}_{\mathrm{post}}\{g(\theta)\,|\,\mathbf{y}\}$. In general,

$$g\left\{\mathrm{E}_{\mathrm{post}}\left(\theta\right)\right\} \neq \mathrm{E}_{\mathrm{post}}\left\{g\left(\theta\right)\right\}.$$

A notable exception to this inequality arises when g is a linear function; then the applications of g and $\mathrm{E}_{\mathrm{post}}$ can be interchanged.

For the prior, distributions that represent no or scant information are often used in practice. They are referred to as noninformative. A more effective prior would reflect more closely what is known about θ at the design stage. This requires elicitation that would render the analysis subjective and may introduce some discomforting contention. This statement is intended not as a discouragement but as a characterisation, or diagnosis, of a prevailing attitude in the practice of Bayesian methods. A prior that captures our information about θ, if constructed with integrity, has a great potential to strengthen the inference about θ, by resulting in a smaller posterior variance of θ—in less uncertainty about θ. In applications presented in later chapters, the prior distribution is interpreted as a set of additional (extra-data) observations. This interpretation may be instrumental in eliciting the prior distribution from the client or another source by conveying the effective role of the prior in the analysis. At the same time, it offers a way of incorporating prior information within the frequentist paradigm. Note, however, that not all priors can be (easily) expressed in terms of additional observations.

Bayesian analysis is said to be *coherent*—it involves a clear and transparent process with no improvisation, and the posterior distribution is a complete and unambiguous statement about θ, conditional on (that is, exploiting fully) what was observed, namely the data $\mathbf{y}$, and what was assumed, namely the model $(\mathbf{y}\,|\,\theta)$. In contrast, frequentist analysis has no principled recipe for estimation, although the maximum likelihood, supported by a comprehensive asymptotic theory and numerical methods for locating the extremes of a real function, is constructive in many settings. A model is common to both paradigms.

Bayesian analysis with a noninformative prior often reproduces the result of the (frequentist) maximum likelihood analysis that uses the same model, so the two paradigms implement essentially different approaches that lead to similar results. Having negligible or no prior information about a parameter of interest is a very rare situation. It is pretended in practice to avoid issues of integrity at the expense of reduced efficiency, and to spare the analyst having to expend the effort and engage other parties in collecting and processing the prior information. The other extreme, settings in which the information in the data is dwarfed by prior information, is also exceedingly rare.

An analyst may prefer one paradigm over another because it is easier to devise an appropriate analysis and implement it, but this advantage is in most problems subjective. Analysis in both paradigms is second-rate if it fails to draw on all the relevant information. The Bayesian paradigm has the device of the prior distribution for this purpose, and requires no improvisation. The frequentist paradigm requires some improvisation, such as the appeal to extra-data observations. Much of the current practice in both paradigms is remiss by analysing the 'data' only, with noninformative priors or no extra-data observations, shying away from the detective work of collecting, appraising and quantifying the information available before the data is inspected, that is, at the design stage.

Metaanalysis and research synthesis are concerned with pooling the information across studies that have, or had, the same or closely related targets. It can be loosely interpreted as an analysis in which studies are the units. Most of its methods assume that the studies are independent. This assumption fails when studies share some information that they use or indeed when one or a few studies provide prior information for another study.

2.3   Computer-based replications

Whilst references to hypothetical replications may appear to be too abstract, as they cannot be realised in practice, the concept is very useful in computer simulations. A simulation comprises replications of a data-generating process (that is, a model), which involves some randomness. Each replication yields an outcome called a *replicate*, such as an estimate. The outcome can be regarded as random, defined by the process, or as fixed, the result of a particular replication. A collection of many replicates defines the *empirical distribution* of the random version of the outcome (the estimator). An empirical distribution of an estimator $\hat{\theta}$ approximates the sampling distribution of $\hat{\theta}$. This approximation is close when sufficiently many replications are generated.

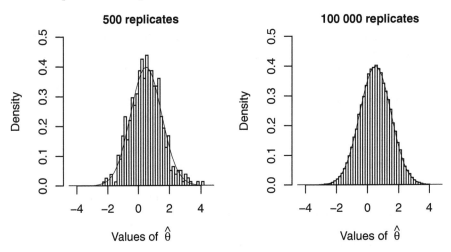

FIGURE 2.1
Empirical and sampling distributions of a normally distributed estimator; an illustration. Based on 500 replications (left-hand panel) and 100 000 replications (right-hand panel).

A simple, although artificial, example is presented in Figure 2.1. Suppose the estimator $\hat{\theta}$ has, according to the adopted model, the normal distribution $\mathcal{N}(\theta, 1.0)$; θ is the target. In simulations on the computer, we assume that $\theta = 0.5$. The left-hand panel of Figure 2.1 displays a histogram (the empirical distribution) based on 500 replicates and the right-hand panel is based on 100 000 replicates. The density of the (theoretical) sampling distribution of $\hat{\theta}$, $\mathcal{N}(0.5, 1.0)$, is drawn in each panel by a solid line. In the left-hand panel, the empirical distribution does not approximate the sampling distribution well, although it matches it quite closely in its location (e.g., mean and median) and dispersion (variance and interquantile range). In the right-hand panel, the agreement is impeccable. Note the logical error one would commit by concluding from the poor agreement of the histogram with the density in the left-hand panel that the estimator $\hat{\theta}$ is not normally distributed, arguing that the sample size of 500 is quite large.

In summary, the frequentist paradigm is centred on conditioning on the unknown parameter, as epitomised by the bias $b(\theta) = \mathrm{B}(\hat{\theta}; \theta)$ and MSE $m(\theta) = \mathrm{E}\{(\hat{\theta}-\theta)^2; \theta\}$, evaluated as functions of θ. This is greatly simplified when these functions do not depend on θ, as when there is no bias, $b(\theta) = 0$ for all θ, and $m(\theta)$ is a constant. Otherwise $b(\hat{\theta})$ and $m(\hat{\theta})$ may be used as the respective estimators of $b(\theta)$ and $m(\theta)$, but this is problematic when b or m is a distinctly nonconstant function. An extreme example of this problem arises when $m(\theta)$ is proportional to θ. The error in estimating θ by $\hat{\theta}$ is repeated in estimating $m(\theta)$ by $m(\hat{\theta})$.

In the Bayesian paradigm, the posterior distribution has a central role. Its summaries, such as expectation and variance, are evaluated as functions of the dataset $\mathbf{y}$ after its realisation. For a theoretically minded student, the integral in the denominator in equation (2.1) may present a difficulty or hindrance. However, there are numerous methods for its approximation. Moreover, its evaluation can often be avoided when the numerator is recognised as a scalar multiple of a density. In this case, the posterior distribution is given by this density.

An undoubted advantage of the Bayesian paradigm is that $\mathbf{y}$ is already realised when the posterior distribution is evaluated, so the posterior distribution is well identified as a function of the prior distribution. Its drawback is that the posterior distribution is a complicated function at the design stage, before $\mathbf{y}$ is realised. This problem diminishes when the posterior distribution depends on $\mathbf{y}$ only through a sufficient statistic, especially when this statistic has a tractable distribution under the assumed model. Sufficient statistics are equally useful in both paradigms.

2.4   Design and estimation

In both paradigms, data (or a dataset) is collected according to a specified design (protocol), the purpose of which is to ensure that the data would be suitable for estimation or a similar purpose, or be (nearly) optimal, as would be appraised by a specified criterion. Such a criterion considers the expense of implementing the design (collecting the data) and the precision of the resulting estimator $\hat{\theta}$. The expense is usually interpreted as a quantity proportional to sample size, regarding the observation of a single subject, or the generation of a single data record, as involving a unit of expense (cost or effort). More complex schemes, involving a setup cost and economies of scale, are often more realistic.

A general consideration is to obtain the highest precision for a parameter of interest for least resources expended—to 'purchase' precision at the lowest relative cost, that is, to get the best deal of all combinations of design and analysis. This goal is sometimes modified by fixing the resources that are available and finding the design and estimator which in tandem conclude with least uncertainty about θ. As an alternative, a level of uncertainty may be specified and a design that requires minimum resources is sought for which θ would be estimated with the prescribed (or higher) precision.

In studies with more than one target, when several parameters are to be estimated, some compromise may have to be struck between these goals. However, these goals are often in accord; greater sample size promotes every one of them. An example to the contrary arises with grouped data, such as households and their members. For some parameters, it is advantageous to collect data from one (adult) member of each household, and contact more households. For others, it is advantageous to collect data from every member of a contacted household. The (average) expense involved in contacting a household is an important consideration in this context.

In another example, the targets are quantities, such as the rate of unemployment, in the regions of a country. In this setting of a national survey, the regions in effect compete for the available resources, assumed for this illustration to be fixed. A compromise has to be struck by 'trading' the precisions of the regions' estimators. In such a trade, the relative importance of the regions, which may be related to their population sizes, is one factor, and the estimators applied are another.

2.5 Likelihood and fiducial distribution

The likelihood for a vector of outcomes $\mathbf{y}$ is defined as the joint density (or probability) of $\mathbf{y}$, that is, the model, with the roles of the observations $\mathbf{y}$ and the parameter θ (or a set of parameters $\boldsymbol{\theta}$) interchanged. For a joint density of the outcomes, we use the notation $f_{\text{mod}}(\mathbf{y}; \theta)$. The corresponding likelihood is defined as $L(\theta; \mathbf{y}) = f_{\text{mod}}(\mathbf{y}; \theta)$. Note the convention for the arguments of a function, which departs slightly from the notation used in equation (2.1), but is in accord with the notation for bias and MSE introduced in Section 2.1.1. In parentheses we list first the arguments with variable or uncertain values (θ for L), followed by the arguments with fixed values ($\mathbf{y}$ for L). The notation implies an interest in the behaviour of the function with respect to the 'variable' arguments.

The joint density is a function of $\mathbf{y}$, parametrised by θ. In contrast, the likelihood is a function of θ, parametrised by $\mathbf{y}$. Thus, the likelihood serves a purpose similar to the Bayes theorem—to turn the focus from $\mathbf{y}$ at the design stage to the target θ in the analysis (estimation) stage.

The likelihood is instrumental in estimation. The maximum likelihood estimator and estimate are defined as the value $\hat{\theta}$ of θ for which $L(\theta; \mathbf{y})$ attains its maximum. The estimate relates to $\mathbf{y}$ as a fixed vector, as the realised data. In contrast, the estimator relates to $\mathbf{y}$ prior to its realisation, when it is (or was) a random vector.

The fiducial distribution of an estimator is derived by inverting the association of the estimator to the parameter. For example, for a normally distributed unbiased estimator of θ, we write

$$\hat{\theta} = \theta + \varepsilon, \qquad (2.2)$$

where $\varepsilon \sim \mathcal{N}(0, \tau^2)$; τ^2 is the sampling variance of $\hat{\theta}$. Inversion of the identity in (2.2) is simple,

$$\theta = \hat{\theta} - \varepsilon.$$

Since the normal distribution is symmetric, this implies that after $\hat{\theta}$ is evaluated and becomes a constant, $\theta \sim \mathcal{N}(\hat{\theta}, \tau^2)$. We refer to this distribution as *fiducial*, to distinguish it from sampling distribution of $\hat{\theta}$ associated with equation (2.2).

Although obvious in this example, such inversion is in general not unique, nor is it additive, and the result may involve some contradictions, as in the following example. For an estimator $\hat{\theta}$ of a parameter $\theta \in (0, 2)$, with estimation error $\varepsilon = \hat{\theta} - \theta$ uniformly distributed on $(-1, 1)$, the fiducial distribution of $\theta = \hat{\theta} - \varepsilon$ has a positive probability outside the interval $(0, 2)$ whenever $\hat{\theta} \neq 1$. Admittedly, this is an artificial example, but the lack of an unambiguous way of deriving a fiducial distribution may be disconcerting. In contrast, the (Bayesian) posterior distribution is unique and without any controversy, except for the specification of the prior distribution for θ, which in practice often lacks any profundity or integrity.

2.5.1 Example. Variance estimation.

The variance σ^2 of a normal distribution is estimated without bias from a random sample $\mathbf{Y} = (Y_1, Y_2, \ldots, Y_n)$ from this distribution by

$$\hat{\sigma}^2 = \frac{1}{n-1} \sum_{i=1}^{n} (Y_i - \bar{Y})^2 \, ,$$

where $\bar{Y} = \frac{1}{n}(Y_1 + \cdots + Y_n)$ is the sample mean. The estimator $\hat{\sigma}^2$ has a scaled chi-squared distribution. Its number of degrees of freedom is $k = n - 1$. The sampling distribution of $\hat{\sigma}^2$ is such that

$$\frac{k\,\hat{\sigma}^2}{\sigma^2}$$

has chi-squared distribution with k degrees of freedom, denoted by χ_k^2. Let X be a random variable with this distribution; $X \sim \chi_k^2$. Then the obvious way of inverting the distributional identity leads to the fiducial identity

$$\sigma^2 = \frac{k\,\hat{\sigma}^2}{X} \, ,$$

so the fiducial distribution of σ^2 is scaled inverse chi-squared. The multiplicative scale implied by this operation is natural for an estimator related to a chi-squared distribution.

The χ_k^2 distribution has the density

$$f_k(x) = \frac{1}{\Gamma_2(k)} \frac{x^{\frac{k}{2}-1}}{2^{\frac{k}{2}}} \exp\left(-\tfrac{1}{2} x\right) \, , \tag{2.3}$$

where $\Gamma_2(k) = \Gamma(\tfrac{1}{2} k)$, called the half-gamma function, is introduced solely for typographical reasons. The inverse-χ_k^2 distribution has the density

$$g_k(y) = \frac{1}{\Gamma_2(k)} \frac{1}{2^{\frac{k}{2}}} \frac{1}{y^{\frac{k}{2}+1}} \exp\left(-\frac{1}{2y}\right),$$

obtained by the transformation $y = 1/x$ in f_k.

In a Bayesian setting, suppose the prior for σ^2 is scaled inverse chi-squared, such that for a given prior degrees of freedom $h > 0$ and prior variance σ_0^2, $h\sigma_0^2/\sigma^2$ has χ_h^2 distribution. Then the numerator in the expression for the posterior distribution of σ^2 in the Bayes theorem in (2.1) is

$$\frac{1}{\Gamma_2(k)\,\Gamma_2(h)} \frac{1}{2^{\frac{1}{2}(k+h)}} \left(h\sigma_0^2\right)^{\frac{1}{2}h} \frac{k^{\frac{1}{2}k}\,y^{\frac{1}{2}k-1}}{(\sigma^2)^{\frac{1}{2}(k+h)+1}} \exp\left(-\frac{ky+h\sigma_0^2}{2\sigma^2}\right). \qquad (2.4)$$

This is a function of σ^2. We do not have to evaluate the denominator in the Bayes theorem because it is equal to the divisor for which the result is a density. In the expression for the numerator in (2.4) we recognise that, as a function of σ^2, it is a scalar multiple of the density of the inverse χ_{k+h}^2 distribution, with scaling $ky + h\sigma_0^2$.

This result has an attractive interpretation. The prior can be related to an estimate σ_0^2 of σ^2 based on h degrees of freedom. The corresponding estimator is independent of $\hat{\sigma}^2$, so these two estimators can be combined as

$$\tilde{\sigma}^2 = \frac{k\hat{\sigma}^2 + h\sigma_0^2}{k + h}$$

to form an estimator with scaled χ_{k+h}^2 distribution. The fiducial distribution coincides with this distribution when h is set to zero, ignoring (or having no) prior information. We shall proceed by issuing a verdict (making a decision) about σ^2 using this distribution without any regard for its origin—whether it is a fiducial or a posterior distribution.

2.6   From estimate to decision

We return to the problem of using an unbiased normally distributed estimator $\hat{\theta}$ to assign its target θ to one of the (real) intervals separated by thresholds $T_1, \ldots, T_{K-1}$. By inversion, we obtain the normal fiducial (or posterior) distribution of θ. With no regard for the consequences of the various kinds of incorrect assignment, one way to proceed would be to substitute $\hat{\theta}$ for θ: assert (or, issue the verdict) that θ belongs to the same interval as $\hat{\theta}$. When $K = 2$, hypothesis testing provides a different solution, which can be adapted, with some difficulty, for $K > 2$.

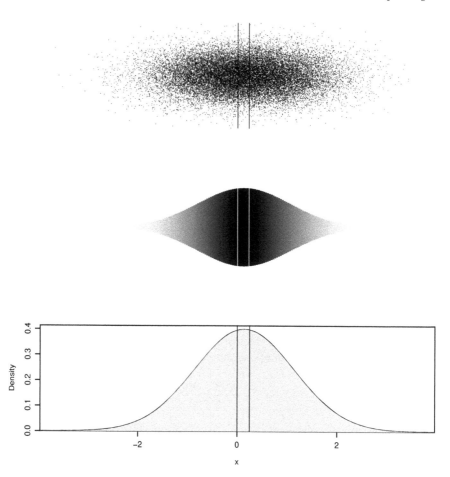

FIGURE 2.2

Representations of the fiducial distribution of a parameter; $\theta \sim \mathcal{N}(0.14, 1)$. The vertical bars mark the cutpoints $T_1 = 0$ and $T_2 = 0.25$ for three states.

The two steps, estimation and decision, would yield the right answer if estimation in step 1 were perfect and $\hat{\theta}$ would not differ from θ. Alas, with uncertainty about θ we do not get the right answer, but the two-step procedure can be improved. To see this, consider the following example. Suppose $\Theta = (-\infty, +\infty)$ and there are three states, separated by $T_1 = 0$ and $T_2 = 0.25$. We evaluate $\hat{\theta}$ and obtain 0.14, but with a lot of uncertainty, so that it is quite plausible that $\theta < 0$, and $\theta > 0.25$ is also plausible. If failing to discover that $\theta < 0$ has catastrophic consequences, then verdict V_1 $(\theta < T_1)$ is appropriate, even though $\hat{\theta} > 0$.

An illustration is given in Figure 2.2. The fiducial distribution of the parameter θ, $\mathcal{N}(0.14, 1)$, is presented in three distinct ways. At the top, the distribution is represented by a large set of replicates simulated on a computer. The vertical coordinate of a point is immaterial; random noise is added in the vertical direction to avoid overprinting and to better represent the frequency (density) of the values. Below, the density of the distribution is represented by the vertical thickness and shade of grey used for filling. At the bottom, the distribution is represented by its density. Histogram and box plot are further alternatives. The two vertical lines in each panel mark the cutpoints $T_1 = 0$ and $T_2 = 0.25$, to illustrate that verdict V_2 ($T_1 < \theta < T_2$) is poorly supported, even when $T_1 < \hat{\theta} < T_2$, because the interval (T_1, T_2) is so narrow.

Suppose we have only two options, Θ_1 and Θ_2, separated by cutpoint T_1, and have settled on an unbiased estimator $\hat{\theta}$ with positive and finite variance $\text{var}(\hat{\theta})$. Which option should be chosen? A simple but poorly qualified solution is to choose Θ_1 if $\hat{\theta} < T_1$ (when $\hat{\theta} \in \Theta_1$) and Θ_2 otherwise (when $\hat{\theta} \in \Theta_2$). Its deficiency stems from the possibility that θ and $\hat{\theta}$ may be separated by T_1; we cannot rule out the scenarios $\hat{\theta} < T_1 < \theta$ and $\theta < T_1 < \hat{\theta}$. After evaluating $\hat{\theta}$ we can eliminate only one of them.

If we are averse to choosing Θ_1 inappropriately, because it would have catastrophic consequences relative to choosing Θ_2 inappropriately, then Θ_1 should be chosen only when $\hat{\theta}$ is very small, say, when $\hat{\theta} < T_1 - \Delta$, where $\Delta > 0$ is to be set. If we are averse to choosing Θ_2 inappropriately, then $\hat{\theta} < T_1 - \Delta$ is a reasonable rule for choosing Θ_1, but with a negative value of Δ.

Thus, all we have to do is set the value of Δ. This is equivalent to adjusting the estimator $\hat{\theta}$ to $\hat{\theta} + \Delta$, and comparing it with cutpoint T_1. This suggests that unbiasedness is possibly a valuable property only when $\Delta = 0$, when the two kinds of error have equally severe consequences. Apart from the sign of the error $e = \theta - T_1$, which determines the type of the error, its magnitude, $|e|$, should also be considered. Adopting MSE as the scale for the quality of an estimator is similar to assessing the gravity of the error e by e^2. The principal difference is that an estimation error is innocuous if it leads to the appropriate decision (action). That is, we are concerned only with the signs of $\hat{\theta} + \Delta - T_1$ and $\theta - T_1$. When these two signs coincide, the magnitude of the error $\hat{\theta} - \theta$ or $\hat{\theta} - \theta + \Delta$ is immaterial.

2.7 Hypothesis testing

In a conventional mindset, the problem of deciding between the two signs, positive (+) and negative (−), for a parameter θ, or for $\theta - T_1$, can be fitted into the framework of hypothesis testing. Here we describe this approach in neutral terms, and offer a criticism in the next section.

We assign one of the options to the hypothesis and the other to the alternative. There is no ready prescription as to which option should be the hypothesis and which the alternative, but the choice turns out to be far from

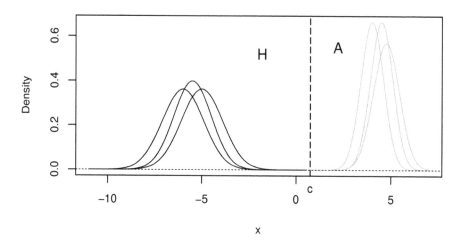

FIGURE 2.3
Example of two perfectly separated sets of continuous distributions.

innocuous. We choose a (test) statistic, such as an estimator $\hat{\theta}$. For each value of θ in its parameter space Θ, this statistic has a particular distribution, denoted by $\mathcal{D}_\theta$. These distributions, $\mathcal{D}_\theta$, $\theta \in \Theta$, are split into two groups, one for $\theta \in \Theta_H$, and the other for $\theta \in \Theta_A$. The indices stand for the hypothesis (H) and alternative (A). Based on the realised value of $\hat{\theta}$ we want to infer the group, H or A, to which this distribution belongs.

The choice of the statistic $\hat{\theta}$ is guided by the desire to make this inference as decisive as possible. The task would be trivial if the two sets of distributions were perfectly separated, if they satisfied the identities $P(\hat{\theta} < c; \theta \in \Theta_H) = 1$ and $P(\hat{\theta} > c; \theta \in \Theta_A) = 1$ for a constant c. In this case, we could readily establish from the value of $\hat{\theta}$ whether θ is in the hypothesis or in the alternative; the inference would be without any uncertainty. Figure 2.3 gives an illustration of two groups of three (continuous) distributions each, represented by their densities. The separator c is marked by thick vertical dashes. One set of densities, drawn by black colour, is concentrated entirely to the left of c, and another, drawn by grey colour, entirely to the right of c.

This is rarely a realistic proposition. First, when groups of distributions, are considered in practice, usually one or both of them have infinite supports. Typically, they are specified by disjoint subspaces of the parameter space. And second, the two sets of densities overlap. That is, a density can be found in one set, and another in the other set, so that their supports overlap, and often do so on a wide interval, or even the entire axis $(-\infty, +\infty)$. The support of any normal distribution is $(-\infty, +\infty)$, and so any subsets of normal distributions have this property.

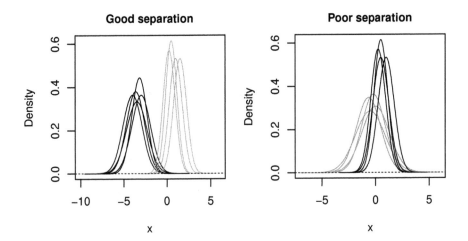

FIGURE 2.4
Examples of good and poor separation of sets of continuous distributions.

It is more meaningful to search for a statistic $\hat{\theta}$ and a constant c for which the probabilities $P(\hat{\theta} < c; \theta \in \Theta_H)$ and $P(\hat{\theta} > c; \theta \in \Theta_A)$ are both large, in a sense that is left to be defined. Figure 2.4 gives an illustration of (relatively) good separation in the left-hand panel and of poor separation in the right-hand panel. With good separation, we can identify from $\hat{\theta}$ the group (H or A) to which θ belongs with near certainty for all but a narrow range of values of $\hat{\theta}$. With poorer separation, $\hat{\theta}$ is less useful for this task.

We are at liberty to choose the statistic $\hat{\theta}$, and it is obvious that good separation is a desirable property for this task. An efficient estimator of θ is often a good choice but there are many other choices. For example, if a statistic $\hat{\theta}$ is good for this purpose, then $d\hat{\theta}$ is equally good for any constant $d \neq 0$. Delving into the details of other choices is beyond our scope.

In the established way of testing a hypothesis, we specify a small probability α, called the size of the test; $1 - \alpha$ is referred to as the level of significance. The size of the test is set conventionally to 0.05, and less commonly to 0.01 or 0.10. For the specified α, we find the critical value c for which the inequality

$$P\left(\hat{\theta} > c; \theta\right) \leq \alpha \tag{2.5}$$

holds for all $\theta \in \Theta_H$. So, for any θ in the hypothesis, the probability of exceeding the critical value c is small, more precisely, not greater than α. The critical value is not unique. If the inequality in (2.5) is satisfied for a particular c, then it is satisfied also for every $c' > c$ because the probability is a decreasing function of c for any θ. However, c is in most settings unique if we add the condition that equality in (2.5) has to hold for (at least) one value of $\theta \in \Theta_H$. That is, we want to select the smallest value c that satisfies (2.5).

The critical value c separates the possible values of $\hat{\theta}$ to those that are regarded as not contradicting the choice of Θ_H ($\hat{\theta} < c$), and those that contradict it. In this statement, 'no contradiction' is interpreted as small probability of $\hat{\theta}$ exceeding c. Thus, if $\hat{\theta} > c$, we conclude that we have evidence against the hypothesis, because such values of $\hat{\theta}$ are atypical when $\theta \in \Theta_H$. However, they are not implausible; we permit a small probability for the verdict (statement or claim) that $\theta \in \Theta_H$ to be incorrect. If $\hat{\theta} < c$, we have a vacuous conclusion; we have support for neither the hypothesis nor the alternative. We tried to find a contradiction with the hypothesis, but we failed.

Paraphrased, using the setting of a family idyll, consider the venture of having a look at the sky at night with your small niece or granddaughter. If you see the Moon ($\hat{\theta} > c$), that proves once and for all that there is a Moon. But if you do not see the Moon ($\hat{\theta} < c$), the conclusion that there is no Moon is far fetched; the Moon may be hiding behind the clouds, or beyond the horizon, and may appear on another night.

2.8 Hypothesis test and decision

The hypothesis test is poorly suited for our task of selecting one of the available options for several reasons. First, the hypothesis and the alternative have asymmetric roles in the test; switching the roles of Θ_H and Θ_A results in a different test. The critical value c is set with no regard for Θ_A and, after the switch, with no regard for Θ_H.

Second, no support can be found for the hypothesis H. Support for A is found when H is rejected. 'No rejection' of H does not amount to any support for H. There are no circumstances in which A could be rejected. The definition of c in equation (2.5) can be motivated as an exploration of what would happen if θ were in the hypothesis. The interval $(c, +\infty)$ is reserved for the conclusion that the hypothesis H is unlikely, and thus a verdict for its complement, the alternative A, that is, rejection of H. The hypothesis cannot be ruled out, but the probability of incorrectly rejecting it, conditional on any particular value of θ, is small, bounded by α. We say that this type of error, false rejection, is under control, by the size of the test, α. An example in which Θ_H and Θ_A comprise a single distribution each, $\mathcal{N}(-1, 1.1^2)$ and $\mathcal{N}(1, 0.9^2)$, respectively, is given in Figure 2.5. The density under the hypothesis is marked as f_H (black line) and the density under the alternative as f_A (grey line). Darker shading indicates the critical region, the right-hand tail of $\mathcal{D}_H$, covering $\alpha = 0.05$ of the area under the density f_H. It is the probability of inappropriate rejection (under the assumption of H), which was set by design (and convention). The critical value is $c = 0.81$. The lighter shade covers the probability of failing to reject the hypothesis (under the assumption of A). This probability is $P(\hat{\theta} < c; A) = 0.42$. Since we explore below alternative settings of c, it is

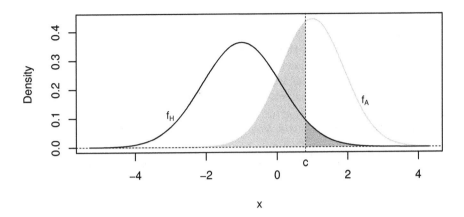

FIGURE 2.5
The size of the test and its power.

useful to adopt the notation c_α, indicating the test size on which the critical value is based.

If we increase the test size α, we reduce the critical value c_α. The dark-shaded area is increased at the expense of the light-shaded area. The probability of failing to reject the hypothesis when the alternative holds is reduced. We could have selected α with a view towards a balance of the two probabilities. This is achieved for $\alpha = 0.16$, when $c = 0.10$. Another well-motivated choice is $c = 0.002$, where the two densities intersect; $f_H(0.002) = f_A(0.002) \doteq 0.240$, permitting errors when one density is smaller and the other is greater than the threshold of 0.240. This corresponds to $\alpha = 0.181$.

We are left in an inferential limbo as to the choice of the critical value c because we have no principle or guidance to follow. We propose to fill this void by specifying and taking into account the consequences (ramifications) of the two kinds of error, expressed in numerical terms as losses that have the properties set out in Section 1.2.1. With these losses, we have an unequivocal principle—to minimise the expected loss. In the setting of Figure 2.5, this can be implemented straightforwardly, by evaluating the probabilities of the two kinds of error for values of c on a fine grid of points. For each pair of probabilities, $p_H(c) = P(\hat{\theta} > c; H)$ and $p_A(c) = P(\hat{\theta} < c; A)$, we have to decide what is preferable: to inappropriately reject the hypothesis with probability $p_H(c)$, or to inappropriately adopt the hypothesis with probability $p_A(c)$. The probability $p_H(c)$ decreases from unity to zero as a function of $c \in (-\infty, +\infty)$, and $p_A(c)$ increases from zero to unity. So, the alternative A gradually becomes less attractive and the attraction of the hypothesis H is increased with increasing c. We have to find the equilibrium c for which the choice (the dilemma) between H and A is in a balance and would cause least loss (harm or damage).

This scheme has not been taken up because it cannot be adapted to settings in which H or A comprise a continuum of distributions that are poorly separated. By way of an example, consider the distribution $\mathcal{N}(0,1)$ as the hypothesis and $\mathcal{N}(\theta,1)$ for $\theta > 0$ as the alternative. The latter contains innumerably many distributions that are arbitrarily poorly separated from $\mathcal{N}(0,1)$. With the approach we outlined in the discussion of Figure 2.5, it is impossible to set a single value of c that would be suitable for all the distributions in A.

If with a pre-specified test size α we conclude by supporting the hypothesis when $\hat{\theta} < c_\alpha$, we control the error of the first kind (false discovery of A). But we have no comparable control over the second kind (failure to discover A), which has the probability $\mathrm{P}(\hat{\theta} < c; \mathrm{A})$. In most applications, this probability is a continuous function of both c and θ. Then the probability of not rejecting the hypothesis for a θ in Θ_A in the vicinity of the cutpoint T_1, which separates Θ_H and Θ_A, is close to $1 - \alpha$, and this is far too large in most cases.

Hypothesis test operates with *hypothetical* probabilities, $\mathrm{P}(\hat{\theta} < c; \theta)$, for $\theta \in \Theta_\mathrm{H}$ and Θ_A, evaluated under the false presumption of a particular value of θ. It imposes a limit on the probability of one kind of error, and leaves us in the lurch with regard to the other kind of error. The probability of rejecting the hypothesis when the alternative is valid is called the *power* of the test. The power is a function of θ. Values of the power close to unity are desirable.

In the design of a study, some control can be imposed on the power, but not for values of θ close to the cutpoint T_1. The sample size and other aspects of the study are set by the following procedure. A value $\theta_M \in \Theta_\mathrm{A}$ is specified. It is referred to as the smallest important value of θ. The sample size and other aspects of the study are then set so as to ensure that the power exceeds a set bound under the presumption that $\theta \geq \theta_M$. So, the 'lurch' is only partial, not resolved in the interval (T_1, θ_M), which is associated with low power—large probability of failing to reject the hypothesis, even though the alternative holds.

To summarise, the analyst is placed into a mindset of the hypothesis, and this hypothesis is rejected if the data make this mindset (hypothesis) appear implausible—the hypothesis is contradicted. This amounts to evidence against the hypothesis, and therefore for the alternative, the hypothesis and the alternative cover all options. If the hypothesis is not rejected, then we end up with no contradiction with the adopted position. That makes a very poor case for supporting the hypothesis because a contradiction has been found with neither the hypothesis nor the alternative.

However, we dwell on a different argument, developed in Chapter 3. Hypothesis testing is ill-suited for making decisions because it operates solely with probabilities and has no means of incorporating the consequences (ramifications) of the two kinds of erroneous choices. No method of choosing between H and A deserves any respect if it is oblivious to these consequences. If we adopted the rule for crossing the road when the probability of a vehicle coming at the same time was only 0.05, or when it was 0.01, unacceptably many of us would soon come to grief.

2.9 Combining values and probabilities—Additivity

The calculus of probabilities in conducting several hypothesis tests is far from trivial, especially when the test statistics are dependent (correlated) and the tests are conducted conditionally, one test depending on the outcome of some others. In contrast, if we constructed a currency for error, akin to a monetary currency, it would be additive.

In a hypothesis test, certain conditional probabilities are under control, but values, such as gains or losses, are ignored. Expectation combines these two aspects of uncertainty, the chance and the stake. In its simplest form, suppose we lose $L_1, L_2, \ldots, L_K$ units of a currency, called lossiles), with respective probabilities $p_1, p_2, \ldots, p_K$. With complementary probabilities, $1 - p_1, 1 - p_2, \ldots, 1 - p_K$, we incur no loss. Then the expected loss, as interpreted in the frequentist paradigm, is the weighted average

$$\mathrm{E}(L) \; = \; \sum_{k=1}^{K} p_k L_k \, .$$

It stands to reason that scenarios that are more likely are given greater weight. The expectation $\mathrm{E}(L)$ has another important interpretation. Let $L^{(1)}, L^{(2)}, \ldots, L^{(n)}$ be a sequence of replicates of the loss L, realised according to the probabilities $p_1, \ldots, p_K$. Then the average loss

$$\bar{L}^{(n)} \; = \; \frac{1}{n} \left(L^{(1)} + L^{(2)} + \cdots + L^{(n)} \right)$$

converges to $\mathrm{E}(L)$ as $n \to +\infty$.

The expression for $\bar{L}^{(n)}$ indicates an assumption that should not be glossed over. The currency for loss has to be such that addition, or averaging, of losses is a meaningful operation. That is, the loss of L_1 lossiles in one instance, and of L_2 lossiles in another, is for all intents and purposes equivalent to the loss of $L_1 + L_2$ lossiles in a single instance. We refer to this property as *additivity*. In many contexts related to rational and responsible conduct, a monetary currency has the same property.

It should be understood that the monetary funds at our disposal are sometimes treated as not additive, by both individuals (and households) and institutions. One reason is the singularity of zero funds. Being in debt (to a bank or another financial institution) is expensive, associated with social or corporate stigma and constraints, and so an expenditure of 100 units has different consequences when it is a small fraction of our disposable assets and when we have only 50 units at our disposal. At the other extreme, some of us spend liberally on a lottery well knowing that in expectation it is a losing proposition. However, some enjoy the thrill of betting and associate winning a big prize with a kudos in excess of its actual monetary value.

Any loss or loss function has to have the property of additivity. There is no clear-cut analytical way of checking that a particular loss has this property, because an appropriate loss reflects the client's perspective which usually defies any codification. A simple way to deal with this problem is to consider a range of plausible losses. Such a range has the property that the client is satisfied that it contains the appropriate loss, even if this loss has not been identified (pin-pointed). The analysis then proceeds by separate evaluations for each plausible loss, although shortcuts can often be found.

This device has to be used sparingly because a plausible range can be defined for any aspect of the analysis. If it is used for several aspects, then the analysis may get out of hand by its complexity. However, we want to proceed with honesty and integrity, without pretending certainty when it is absent. At the same time, we want to exploit all the relevant information that is available. Discarding it with the excuse that it is incomplete and that it still leaves considerable uncertainty is beyond the statistical pale.

2.10 Further reading

The fiducial distribution was invented and applied for the first time by Fisher (1955). Lindley (1958) established its connection with the Bayes theorem. For a review of the method and its generalisations, see Seidenfeld (1992) and Hannig (2009). Lehmann and Romano (2005) is a comprehensive volume on hypothesis testing. Hedges and Olkin (1985) spawned an extensive literature on metaanalysis, in medical sciences in particular.

2.11 Exercises

2.1. Review the definitions of the terms 'estimator' and 'estimate'. In a study that comprises planning, data collection and analysis, when are they set and when are they evaluated?
   In what circumstances could the value of an estimation error be established? (Its value known with precision?) When could suspicions be raised that an estimation error is excessive?

2.2. Review the differences between the frequentist and Bayesian paradigms. Recall your earlier courses in statistics and summarise them according to the paradigm used in them. State which paradigm you prefer, if any, and give your reasons.

2.3. I do my shopping every Saturday morning and record the bill incurred. Are the recorded amounts replications of a process? What would you like to know, so that you could answer this question with greater confidence?

2.4. I walk (drive a car, cycle or take a train) to work every working day at around the same time. Can the times taken be regarded as a set of replications? Under what circumstances?

2.5. Suppose a study involves replications of an experiment that concludes with either success or failure. The outcome of the study is one success and nine failures. Do you need to know more about the details of the study, its design in particular, to conduct a competent analysis?

2.6. Suppose we have the task of estimating a parameter θ. We have two choices for the estimator. One has a binary distribution with its support (that is, its possible outcomes) on the values $\theta - 1$ and $\theta + 1$, each with probability $\frac{1}{2}$. The other estimator is distributed uniformly on the interval $(\theta - 0.1, \theta + 0.4)$. Which estimator would you prefer and why (and in what circumstances)?
Evaluate the biases, variances and mean squared errors of the two estimators.

2.7. Recall (or find in your lecture notes or in the literature) the densities and the distribution functions of the exponential and (continuous) uniform distributions and verify that the differential of the distribution function is indeed equal to the density, and the integral of the density is equal to the distribution function. Explore the difficulties in integrating the density of the (standard) normal distribution.

2.8. Generate on the computer a random sample of size m, of your choice, from the binomial distribution with probability $p = 0.35$ and number of trials $n = 250$. Demonstrate the convergence of the binomial to the normal distribution. By trial and error, find how large the sample size m has to be for this demonstration to be convincing. Repeat this exercise with the standard normal distribution. That is, generate a (large) random sample from $\mathcal{N}(0, 1)$, and draw a histogram of these values. How large a sample is needed for the histogram to have the shape of the normal density?

2.9. In the setting of Exercise 2.8, explore the options available in the software you use for drawing a histogram: the width of a bin and the location of the cutpoints of the bins. Explore what alternatives there are for the histogram, or devise some yourself. How would they have affected your conclusions in Exercise 2.8?

2.10. Discuss how the general perception of additivity of personal or corporate assets might be undermined in the following circumstances: (a) the threat of insolvency; (b) having to borrow money; (c) the kudos of wealth acquired suddenly and effortlessly.

2.11. An opinion poll is planned among the students of University U. It has the form of a referendum, containing a single question: 'Would you have any objections if University U amalgamated with University V in the next academic year?' What is the population and the sample in this case? Discuss what might be the key parameter of interest in this setting and how it might be estimated. What is the estimator in this setting? What is the estimate and when is it established? How would you go about deciding whether a majority of students supports the merger?

2.12. Demonstrate on the computer, that if $X_1, X_2, \ldots, X_n$ is a random sample from $\mathcal{N}(\mu, \sigma^2)$, then the mean $\frac{1}{n}\{\exp(X_1) + \exp(X_2) + \cdots + \exp(X_n)\}$ is biased for $\exp(\mu)$, even though $\frac{1}{n}(X_1 + X_2 + \cdots + X_n)$ is unbiased for μ. Choose arbitrary values for μ and $\sigma^2 > 0$. Explore what happens if you set σ^2 to a very large value, such as 1000. Show that this demonstration is sufficient for $\mu = 0$. The results for any other expectation μ can be deduced (derived analytically) from the results for $\mathcal{N}(0, \sigma^2)$.

2.13. A random sample $X_1, X_2, \ldots, X_n$ from a normal distribution with unknown mean μ and variance $\sigma^2 = 1$ is drawn. The prior distribution for μ is normal with mean θ and variance τ^2. Derive the posterior distribution of μ.

2.14. Repeat Exercise 2.13 with a random sample from a binomial distribution (p, n) with a beta distributed prior for the probability p.

2.15. Review the steps in deriving the fiducial distribution for a variance parameter based on a chi-squared statistic.

2.16. Derive the density of the χ_1^2 distribution by basic principles from the density of the normal distribution. For a bonus, derive the density of the χ_2^2 distribution from the density of χ_1^2.

2.17. The manufacturer of a particular product has an order-book for 200 of its products. The products undergo a stringent quality control, and about 20% of the inspected items are rejected. The rejected items are transported to a 'repair shop' where, after fixing their defects, they are offered on the market at a discount. How many items should the manufacturer plan to make to fulfil the order? Consider the consequences of (a) failing to fulfil the order and (b) having unsold items that have to be stored. After weighing the consequences, how would you adjust your answer (if at all)?

2.18. Review the steps of the test of the hypothesis that two normally distributed random samples of unequal sizes have identical expectations. (Assume that the two distributions have identical variances.)

2.19. In the setting of Exercise 2.18, suppose the conclusion is 'not significant'. Discuss the validity or appropriateness of the following interpretations:

- There is no difference between/among the studied groups.
- The expectations of the groups do not differ substantially.
- There is no evidence of any difference between/among the expectations of the groups.
- The sizes of (some of) the groups are not sufficiently large.
- The differences between/among the within-group expectations, even if there are some, can for all reasonable purposes be ignored.

Feel free to add to this list any other interpretations, with specific (data) examples.

2.20. Suppose the result of testing a hypothesis that the expectations of two (unspecified) distributions coincide with test size $\alpha = 0.05$ is 'significant'. Discuss the validity of the following statements:

- The probability that this result is incorrect is 0.05.
- If the two expectations do not differ, then the probability that the conclusion is correct is 0.95.
- If the two expectations do not differ, then the conclusion is correct.
- The hypothesis is contradicted by the text.

Feel free to add further statements to this list, for example, for a classroom discussion.

2.21. Compile the code for simulating the comparison of the expectations of two normal random samples. In R, this can have the form of a function in which the arguments (inputs) are the two sample sizes, means and variances and the level of significance. The output of the function is the verdict of significance (T or F). Use the R function `replicate` for a large number of replications.

For various configurations of the arguments of this function, evaluate the rate (percentage) of correct verdicts.

2.22. Draw the densities of two distributions with different expectations, such as $X_1 \sim \mathcal{N}(1,3)$ and $X_2 \sim \mathcal{N}(5,2)$. Find by trial and error the point T for which the sum of the tail probabilities,

$$P(X_1 > T) + P(X_2 < T),$$

is minimised. One route is by evaluating this probability for values of T on a regular grid defined in the range $(1,5)$, or narrower. After identifying two points on this grid within which the minimum is bound to lie, define a finer grid, and repeat this process a few times if necessary.

By a similar process, find the point where the two densities intersect. This problem can be solved analytically, concluding that there are in fact two such points. Explain the apparent contradiction.

2.23. Discuss the following proposition:

> Taking part in games of chance is not rational because the house has to take a fraction of the bets to cover its running costs and make a profit. By design, no gambler can recover by wins a greater fraction of the amount placed in bets than any other gambler. Therefore, in the long run, every gambler loses.

2.24. Find or construct an example of estimating a parameter, in which errors of the same magnitude but with different signs, say e and $-e$, have different consequences.

3

Positive or negative?

The focus of this chapter is on the simplest of all the statistical problems, arbitrating about the sign of a parameter—whether it is positive or negative. This subsumes the problem of deciding whether the mean of a variable is positive or negative and whether one mean is greater or smaller than another. Our attention is narrowed down further to estimators that are normally distributed, invoking the established theory that proves asymptotic normality of a variety of common statistics, including the mean of a random sample from a distribution with finite variance.

Having dismissed hypothesis testing, we devise a solution for the simplest version of the problem with $K = 2$ options. We have to decide whether an unknown quantity θ is greater or smaller than a given cutpoint T_1 which separates the two options, $\Theta_1 = (-\infty, T_1)$ and $\Theta_2 = (T_1, +\infty)$. We have an unbiased estimator $\hat{\theta}$ of θ. Its sampling distribution is normal, $\hat{\theta} \sim \mathcal{N}(\theta, \tau^2)$, and we assume that the variance τ^2 is known. No generality is lost by assuming that $T_1 = 0$ and $\tau = 1$, because we could consider $\theta' = (\theta - T_1)/\tau$, estimated without bias by $\hat{\theta}' = (\hat{\theta} - T_1)/\tau$.

3.1 Constant loss

Suppose the losses for the two kinds of inappropriate verdict, false negative and false positive, are $L_{12} = 1$ and $L_{21} = R$, respectively; R is called the loss ratio. When $T_1 = 0$, the indices 1 and 2 may be changed to $-$ and $+$, respectively, so that the losses are denoted by L_{-+} and L_{+-}, and the expected losses by Q_- and Q_+. These expectations are evaluated over the fiducial (or posterior) distribution of θ.

Denote by $\Phi(x; \mu, \sigma)$ the distribution function of $\mathcal{N}(\mu, \sigma^2)$ and by $\varphi(x; \mu, \sigma)$ the corresponding density. We drop the arguments μ and σ^2 when $\mu = 0$ and

$\sigma^2 = 1$. Thus,

$$\varphi\left(\theta; \hat{\theta}, \tau\right) \;=\; \frac{1}{\tau}\varphi\left(\frac{\theta - \hat{\theta}}{\tau}\right) \;=\; \frac{1}{\tau\sqrt{2\pi}}\exp\left\{-\frac{1}{2\tau^2}\left(\theta - \hat{\theta}\right)^2\right\}$$

$$\Phi\left(\theta; \hat{\theta}, \tau\right) \;=\; \Phi\left(\frac{\theta - \hat{\theta}}{\tau}\right),$$

and

$$\Phi(\theta) \;=\; \int_{-\infty}^{\theta} \varphi(x)\,\mathrm{d}x.$$

We shall study the expected losses associated with the two available verdicts and will issue the verdict for which the expected loss is smaller. This way of electing a verdict is known as the Bayes rule. The solution will turn out to have the form of issuing verdict V_1 when $\hat{\theta} < E$, and verdict V_2 otherwise, for a suitable constant E called the *equilibrium*.

With verdict V_1 we are in error if $\theta > T_1$. The fiducial density of θ is $\varphi(x; \hat{\theta}, \tau)$ and the fiducial probability of such error is

$$Q_1 \;=\; \int_{T_1}^{+\infty} \varphi\left(x; \hat{\theta}, \tau\right)\mathrm{d}x.$$

This is equal to the expected loss because $L_{12} = 1$. Further,

$$Q_1 \;=\; 1 - \Phi\left(T_1; \hat{\theta}, \tau\right) \;=\; \Phi\left(\frac{\hat{\theta} - T_1}{\tau}\right),$$

owing to the symmetry of the normal distribution; $\varphi(x) = \varphi(-x)$ and $\Phi(x) = 1 - \Phi(-x)$. The expected loss associated with verdict V_2 is

$$Q_2 \;=\; R\int_{-\infty}^{T_1} \varphi\left(x; \hat{\theta}, \tau\right)\mathrm{d}x$$

$$\;=\; R\left\{1 - \Phi\left(\frac{\hat{\theta} - T_1}{\tau}\right)\right\}. \qquad (3.1)$$

Preferring to lose less in expectation, we issue verdict V_1 when $Q_1 < Q_2$, that is, when

$$\Phi\left(\frac{\hat{\theta} - T_1}{\tau}\right) \;<\; \frac{R}{R+1},$$

or equivalently, when $\hat{\theta} < E$, where

$$E \;=\; T_1 + \tau\Phi^{-1}\left(\frac{R}{R+1}\right). \qquad (3.2)$$

This implies a simple rule, to issue

- verdict V_1, asserting that $\theta < T_1$, when $\hat{\theta} < E$;

- verdict V_2 $(\theta > T_1)$ otherwise.

The equilibrium E is an increasing function of R; with greater R we are better disposed towards V_1 because higher loss ratio R $(= L_{21}/L_{12})$ amounts to stronger aversion to electing V_2. The equilibrium diverges to $-\infty$ as R converges to zero, marginalising verdict V_1, and it diverges to $+\infty$ as R diverges to $+\infty$, when verdict V_2 is marginalised.

The expected losses Q_1 and Q_2 are monotone functions of $\hat{\theta}$; Q_1 increases with $\hat{\theta}$ from zero in the vicinity of $-\infty$ to $L_{12} = 1$ in the vicinity of $+\infty$. In contrast, Q_2 decreases from $L_{21} = R$ in the vicinity of $-\infty$ to zero at $+\infty$. The two functions intersect at E, where

$$Q_1(E) = Q_2(E) = \frac{R}{R+1}.$$

For $\hat{\theta} \neq E$ one of the expected losses is smaller and the other is greater than $R/(1+R)$. So, if we adopt the (Bayes) rule of minimising the expected loss, $R/(1+R)$ is the largest possible expected loss that could be incurred.

The loss incurred, $Q(\hat{\theta}) = \min\{Q_1(\hat{\theta}), Q_2(\hat{\theta})\}$, depends on τ, except at $\hat{\theta} = E$. This is illustrated in Figure 3.1, where functions Q_1 and Q_2 are drawn for $R = 4$ and $\tau = 0.1$, 0.25 and 0.75. The equilibria, where the verdict switches from V_1 to V_2, are 0.084, 0.210 and 0.631 for $\tau = 0.1$, 0.25 and 0.75, respectively. At these points $Q_1 = Q_2 = 0.8$.

The triplets of functions Q_1 and Q_2 in the diagram each intersect at $\hat{\theta} = 0$, where $Q_1(0) = 0.5$ and $Q_2(0) = \frac{1}{2}R = 2$. For smaller τ, the functions are steeper at zero and converge faster to their limits as $\hat{\theta} \to \pm\infty$. This can be interpreted as decision being easier to make (us being more decisive), because the values of Q_1 and Q_2 are similar, and close to $R/(R+1)$, in narrower ranges around the equilibrium.

3.1.1 Equilibrium and critical value

The rule defined by the equilibrium E in equation (3.2) has the same form as the test of the hypothesis that $\theta \leq T_1$, given by the critical value c in equation (2.5). Its solution is

$$c^* = T_1 + \tau \Phi^{-1}(1 - \alpha),$$

so the two rules, or verdicts, coincide after setting $\alpha = 1/(R+1)$. It would seem that we have achieved nothing more than a reparametrisation of the problem from α to R; $\alpha = 0.05$ corresponds to $R = 19$. Further, α converges to zero for $R \to +\infty$ and it converges to unity for $R \to 0$. However, we now have a strong motivation to set the value of R, based on the specifics of each problem, whereas in a hypothesis test we would rarely consider any alternative to $\alpha = 0.05$, and hardly ever a value also different from 0.01 and 0.10.

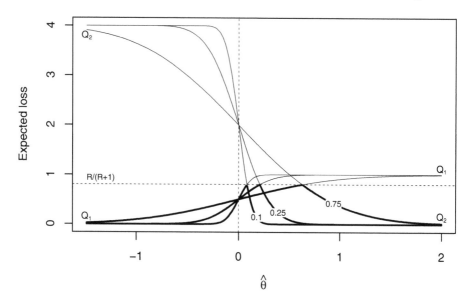

FIGURE 3.1
Expected losses as functions of the estimate $\hat{\theta}$ with $R = 4$, for standard errors $\tau = 0.1$, 0.25 and 0.75, marked on the curves; Q_1 is increasing and Q_2 is decreasing. The incurred expected loss functions $Q = \min(Q_1, Q_2)$ are drawn by full solid lines.

For a practically oriented analyst, this may not be enough to write off hypothesis testing altogether. There are estimators and other statistics with distributions different from the normal, and hypotheses other than 'negative vs. positive' are often tested. We chip away at these commonly perceived advantages in the next two chapters, and in the remainder of this chapter overlay our assault on hypothesis testing by introducing loss structures that depend not only on the type of the error but also on its magnitude. In this process, we come to consider a general problem of classification. In this problem, there is a finite number of classes and, based on a statistic $\hat{\theta}$, we want to place the parameter θ in the class that is expected to cause our client least damage, harm or loss; or equivalently, the greatest profit, benefit or gain.

3.2   The margin of error

The loss introduced in the previous section depends only on the type of the error made, false negative or false positive. Errors by a narrow and a wide margin are assumed to be equally grave, to have the same consequences, so long as they are of the same type. We refer to such loss functions (or loss structure) as piecewise constant. This term is better motivated by specifying the loss as a function of the error, defined on a continuous scale.

For problems with two states, Θ_1 and Θ_2, separated by a cutpoint T_1, the magnitude of the error is $e = \theta - T_1$ when we issue verdict V_1 even though $\theta > T_1$, and $e = T_1 - \theta$ when we issue V_2 even though $\theta < T_1$. For the appropriate verdicts no error is committed, even when $\hat{\theta} \neq \theta$. In these cases, we set $e = 0$. Note that the error e does not involve $\hat{\theta}$ directly; it is a function of the verdict V and parameter θ.

To conform with the term 'error' as used in estimation for the deviation of the estimate from the target, $\Delta\theta = \hat{\theta} - \theta$, the error may be defined as $e = (\theta - T_1) I_{V \neq \Theta}$. Here, I is the indicator function; its value is unity if its (logical) argument is true and zero if it is false. Thus, the 'original' error is wiped out if the verdict is appropriate ($V_1 = \Theta_1$ or $V_2 = \Theta_2$), and otherwise its sign indicates the verdict.

Linear loss for a false negative is defined as $L_{12} = R_2 + S_2(\theta - T_1)$ when we issue verdict V_1 even though we are in state Θ_2, where $\theta > T_1$. Linear loss for a false positive is defined similarly, as $L_{21} = R_1 + S_1(T_1 - \theta)$, when we issue verdict V_2 inappropriately because we are in state Θ_1. Since loss is always nonnegative and nondecreasing with distance from T_1, the coefficients R_1, R_2, S_1 and S_2 have to be nonnegative. Setting R_1 or R_2 to zero (or to a small positive value) is appropriate only when certain small errors are associated with small losses. In such a setting, there may be a narrow range in which either verdict is (almost) equally appropriate and the choice between V_1 and V_2 is almost inconsequential.

The loss functions L_{12} and L_{21} do not have to have the same form. For example, L_{21} may be linear and L_{12} constant. In this case, the magnitude matters only for one type of error. If $R_1 > 0$, then no generality is lost by setting $R_1 = 1$, because it corresponds to a (multiplicative) conversion of the currency for error, akin to a conversion from one monetary currency to another.

The expected loss for a false negative (verdict V_1) is

$$Q_1 = \int_{T_1}^{+\infty} \{R_2 + S_2(x - T_1)\} \varphi\left(x; \hat{\theta}, \tau\right) dx.$$

Denote $z_1 = (T_1 - \hat{\theta})/\tau$. After the transformation $y = -(x - \hat{\theta})/\tau$, the integral becomes

$$Q_1 = \int_{-\infty}^{-z_1} \{R_2 - S_2\tau(y + z_1)\} \varphi(y) dy.$$

Integration by parts, differentiating $R_2 - S_2\tau(y + z_1)$ and integrating $\varphi(y)$, yields

$$Q_1 = \left[\{R_2 - S_2\tau(y + z_1)\} \Phi(y)\right]_{-\infty}^{-z_1} + S_2\tau \int_{-\infty}^{-z_1} \Phi(y) dy$$

$$= R_2\{1 - \Phi(z_1)\} + S_2\tau \int_{-\infty}^{-z_1} \Phi(y) dy.$$

In this derivation we used the identity $\lim y\, \Phi(y) = 0$ as $y \to -\infty$. It follows from the inequalities

$$0 > y \int_{-\infty}^{y} \varphi(x)\, dx > \int_{-\infty}^{y} x\, \varphi(x)\, dx = -\varphi(y) \to 0 \qquad (3.3)$$

for $y < 0$; it is easy to check that $\varphi'(y) = -y\, \varphi(y)$.

The primitive function for $\Phi(y)$ is $\Phi_1(y) = y\Phi(y) + \varphi(y)$; this is checked by differentiation. Owing to symmetry of the normal distribution, this function has the property $\Phi_1(-y) = -y + \Phi_1(y)$. By substituting this identity we obtain

$$\begin{aligned}
Q_1 &= R_2\{1 - \Phi(z_1)\} + S_2\tau \Phi_1(-z_1) \\
&= R_2\{1 - \Phi(z_1)\} + S_2\tau\{-z_1 + z_1 \Phi_1(z_1)\}
\end{aligned}$$

It is left for an exercise to show that

$$Q_2 = R_1\, \Phi(z_1) + S_1\tau \Phi_1(z_1)\ .$$

Both expected losses Q_1 and Q_2 are totals of the constant part of the loss, $R_2\Phi(-z_1)$ and $R_1\Phi(z_1)$ respectively, and the part that is proportional to the error, $S_2\tau\Phi_1(-z_1)$ and $S_1\tau\Phi_1(z_1)$. This confirms that when losses have additive components, such as R_2 and $S_2(\theta - T_1)$ for verdict V_1, the expected losses have the corresponding additive components. Further, the expressions for the proportional loss mirror their counterparts for constant loss; the only change involved is the replacement of $\Phi(z_1)$ with its primitive function $\Phi_1(z_1)$.

3.3   Quadratic loss

The quadratic loss function for a false positive is defined as

$$L_{21} = R_1 + S_1\, (\theta - T_1)^2\ ,$$

where R_1 and S_1 are positive constants. The corresponding expected loss is

$$\begin{aligned}
Q_2 &= \int_{-\infty}^{T_1} \left\{R_1 + S_1\, (x - T_1)^2\right\} \varphi\left(x; \hat{\theta}, \tau\right) dx \\
&= R_1 \int_{-\infty}^{z_1} \varphi(y)\, dy + S_1 \int_{-\infty}^{z_1} \left(\hat{\theta} - T_1 + y\tau\right)^2 \varphi(y)\, dy \\
&= R_1\, \Phi(z_1) + S_1\tau^2 \int_{-\infty}^{z_1} (y - z_1)^2 \varphi(y)\, dy\ ,
\end{aligned}$$

after applying the linear transformation $y = (x - \hat{\theta})/\tau$ and introducing $z_1 = (T_1 - \hat{\theta})/\tau$. The first term is also obtained from equation (3.1). For integrating by parts, we require the primitive function for $\Phi_1(y)$. This function is

$$\Phi_2(y) = \tfrac{1}{2}\left\{\left(1+y^2\right)\Phi(y) + y\varphi(y)\right\};$$

check it by differentiation, although the function can be 'constructed'; see Exercise 3.9. The function has the property

$$2\Phi_2(-z) = 1 + z^2 - 2\Phi_2(z).$$

The integral in the expression for Q_2 is equal to

$$
\begin{aligned}
\int_{-\infty}^{z_1} (z_1 - y)^2\,\varphi(y)\,dy &= 2\int_{-\infty}^{z_1} (z_1 - y)\,\Phi(y)\,dy \\
&= 2\int_{-\infty}^{z_1} \Phi_1(y)\,dy \\
&= 2\Phi_2(z_1),
\end{aligned}
$$

obtained by applying two-fold integration by parts. Hence,

$$Q_2 = R_1\,\Phi(z_1) + 2S_1\,\tau^2\,\Phi_2(z_1).$$

For the quadratic loss $L_{12} = R_2 + S_2(\theta - T_1)^2$, similar operations yield the identity

$$
\begin{aligned}
Q_1 &= R_2\,\Phi(-z_1) + 2S_2\,\tau^2\,\Phi_2(-z_1) \\
&= R_2\left\{1 - \Phi(z_1)\right\} + S_2\,\tau^2\left\{1 + z_1^2 - 2\Phi_2(z_1)\right\}.
\end{aligned}
$$

The concluding comments from Section 3.2 about the linear expected loss carry over to quadratic loss directly. The role of $\Phi(z_1)$ in constant loss and $\Phi_1(z_1)$ in proportional loss is taken over by $2\Phi_2(z_1)$ in quadratic loss. The wider variety of loss functions, expanded from constant to quadratic, comes with only modest computational complexity.

3.4 Combining loss functions

Any pair of loss functions L_{12} and L_{21} can be re-defined as a single loss function $L(\theta)$; for $\theta < T_1$, $L(\theta) = L_{21}(\theta)$ and for $\theta > T_1$, $L(\theta) = L_{12}(\theta)$, if the inappropriate verdict is issued; otherwise $L(\theta) = 0$. See Figure 3.2 for an illustration. Here the loss function is constant in $\Theta_1 = (-\infty, 0)$ and linear in $\Theta_2 = (0, +\infty)$. The discontinuity at $T_1 = 0$ is of no concern if two slightly different values of θ on either side of T_1 are indeed associated with substantially different losses. Discontinuity at T_1 arises also for the piecewise constant loss with loss ratio $R \neq 1$.

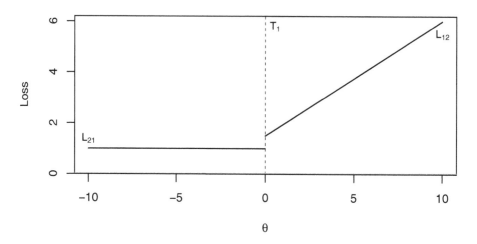

FIGURE 3.2
Loss function defined by its two components, linear for V_1 and constant for V_2.

Any convex combination of loss functions is also a loss function. For example, the quadratic loss $L_{21} = R_1 + S_1 (\theta - T_1)^2$ is a convex combination of the constant loss $2R_1$ and the squared loss $2S_1 (\theta - T_1)^2$, with coefficients (weights) $\frac{1}{2}$ each. Note that if L is a loss function, then for any scalar $d > 0$, dL is also a loss function. However, this expansion of the class of loss functions is vacuous because L and dL are effectively identical loss functions. With a given currency, the lossile, more is at stake for $d > 1$ than for $d = 1$, but the verdict does not depend on d. For any loss function L, the functions $\{dL; d > 0\}$ form a class of equivalence.

Convex combinations expand our horizon of loss functions somewhat, but not substantially because we are limited by the relatively small basis from which we can generate loss functions. At present, this basis comprises constant, linear and quadratic functions. An extension to polynomials of higher degree is straightforward but an example in which, say, a cubic loss function would be suitable is difficult to construe. Nevertheless, analytical evaluation of the expected loss with such a loss function is not demanding. Details for an exponential loss function are left for an exercise.

3.5   Equilibrium function

When there are two options the verdict is obtained by comparing the expected losses Q_1 and Q_2. We refer to the difference $B = Q_1 - Q_2$ as the *imbalance*, to B, as a function of $\hat{\theta}$, as the balance function, and to the equation $B = 0$ as the balance equation. Apart from $\hat{\theta}$, B depends also on the parameters involved in the loss function, such as the loss ratio R for piecewise constant loss.

In all the settings we explored thus far, Q_1 and Q_2 are monotone functions of $\hat{\theta}$, one increasing and the other decreasing. Further, they are positive and converge to zero at the opposite limits of the support of $\hat{\theta}$. Thus, for fixed loss functions, the balance function is strictly monotone and the balance equation has a unique solution in $\hat{\theta}$. This solution is the equilibrium E. It separates the domains of the two verdicts: for $\hat{\theta} < E$ one verdict and for $\hat{\theta} > E$ the other verdict is issued.

When there is uncertainty about one or several parameters involved in one or both loss functions L_{12} and L_{21}, it is useful to explore how the equilibrium depends on such a parameter R. This function, $E(R)$, is called the *equilibrium function*. The following general result gives a condition for when the equilibrium function is strictly monotone (increasing or decreasing).

If the balance function $B(\hat{\theta}, R)$ is differentiable in both $\hat{\theta}$ and R, and each partial differential is either positive or negative throughout, then the equilibrium function $E(R)$ is also monotone and its differential has the same sign throughout. This result follows immediately by applying the implicit function theorem,

$$\frac{\partial E}{\partial R} = -\frac{\partial B\left(\hat{\theta}, R\right)}{\partial R} \Big/ \frac{\partial B\left(\hat{\theta}, R\right)}{\partial \hat{\theta}},$$

assuming that two of these three differentials are well defined and the denominator on the right-hand side is never equal to zero.

Cases in which $B(\hat{\theta}, R)$ is not a monotone function of $\hat{\theta}$ are very unusual. An increase in R usually amounts to being more averse to the verdict associated with small (or large) values of θ, and then $B(\hat{\theta}, R)$ is also a strictly monotone function of R.

Example 2

Suppose the sampling distribution of $\hat{\theta}$ is normal with expectation θ (i.e., with no bias) and variance $\tau^2 = 0.12$. The available courses of action correspond to states $\Theta_1 = (-\infty, 0)$ and $\Theta_2 = (0, +\infty)$, so $T_1 = 0$. The losses are piecewise constant and we consider loss ratios in the range $R \in (1, 10)$. The balance equation has a closed-form solution, derived in (3.2),

$$E(R) = \tau \Phi^{-1}\left(\frac{R}{R+1}\right).$$

This equilibrium function is drawn in Figure 3.3. It is increasing, as can also be inferred from its expression directly.

In the right-hand panel, $E(R)$ is reproduced with the horizontal axis on the log scale. On this scale, $E(R)$ is very close to linearity. The log (or

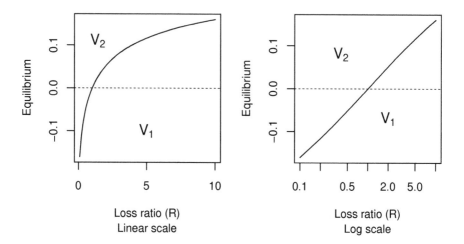

FIGURE 3.3
Equilibrium function $E(R)$ for piecewise constant loss with loss ratio R. The
horizontal axis in the right-hand panel is on the log scale.

multiplicative) scale is much more natural for R because the range $(0,1)$ is the
natural counterpart of $(1,+\infty)$; a value R^* of R has its counterpart in $1/R^*$.
The difference of one unit on the scale of R has a very different meaning for
$R = 1$ and 2 than, say, for $R = 100$ and 101.

Why is the equilibrium function so close to linearity on the log scale? After
substituting $R = e^r$, the function plotted is

$$E(r) = \tau \Phi^{-1}\left(\frac{e^r}{1 + e^r}\right). \tag{3.4}$$

Apart from the factor τ, it is a composition of the inverse of the logit, $\log(R) -
\log(R + 1)$, and the inverse of the normal distribution function, known as the
probit function. The logit and probit are known to be very similar in a wide
range of values R distant from both zero and $+\infty$, that is, of values r distant
from $\pm\infty$. Hence the approximate linearity of the equilibrium function on the
log scale, as a function of $r = \log(R)$. □

Not all equilibrium functions have explicit forms, not even when the balance function has a closed form. The equilibrium can be found by the Newton
method or another (simple) algorithm for solving a nonlinear equation.

To find the root of a continuous function $B(\hat\theta)$, the Newton method starts
with a pair of initial guesses E_0 and E_0'. A provisional solution E_1 is found
by linear interpolation or extrapolation, as may be, solving the equation

$$\frac{B(E_0)}{E_0 - E_1} = \frac{B(E_0')}{E_0' - E_1}.$$

The solution is

$$E_1 = \frac{E_0' \, B(E_0) - E_0 \, B(E_0')}{B(E_0) - B(E_0')}.$$

In the next iteration, E_0' and E_1 are placed in the respective roles of E_0 and E_0', yielding a new provisional (updated) solution E_1. Iterations are stopped when the consecutive solutions E_1 are only slightly apart and $B(E_1)$ is sufficiently close to zero.

The Newton-Raphson algorithm proceeds from an initial solution E_0, which may be based on a guess, by iterations

$$E_t = E_{t-1} - \frac{B(E_{t-1}, R)}{\dfrac{\partial B(\hat{\theta}, R)}{\partial \hat{\theta}} \Big|_{\hat{\theta} = E_{t-1}}}, \tag{3.5}$$

$t = 1, 2, \ldots$, until convergence is reached. (The differential in the denominator is evaluated at $\hat{\theta}$ set to the current solution E_{t-1}.) The iterations are stopped when $|E_t - E_{t-1}|$ or $|B(E_t, R)|$ is sufficiently small. These two criteria can be combined by insisting that their maximum, or total, be sufficiently small. Convergence is slow when the differential in the denominator is small in absolute value in the neighbourhood of the solution E or if an intermediate solution happens to fall in such a region, because there the function is 'flat'—very close to a constant.

Example 3

Suppose the two available options are separated by cutpoint $T_1 = 0$ and the loss function for a false negative is linear, $L_{12} = R_2 + S_2 \theta$ when $\theta > 0$, and the loss for a false positive is constant, $L_{21} = 1$. The (normally distributed) unbiased estimator of θ has standard error $\tau = 0.12$. The expected losses are

$$Q_1 = R_2 \{1 - \Phi(z_1)\} + S_2 \tau \{-z_1 + \Phi_1(z_1)\}$$
$$Q_2 = \Phi(z_1),$$

where $z_1 = -\hat{\theta}/\tau$. The equilibrium is the root of the balance function

$$B(R_2, S_2) = R_2 - (R_2 + 1)\,\Phi(z_1) + S_2 \tau \{-z_1 + \Phi_1(z_1)\}, \tag{3.6}$$

solved for $E = -z_1 \tau \; (= \hat{\theta})$, or for z_1. For the solution, obtained by the Newton-Raphson algorithm, we use the expression for the partial differential,

$$\frac{\partial B}{\partial z_1} = -(R_2 + 1)\,\varphi(z_1) - S_2 \tau \{1 - \Phi(z_1)\}. \tag{3.7}$$

In an implementation in R, the solution is obtained instantly. The Newton-Raphson algorithm requires only a handful of iterations ($5 - 10$) in a wide range of values of R_2 and S_2. Figure 3.4 presents the equilibrium function

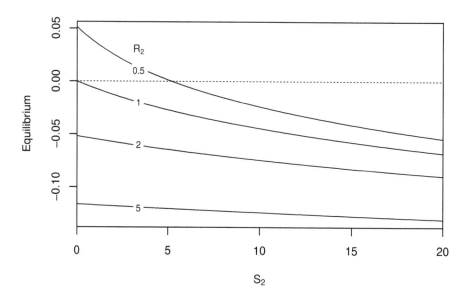

FIGURE 3.4
Equilibrium functions for linear loss L_{12} and unit loss L_{21} of the linear coefficient S_2 for values of R_2 printed on the curves.

for $S_2 \in (0, 20)$ and the values of R_2 indicated on the curves. Each function (curve) is drawn as a piecewise linear function connecting the (101) equilibria for $S_2 = 0, 0.2, \ldots, 20.0$. Convergence is slow only for $R_2 = 5$ and S_2 close to 20.0, when 51 iterations are required to achieve precision to eight decimal places. In this region, the balance function is very flat.

The bivariate equilibrium function $E(R_2, S_2)$ is decreasing in both R_2 and S_2. The two parameters can be interpreted as characterising aversion to false negatives; R_2 is its uniform component, applicable equally to small and large errors involved in false negatives, and S_2 for differential aversion to large errors over small errors of the same kind. The differences among the univariate equilibrium functions $E(S_2; R_2)$ decrease with S_2.

The equilibrium for $R_2 = 1$ is at $S_2^* = 0$. In this case, the loss is equal to one lossile for any inappropriate verdict. For $R_2 > R_1 = 1$, the equilibrium is negative because the loss for any error $e > 0$, equal to $R_2 + eS_2$, is greater than $R_1 = 1$, the loss for $-e$. Conversely, $R_2 = 0$ is not compatible with very small S_2 because V_2 would then be appropriate even for large positive values of $\hat{\theta}$. For $R_2 = 0$ and $S_2 = 0$ there is no equilibrium because V_2 is associated with uniformly smaller loss (0 vs. 1); in this case, V_2 is a safe choice. As $R_2 \to +\infty$ or $S_2 \to +\infty$ (and the other parameter is held constant), the equilibrium diverges to $-\infty$, although the divergence is very slow.

There is no inherent contradiction or error in choosing pairs (R_2, S_2) which we regard here as incompatible. Simply, the associated problem is trivial; one verdict is associated with smaller loss than the other for almost any data.

3.6 Plausible values and impasse

If the loss function is identified, so that its functional forms for the two kinds of error and the parameters they involve are set, comparing two normal samples would seem to be straightforward. This scheme is unrealistic on two counts. First, we assume that the sampling variance $\text{var}(\hat{\theta}) = \tau^2$ is known, and second, any setting of the parameters of a loss function entails at least a modicum of arbitrariness. This is only partly because the elicitation exercise is unusual, with little accumulated experience or sound advice, or a well-established pro-tocol to be followed by the parties involved. The value of any parameter is unlikely to be specified clinically, so that any other value, however close to the declared one, could be ruled out. Simply, we have to contend with the mix of uncertainty and ambiguity, as well the unwillingness of any professional to commit themselves to a single numerical value that would characterise a concept as abstract as the consequences of selecting an inappropriate option.

This problem is resolved by working with plausible values. Instead of a single value of a parameter, say the loss ratio R, we consider a range of values, an interval (R_-, R_+) such that there is no contention that $R \in (R_-, R_+)$, even though the exact value of R is not identified. By this definition, a sufficiently wide range is plausible. However, it is advantageous to declare as narrow a range for R as possible, so long as it is plausible; so long as any value outside the range, $R \notin (R_-, R_+)$, can be ruled out.

Once a plausible range is set, we might solve the problem at hand for a fine grid of values in this range, so that we would establish the verdict (V_1 or V_2) as a function of R. If the verdict is the same throughout the range (R_-, R_+), then the exact value of R is immaterial, so long as it is in the plausible range. We say that such a verdict is *unequivocal*. If for some plausible values of R one verdict and for other plausible values the other verdict would be issued, we reach an *impasse* (stalemate) because either verdict is plausible.

Figure 3.5 presents an example based on the setting of Figure 3.3. In both panels, the equilibrium function is plotted for the piecewise constant loss. The estimate is $\hat{\theta} = -0.08$, with standard error $\tau = 0.12$. The horizontal axis is on the log scale, and so the equilibrium function has very little curvature. In the left-hand panel, the plausible range of R is $(0.15, 0.50)$, marked by the grey strip. The equilibrium function attains the value of $\hat{\theta} = -0.08$ at $R^* = 0.338$. This is within the plausible range, so we have an impasse. For smaller values, $R < R^*$, we would have verdict V_2, and for larger values we would have verdict V_1.

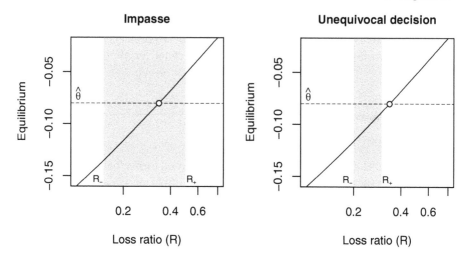

FIGURE 3.5

Equilibrium function $E(R)$ for the piecewise constant loss; $\hat{\theta} = -0.08$ and $\tau = 0.12$. The plausible range of R is marked by the vertical grey strip.

The plausible range of R in the right-hand panel, $(0.2, 0.3)$, is narrower, and now we have the unequivocal verdict V_1. The two panels illustrate the advantage of a tighter plausible range. However, this advantage is illusory if the declared plausible range lacks integrity, if the elicitation of these bounds has gone astray and some values of R outside the declared range (R_-, R_+) are plausible.

When the equilibrium E is a monotone function of R, as is the case in Figure 3.5, the verdict switches from V_2 to V_1, or vice versa, at a single value of R, denoted by R^*. We refer to it as the *borderline* for R. If $R^* \notin (R_-, R_+)$, then we have an unequivocal verdict; otherwise, when $R^* \in (R_-, R_+)$, we reach an impasse. So, when the equilibrium function $E(R)$ is monotone, it suffices to find the verdicts for the limits of the plausible range, R_- and R_+. If they are in accord (both are V_1 or both are V_2) or, in more precise notation, $V(R_+; \hat{\theta}) = V(R_-; \hat{\theta})$, then we have an unequivocal verdict; otherwise we have reached an impasse.

Impasse is undesirable because we cannot point to a course of action without a qualification. An impasse can be ruled out by declaring a single value of R, but such a choice may be neither credible nor agreeable to the client or its representatives. An effort to agree on as narrow a plausible range as possible may be rewarded by an unequivocal verdict that is credible. At the same time, it is essential to resist any false incentives to declare a plausible range that is narrower than what is warranted by the client's perspective.

3.7 Elicitation

The discussion, negotiation or deliberation to set a plausible range of values of a parameter is referred to as *elicitation*. In principle, this range could shrink to a single value, but that may be an unrealistic goal. The participants in elicitation are the analyst and the client. A moderator may also take part, with the remit to assist in making the process constructive and effective. The client may comprise several individuals, either forming a committee, or representing distinct parties. The client may be represented in the elicitation by a party with a specific remit. For instance, a business enterprise may be represented by one of its executives.

Elicitation may start by defining a very wide range of values of the parameter, say R, so that all parties would readily agree that this range is plausible. Elicitation then proceeds by reducing this range, by increasing its lower limit R_- and decreasing its upper limit R_+ in small steps, until objections are raised that the resulting range is no longer plausible. A reduction is accepted only when the client confirms (is convinced) that the proposed range is still plausible. When one limit cannot be improved the other may still be sharpened.

The client may be a heterogenous (disunited) group with differing views and perspectives that correspond to different values of R. The plausible range has to cover all these values. Further, there may be uncertainty about this 'overall' range. In an elicitation exercise, the values that correspond to the individual perspectives should be regarded as untouchable, not subject to negotiation. But the overall plausible range may be reduced by a well organised and structured discussion. There may be a single perspective, such as a national government's programme in which the country's public is the client, but there may be considerable variation in its interpretation among the representatives participating in the elicitation. Dealing with such grey area is beyond our scope; we assume throughout that the client's perspective is accurately and faithfully represented, or that the inevitable ambiguities are covered by a plausible range. But in general, imperfect representation is preferred to no representation.

Elicitation may also start by a single value that would not be rejected by the client as implausible. Then a narrow interval around this value is proposed as a plausible range, and it is expanded in small steps, at one limit at a time, until the client is satisfied that a plausible range has been obtained. A distinct drawback of this approach is that the initial value may function as an *anchor*, subconsciously influencing the choices and calls made by the experts taking part in elicitation.

Elicitation involves two counteracting principles: integrity (inclusiveness)—that the agreed range is plausible—and tightness—that the range is as narrow as possible. Tightness can be interpreted as richness (exclusivity) of information. After all, $(-\infty, +\infty)$ is a plausible range for any real parameter, but it is not useful; for our purposes it is vacuous.

Another meaning of integrity in elicitation is the absence of any purpose or agenda other than the faithful (impartial and detached) representation of the client's perspective. Elicitation relies on the integrity of all the parties involved and on a genuine desire to obtain conclusions that best serve the interests of the client. Such conclusions, or processes that yield them, are commonly referred to as *valid*. We refrain from using this qualifier because we regard all processes that do not serve this purpose unconditionally as inappropriate. In the process of arriving at a verdict, we may make some assumptions that are or may be invalid. We regard the process as appropriate if such lack of validity is clearly stated or delineated, together with the reason for its adoption: illustration (e.g., as part of training or instruction), exploration (sensitivity analysis), our failure to identify the value of a parameter, and using a guess or values in a plausible range instead, and the like.

The outcomes of elicitation cannot be externally verified. In principle, the elicitation process could be replicated, but the client would have to have a different set of representatives, because the original set (the committee) have a memory that could not be wiped out, while leaving the perspectives intact, to ensure independence of the two replicates. Independence would also be undermined if the members of the two committees conferred and discussed their experiences and conclusions.

In summary, we do not regard inaccuracies or errors in elicitation as factors that disqualify an analysis, so long as the subsequent analysis is informed at its outset by the likelihood and nature of such deficiencies. Data should be treated similarly. It can rarely be stripped of all the undesirable features that arise in the process of its collection, such as nonresponse, measurement error and proxy status. The analysis should respond to the challenge of establishing (or estimating) what the results would have been had these data deficiencies been absent. The level of confidence in the analysis should be downgraded to reflect the vagaries of such 'nuisance' processes that reduce the informational content of the data.

3.7.1   Post-analysis elicitation

In ideal circumstances, elicitation is conducted prior to data collection and inspection, so that it is not influenced by the would-be findings of the analysis. The outcomes of elicitation have the status of prior information, independent of the data planned to be collected according to a design.

An analysis may proceed without elicitation, or with only rudimentary (hasty) elicitation, using a wide plausible range for some of the contentious parameters. If the analysis concludes with an impasse, as it is likely, then the client may be requested to narrow down the plausible range. As an alternative, the client may be confronted with the borderline, such as R^*, and asked whether this value of R is plausible. If it is not, then the verdict is unequivocal, otherwise there is an impasse. For example, the borderline in the setting of Figure 3.5 is $R^* = 0.338$. The verdict is unequivocal if $R_+ < R^*$ (the case in the right-hand panel), or $R_- > R^*$. Impasse arises if R^* is a plausible value of R, when $R^* \in (R_-, R_+)$.

Integrity of a post-analysis elicitation is threatened if the clients adjust their conduct in the elicitation with intent to avoid impasse or to pursue another agenda influenced by the analyst's disclosure. The resulting misrepresentation of the client's perspective is referred to as *rigged elicitation*.

3.8 Plausible rectangles

The exploration of the equilibrium function in Example 3 is limited by being univariate, concerned with only one parameter. For each value of the standard deviation τ we find the borderline R^* of R. It is preferable to address the uncertainty about two (or, in general, about all) parameters. We deal first with the setting in which only two parameters, τ and R, are subject to uncertainty. Similarly to the plausible range (R_-, R_+), we consider a plausible range (τ_-, τ_+) for τ. For a given value of $\hat{\theta}$, the estimate of θ, we consider the borderline function $R^*(\tau)$ and relate it to the plausible rectangle $\mathcal{P} = (R_-, R_+) \times (\tau_-, \tau_+)$. If $R^*(\tau)$ intersects this rectangle then we have an impasse because for some plausible pairs (R, τ) one verdict and for their complement in $\mathcal{P}$ the other verdict would be issued.

For piecewise constant loss with loss ratio R, we have equilibrium at the estimate $\hat{\theta}$ when

$$\hat{\theta} = T_1 + \tau \Phi^{-1}\left(\frac{R}{R+1}\right). \tag{3.8}$$

By solving this equation for R and τ we obtain the borderline functions

$$R^*(\tau) = \frac{\Phi(z)}{1 - \Phi(z)}$$

$$\tau^*(R) = \frac{\hat{\theta} - T_1}{\Phi^{-1}\left(\frac{R}{R+1}\right)}, \tag{3.9}$$

where $z = (\hat{\theta} - T_1)/\tau$.

Figure 3.6 presents an example for piecewise constant loss with threshold $T_1 = 0$ and estimate $\hat{\theta} = 0.756$. Both $R^*(\tau)$ and $\tau^*(R)$ are decreasing functions when $\hat{\theta} > T_1$. For $\hat{\theta} < T_1$, they are increasing. Either borderline function partitions the space $R \times \tau$ (or $\tau \times R$) to subsets in which verdict V_1 or V_2 is issued.

The lighter-shaded rectangle with a solid-line border delineates a plausible rectangle. The borderline function intersects this rectangle, so the verdict is

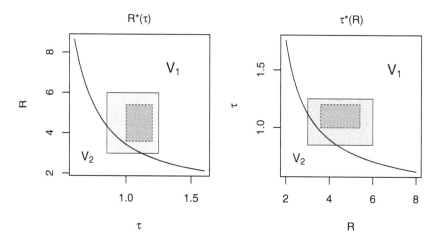

FIGURE 3.6
Borderline functions $R^*(\tau)$ and $\tau^*(R)$ for $\hat\theta = 0.756$ and $T_1 = 0$; piecewise constant loss. The shaded rectangles represent plausible ranges for R and τ.

equivocal—for some plausible pairs (τ, R) verdicts V_1 and for others V_2 is issued. Suppose the smaller rectangle, with darker shading, is also a plausible rectangle for (τ, R). The borderline function does not intersect it, so the verdict V_1 would be unequivocal. This highlights the value of adopting as narrow a plausible range as possible, so long as it does qualify as a plausible range.

Narrowing a plausible range at one or the other of its limits is not equally important. In Figure 3.6, a reduction of the upper limit for R (from 6.0 to 5.4) has no impact, but the increase of the lower limit, from 3.0 to 3.6 is very useful. Similarly, the reduction of the upper limit for τ, from 1.25 to 1.20, is of no consequence, but the increase of the lower limit, from 0.85 to 1.00 is very important. Before realising the study and establishing the value of $\hat\theta$, we do not know which limits will be important, although the form of the equilibrium function implies that it will be either the two upper or the two lower limits of the plausible ranges.

Note how the two plausible ranges 'cooperate' in bringing about an unequivocal verdict. If the plausible range for τ were reduced as indicated in Figure 3.6, but the range for R were not, the verdict would be equivocal; verdict V_2 would be issued, for instance, for $\tau = 1.01$ and $R = 3.1$. Similarly, if the lower limit of the plausible range R were raised, to $R_- = 3.6$, but the lower limit for τ were retained at $\tau_- = 0.85$, the verdict would be equivocal because the borderline function would still intersect the plausible rectangle.

The relationship of the borderline function to the plausible rectangle can be explored even before the study is realised. Simply, we evaluate the function for values of $\hat\theta$ in a (plausible) range, and note the range of values of $\hat\theta$ that lead to an equivocal verdict. An example, using the setting of Figure 3.6 is

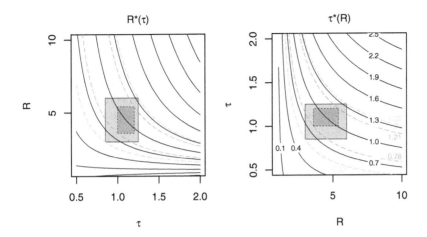

FIGURE 3.7
Borderline functions $R^*(\tau)$ and $t^*(R)$ for a range of values of $\hat{\theta}$ (printed on the curves); the setting of Figure 3.6.

given in Figure 3.7. The borderline is an increasing function of $\hat{\theta}$. The same values of $\hat{\theta}$ are used in both panels but, to reduce the clutter, they are marked only in the right-hand panel.

At the planning stage, the range of values of $\hat{\theta}$ for which there would be an impasse is of interest. The narrower this impasse range the lower the likelihood of an impasse. For the bigger plausible rectangle, the impasse range is $(0.573, 1.335)$; the corresponding borderline functions are drawn by lighter shade of grey, in accord with the shade used for the corresponding plausible rectangle. The limits of the impasse range correspond to the equilibria obtained directly from equation (3.2), with the appropriate vertices of the plausible rectangle substituted for (R, τ). For the smaller plausible rectangle we obtain a narrower impasse range, $(0.781, 1.212)$. This confirms the value of the effort expended in a patient elicitation that (hopefully) yields narrower plausible ranges for at least some of the parameters. We return to this theme in Chapter 6.

The borderline functions in Figure 3.7 motivate the following approach to issuing a verdict without any elicitation prior to conducting the study. We evaluate the estimator $\hat{\theta}$ and draw the borderline function $R^*(\tau; \hat{\theta})$ or $\tau^*(R; \hat{\theta})$. Then we ask the client whether all the plausible values of R and τ are located in one subset of $(0, +\infty)^2$ separated by the borderline function. If the client chooses one of the subsets, then we have an unequivocal verdict. Otherwise there is an impasse.

This approach is distinctly second-rate to elicitation before data collection. A 'wise' client, who appreciates the drawbacks of an impasse, might be reluctant to admit that some values of (R, τ) from either subset are plausible and would want to avoid the responsibility for causing the study to be inconclusive. Of course, when the client resists the temptation and maintains the integrity of the interaction with the analyst, then such post hoc elicitation is not problematic.

The plausible region for (R, τ) need not be a rectangle, although a practical example of this is difficult to construe. Certainly, a case in which it would not be a convex set, or even not a contiguous set (one with separated subsets), would be very strange.

Example 4

The example in Figures 3.6 and 3.7 is based on an equilibrium equation that has a closed-form solution. When the solution can be obtained only by an iterative algorithm, much more computing is involved but this is not barrier to exploring equilibrium functions. We illustrate this on an example with a piecewise linear loss function. We reuse the setting of Example 3, where we derived the balance function $B(R_2, S_2)$ in equation (3.6). We consider the plausible ranges $R_2 \in (0.5, 1.5)$ and $S_2 \in (0.3, 1.5)$. The plausible equilibria are found by the Newton-Raphson algorithm, using equation (3.7). We prefer to search for equilibria for $z_1 = -\hat{\theta}/\tau$ because the balance function is expressed in terms of z_1. Switching to $\hat{\theta}$ is straightforward; z^* is an equilibrium for z_1 if and only if $\hat{\theta}^* = -z_1\tau$ is an equilibrium for $\hat{\theta}$.

The equilibrium functions are drawn in Figure 3.8 for R_2 and S_2 in their plausible rectangle. The Newton-Raphson algorithm converges very rapidly, requiring no more than ten iterations for each of the $51 \times 5 = 255$ evaluations. (Each curve is drawn as a piecewise linear function with 50 connected segments.) The equilibria are increasing functions of both R_2 and S_2. The dependence on S_2 diminishes with increasing R_2.

Impasse arises for $z_1 \in (-0.39, 0.30)$, the extreme values of the bivariate equilibrium function $E(R_2, S_2)$ in the plausible rectangle. They correspond to $\hat{\theta} \in (-0.047, 0.036)$. Before realising the value of $\hat{\theta}$, the client could be informed about this and offered an opportunity for further elicitation, to reduce the plausible ranges of R_2 and S_2. Figure 3.8 indicates that reducing the plausible range for R_2 is more important, especially raising its lower bound R_{2-}, because even a small increase of R_{2-} would result in an appreciably narrower impasse range of $\hat{\theta}$. For example, increasing R_{2-} from 0.5 to 0.6 would raise the lower limit of the impasse range from -0.39 to -0.29, comparable to increasing S_{2-} from 0.3 to 1.2.

Suppose $\hat{\theta} = 0.0216$, so that $z_1 = -0.18$ (horizontal dashes in Figure 3.8). Impasse is then reached because equilibrium arises for pairs (R_2, S_2) on a curve connecting $(0.72, 0.30)$ and $(0.60, 1.50)$. This curve is very close to a straight line. Its exact values are easy to establish because the balance function is linear in both R_2 and S_2.

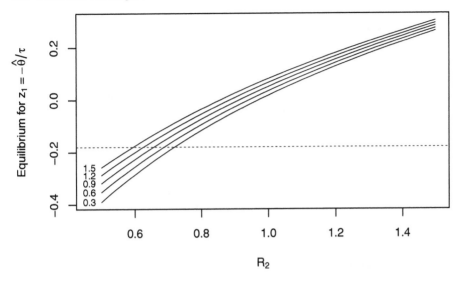

FIGURE 3.8
Equilibrium functions of R_2 (horizontal axis) and S_2 (separate curves) for the setting of Figure 3.6.

3.8.1 Summary

This chapter introduced a method that converts a normally distributed unbiased estimator of a parameter, and its realisation, to the choice between two intervals in which this parameter may lie. The intervals represent the options or states linked to the available courses of action. Key ingredients of the method are the partition of the parameter space to sets that correspond to the available options and the loss function that quantifies the consequences of the inappropriate choices. The loss function is declared after elicitation, usually in the form of functions defined separately for each discordant pair of verdict and state. The loss is zero for each concordant pair. The analysis concludes with a verdict, selected as the option for which the expected loss is smaller. Uncertainty and ambiguity about the loss function and the partition of the parameter space are represented by plausible functions and plausible partitions (cutpoints), and the problem is, in effect, solved for each combination of plausible functions and partitions. The conclusion (verdict) is unequivocal if it is the same for all plausible settings; otherwise an impasse is reached. An unequivocal verdict is always preferred because it provides a straightforward proposal for which course of action to pursue.

3.9 Further reading

Longford (2013) has a more condensed presentation of methods of this Chapter. For the original motivation, see Lindley (1985). An application with exponential loss functions is presented in Varian (1975). Methods for eliciting prior information and converting it to prior distributions are discussed by Garthwaite, Kadane and O'Hagan (2005) and O'Hagan (2019). The latter reference and Hanea *et al.* (2018) emphasise the role of a protocol in a structured elicitation exercise. Accounts of elicitation in practice are given by O'Hagan (1998) and Kadane and Wolfson (1998). They focus on prior distributions but their advice is instructive also for eliciting plausible ranges.

3.10 Exercises

3.1. Suppose the two available courses of action, A and B, correspond to states separated by a cutpoint $T \neq 0$; actions A and B are appropriate when $\theta < T$ and $\theta > T$, respectively. The (normally distributed) estimator $\hat{\theta}$ is unbiased for θ and has variance τ^2. Convert this problem to the problem of deciding whether a different parameter, ξ, is positive or negative, based on an estimator $\hat{\xi}$ that has normal distribution with unit variance.

3.2. Revisit the derivation in equation (3.1) and work out all the intermediate steps.

3.3. Draw a plot of the normal quantile function Φ^{-1}. Derive the relationship between the quantile functions of the standard normal $\mathcal{N}(0, 1)$ and an arbitrary normal distribution $\mathcal{N}(\mu, \sigma^2)$.

3.4. Redraw Figure 3.1 with different settings of the standard errors τ and loss ratio R. Check that the features pointed out in the discussion of Figure 3.1 are present also in your diagram.

3.5. Check that the differential of the standard normal density is related to the density by the identity $\varphi'(x) = -x\varphi(x)$. Derive an expression for the second-order differential of φ.

3.6. Review the conditions necessary for integration by parts, and check that they are satisfied in the derivation of the expected loss in Section 3.2.

3.7. Show that $y^2 \, \Phi(y) \to 0$ as $y \to -\infty$. Using this result, together with equation (3.3), show that $y^q \, \Phi(y) \to 0$ as $y \to -\infty$ for any $q > 0$. Hint: Apply the l'Hôpital's rule and mathematical induction.

3.8. Derive the expression for the integrated distribution function Φ_1 by 'guessing' the form of the function that has its differential equal to Φ.

Hint: By differentiating $x\Phi(x)$ we obtain $\Phi(x)$, the desired function, but with $x\phi(x)$ added. Then refer to Exercise 3.5.

3.9. Review, with no shortcuts, the derivations in Section 3.3. Derive the expression for the function Φ_2 by the method outlined in Exercise 3.8 for Φ_1.

3.10. Suppose L_1 and L_2 are two loss functions defined for the same setting of two courses of action, two corresponding states, and related to the same estimator $\hat{\theta}$. Suppose the respective expectations of L_1 and L_2 (i.e., the expected losses) are Q_1 and Q_2. What is the expected loss for $c_1 L_1 + c_2 L_2$ for $c_1 \geq 0$ and $c_2 \geq 0$?

3.11. Relate the result of Exercise 3.10 to the idea of regarding expected loss like an (additive) monetary currency. Are monetary funds always treated as if they were additive?

3.12. Discuss the merits and drawbacks of characterising an estimator by the transformed MSE, defined as $MSE\{f(\hat{\theta}); f(\theta)\}$ for a suitable (monotone) function f. What about a linear function f?

3.13. What happens to the bias of an estimator $\hat{\theta}$ after a transformation? For example, if $B(\hat{\theta}; \theta) = 0$, what can be said about $B\{f(\hat{\theta}); f(\theta)\}$? Construct some simple examples to accompany your answers or arguments. If you do not want to discuss this generally, focus on a few familiar functions, such as the square, square-root and the exponential.

3.14. Work out the details of the transformation from the χ_k^2 distribution in equation (2.3) to the inverse-χ_k^2 distribution. Derive the density of the fiducial distribution of σ^2—the scaled inverse-χ_k^2 distribution.

3.15. Review the details of the implicit function theorem, including its proof.

3.16. Review the definitions of imbalance, balance function and balance equation, and explain why equilibria are solutions of the balance equation. How are the balance function and equilibrium function related?

3.17. Construct an example of deciding whether a parameter θ is positive or negative, adapted from Example 2, in which an additional loss of R' lossiles is incurred when verdict V_+ is issued, and yet $\theta < -0.25$. Borrow all the other settings from Example 2.

3.18. Adapt the problem from Exercise 3.17 by changing the loss for false negative from constant to linear loss of your choice.

Note: If you compile a general R function for the Newton-Raphson algorithm, in which the objective function is one of the arguments, then you may save a lot of effort later when you have to deal with similar problems.

3.19. Revisit Exercises 3.17 and 3.18, or at least one of them, and solve them for several values of the loss ratio R or another parameter. Summarise the solutions in the form of a plausible range for the parameter concerned. Is the solution unequivocal, or an impasse is reached? If an impasse is reached, consider how the plausible range might be reduced to avoid it. Conversely, if the solution is unequivocal, how an impasse would have been reached had the plausible range been wider.

3.20. Consider the problem of deciding whether a parameter θ is positive or negative, based on an unbiased estimator $\hat{\theta}$ with sampling variance τ^2. Suppose the loss function for false positive is $L_{+-} = 1$ and for false negative is exponential, $L_{-+} = a \exp(b\theta)$, where a and b are positive constants. Derive the balance function for this problem.

3.21. Draw the probit and logit functions in a single plot and make a judgement about the range (interval) in which they are (approximately) linearly related.
Hint: Apply a linear transformation to one of the functions after comparing the differentials of the two functions at $p = 0.5$.

3.22. Discuss why the plausible range for a parameter should not be averaged — collapsed to the centre of the range,

3.23. Compare the merits of the Newton and Newton-Raphson methods for finding the root of a smooth monotone function. Illustrate them on an example of a balance function, such as $B(R_2; S_2)$ in equation (3.6). Consider the programming effort, problems with setting the initial solution(s), speed of convergence, failure to converge, and computing time used.

3.24. Explain the advantage of differentiating the balance function in Example 3, equation (3.6), with respect to z_1 instead of θ. Derive an expression for $\partial B / \partial \theta$.

3.25. Summarise the merits of elicitation at the design stage of a study. What is the advantage of declaring a narrower plausible range? What is the associated risk? Is there any way of discovering that the elicited plausible range is too narrow? How do your conclusions carry over to plausible rectangles?

3.26. Suppose the client is represented for elicitation by a committee of five members. Discuss the merits of

- asking the committee to confer and issue a single statement about a parameter of the loss function;
- asking the committee to confer, clarify the task with the analyst, and then ask each member of the committee to anony-

mously issue a statement; these statements would be later reconciled;

- providing a written or oral explanation by the analyst, asking each member of the committee to anonymously issue a statement, without any conferring;
- removing the anonymity from the previous two proposals;
- reconvening the committee a week later and assigning to them the same task.

Discuss how the disagreements among the members of the committee should be resolved.

3.27. Summarise the problems that might arise in elicitation after the data analysis.

3.28. Discuss the difference between equilibrium and borderline functions and their roles in formulating a verdict.

3.29. Check the expressions for the borderline functions in Section 3.8, equations (3.8) and (3.9).

4

Non-normally distributed estimators

This chapter deals with the problem of selecting one from a few available courses of action in the context of statistics (usually estimators) with non-normal distributions. The Student t and the chi-squared distributions are covered in the next two sections and some other common distributions are dealt with in the following sections.

4.1 Student t distribution

Having to know the sampling variance of a normally distributed estimator $\hat{\theta}$ is a restrictive assumption. Many statistics related to the difference of two means, $\Delta\theta = \theta_2 - \theta_1$, or similar quantities, have non-central t distributions, either exactly or approximately. These statistics have the form

$$
t = \frac{\Delta\hat{\theta}}{\hat{\tau}},
$$

where $\Delta\hat{\theta}$ is an (unbiased) estimator of $\Delta\theta$ and $\hat{\tau}^2$ is an unbiased estimator of $\tau^2 = \text{var}(\hat{\theta})$; $\hat{\tau}^2$ has a scaled chi-squared distribution.

When comparing the expectations of two normal distributions with identical variances, the chi-squared distribution associated with $\hat{\tau}^2$ has $n_1 + n_2 - 2$ degrees of freedom, where n_1 and n_2 are the sample sizes of the respective samples 1 and 2. Let σ^2 be the variance common to the two samples. It is estimated without bias by the pooled sample variance,

$$
\hat{\sigma}^2 = \frac{1}{n_1 + n_2 - 2} \left\{ \sum_{i=1}^{n_1} (y_{i1} - \bar{y}_1)^2 + \sum_{i=1}^{n_2} (y_{i2} - \bar{y}_2)^2 \right\}, \tag{4.1}
$$

where y_{ih} denotes observation i in sample h and $\bar{y}_h$ is the mean of sample $h = 1, 2$.

Matrix notation is much more elegant. Its introduction, with a few definitions and notation, may be tedious for some, but the rewards quickly follow. Denote by $\mathbf{y}_h = (y_{1h}, \ldots, y_{n_h h})^\top$, $h = 1, 2$, the two samples as column vectors, $\mathbf{1}_n$ the column vector of unities of length n, $\mathbf{I}_n$ the $n \times n$ identity matrix, and $\mathbf{J}_n = \mathbf{I}_n - \frac{1}{n}\mathbf{1}_n\mathbf{1}_n^\top$. We have the identities $\mathbf{1}_n^\top\mathbf{1}_n = n$ and $\mathbf{J}_n^2 = \mathbf{J}_n$; $\mathbf{J}_n$ is an *idempotent* matrix.

With this notation, the mean of a vector $\mathbf{y}$ of length n is $\bar{y} = \frac{1}{n}\mathbf{1}_n^\top\mathbf{y}$ and the estimator in equation (4.1) is

$$
\hat{\sigma}^2 = \frac{1}{n_1 + n_2 - 2} \sum_{h=1}^{2} \left(\mathbf{y}_h - \frac{1}{n}\mathbf{1}_{n_h}\mathbf{1}_{n_h}^\top\mathbf{y}_h\right)^\top \left(\mathbf{y}_h - \frac{1}{n}\mathbf{1}_{n_h}\mathbf{1}_{n_h}^\top\mathbf{y}_h\right)
$$

$$
= \frac{1}{n_1 + n_2 - 2} \sum_{h=1}^{2} \mathbf{y}_h^\top\mathbf{J}_{n_h}\mathbf{y}_h .
$$

Estimator $\Delta\hat{\theta}$ has sampling variance $\tau^2 = \sigma^2(1/n_1 + 1/n_2)$, and so $\hat{\tau}^2 = \hat{\sigma}^2(1/n_1 + 1/n_2)$.

4.1.1　Fiducial distribution for the t ratio

We develop methods for the mirror images of the problems addressed in Chapter 3 for normally distributed estimators. We start by inverting the sampling distribution, $(\hat{\theta}\,|\,\theta)$, to the posterior or fiducial distribution, $(\theta\,|\,\hat{\theta})$. To apply the fiducial argument, we express the estimator $\Delta\hat{\theta}$ as $\Delta\theta + \tau X$ and the t statistic as

$$
t = \frac{\Delta\theta + \tau X}{\sqrt{\tau^2 \frac{Y}{k}}} ,
$$

where X and Y are independent random variables, $X \sim \mathcal{N}(0,1)$ and $Y \sim \chi_k^2$, the latter with the chi-squared distribution with k degrees of freedom. The density of the χ_k^2 distribution is given by equation (2.3) in Section 2.5.1.

The fiducial distribution for $\Delta\theta/\tau$ is derived by inverting the expression for t:

$$
\frac{\Delta\theta}{\tau} = t\sqrt{\frac{Y}{k}} - X .
$$

The distribution of the right-hand side is a convolution of a transformation of χ_k^2 and the standard normal. Its variance is greater than unity, although it converges to unity as $k \to +\infty$.

The density of $U = \sqrt{\frac{1}{k}Y}$ is

$$
f(x) = \frac{k}{\Gamma_2(k)}\left(\frac{k}{2}\right)^{\frac{k}{2}-1} x^{k-1} \exp\left(-\frac{1}{2}kx^2\right) , \tag{4.2}
$$

derived by transforming the density of the χ_k^2 distribution. The expectation and variance of U are

$$
\mathrm{E}(U) = \sqrt{\frac{2}{k}}\,\frac{\Gamma_2(k+1)}{\Gamma_2(k)}
$$

$$
\mathrm{var}(U) = 1 - \frac{2\,\Gamma_2(k+1)^2}{k\,\Gamma_2(k)^2} = 1 - \{\mathrm{E}(U)\}^2 . \tag{4.3}
$$

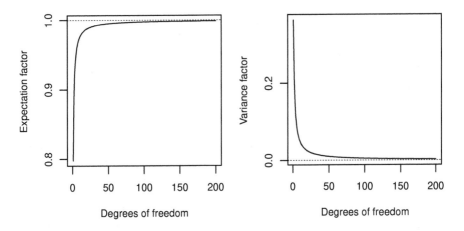

FIGURE 4.1
Expectation and variance factors (4.3) of the fiducial distribution for an estimator with the Student t distribution, as functions of the degrees of freedom.

We refer to them as the expectation and variance factors because the fiducial expectation and variance of $\Delta\theta/\tau$ are $t\,\mathrm{E}(U)$ and $1 + t^2\mathrm{var}(U)$, respectively. The expectation factor converges to unity and the variance factor to zero as k diverges to $+\infty$; see Figure 4.1. For large k we can ignore the uncertainty about τ^2, pretend that τ^2 is equal to $\hat{\tau}^2$, and apply methods for normally distributed estimators with known sampling variance. But bear in mind the presence of the value of t in the fiducial distribution; for large t more degrees of freedom would seem to be necessary for such a simplification.

The expected losses can be evaluated in two ways. The fiducial distribution can be approximated by the normal that has the same expectation and variance. No new derivations are needed for this, and the approximation is problematic only for small numbers of degrees of freedom.

As an alternative to this approximation, the expected losses can be simulated using random samples from the chi-squared and normal distributions. Suppose we want to evaluate the expectation of $f(Z)$, where f is a real function and Z is a random variable with the fiducial distribution for $\Delta\theta/\tau$. The t statistic is realised and has k degrees of freedom. We generate independent random samples $\mathbf{x}$ and $\mathbf{y}$, both of size H, from the respective standard normal and χ_k^2 distributions, and evaluate the vector $\mathbf{z} = t\sqrt{\mathbf{y}/k} - \mathbf{x}$, where the square root is defined elementwise. If $\mathrm{E}\{f(Z)\}$ is well defined, then it is approximated by the mean $\frac{1}{H}(z_1 + \cdots + z_H)$. There is no ready prescription for how large H should be, but modern computers and software present no serious limitations, even with moderately large sets of values of t and degrees of freedom k.

This method of simulation can be applied to a function f and distribution of Z for which the exact (analytical) result is known, such as $f(x) = x^2$ and a χ_k^2 distribution. The values obtained by simulation are close to the analytical result for sufficiently large number of replications H. Extensive evaluations on a grid of values of t and k can be organised in two or more stages. First a cruder approximation, with smaller H, is obtained to narrow down the region of interest, and then more computational effort can be expended, with larger H, in this region. Adopt this method for routine checking of your analytical derivations for any expressions that have the form of an expectation.

Example 5

We consider a study in which the means of a variable in two disjoint subpopulations are compared. The target is the difference of the means, and it is estimated without bias by $\Delta\hat{\theta} = 2.058 - 1.328 = 0.730$, based on samples of sizes 30 and 24. The sampling variance of $\Delta\hat{\theta}$ is estimated by $\hat{\tau}^2 = 1.175$, so the t statistic $\Delta\hat{\theta}/\hat{\tau} = 0.730/\sqrt{1.175} = 0.673$ is associated with $k = 52$ degrees of freedom. As random variables, $\Delta\hat{\theta}$ is normally distributed and the distribution of $k\hat{\tau}^2/\tau^2$ is χ_k^2; the two statistics are independent. Suppose further that piecewise constant loss is applicable and the plausible range for the loss ratio $R = L_{21}/L_{12}$ is $(5, 25)$. We are averse to false positives $(R > 1)$, but there is a considerable uncertainty about the strength of this aversion, characterised by the large ratio of the limits of the plausible range, $R_+/R_- = 5$.

The fiducial distribution of $\Delta\theta/\tau$ has expectation $m = 0.6702$ and variance $v = 1.0043$. These values are obtained by applying the identities in equation (4.3). The (exact) distribution is compared to the normal that has the same moments in Figure 4.2. The histogram is based on a random sample of size $500\,000$ and the density of the normal approximation is drawn by dashes. The approximation is clearly beyond reproach. We therefore proceed by the method for normally distributed estimators, although we apply the adjustment of the mean and variance, and work with θ/τ instead of θ. Note that scaling by an unknown τ, from $\Delta\theta$ to $\Delta\theta/\tau$, does not alter the cutpoint $T_1 = 0$. It would alter the cutpoint if it differed from zero.

The equilibria for $R = 5$ and $R = 25$, $\sqrt{v}\,\Phi^{-1}\{R/(R+1)\}$, are 0.969 and 1.773, respectively. Both values exceed the fiducial expectation of $\Delta\theta/\tau$, so the verdict is unequivocally V_1, that is, $\theta < 0$. We can 'invert' the problem and identify all the values of R for which one or the other verdict is issued. The solution, called the borderline, is given by the value of R that separates the two intervals:

$$R^* = \frac{\Phi\left(\dfrac{m}{\sqrt{v}}\right)}{1 - \Phi\left(\dfrac{m}{\sqrt{v}}\right)}, \tag{4.4}$$

which yields $R^* = 2.97$. If a plausible range for R is not elicited, then we could turn to the client after the analysis and ask whether R^* is a plausible value of R. If it is not, then we have an unequivocal verdict; $\theta > 0$ if R is certainly smaller than R^*, and $\theta < 0$ if R is certainly greater than R^*. In our case, $R^* < R_- = 5 < R$. □

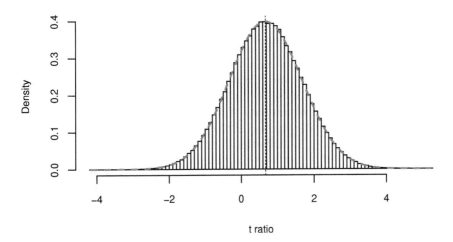

FIGURE 4.2
Fiducial distribution of θ/τ (histogram) and its normal approximation (grey dashes).

Example 6

We consider a setting similar to Example 5 but with two much smaller samples, $n_1 = 3$ and $n_2 = 4$, which provides a sterner test of the approximation to the fiducial distribution. Suppose $\Delta\hat{\theta} = 0.730$ with $\hat{\tau}^2 = 1.368$, and the plausible range of R is $(5, 25)$. The fiducial distribution of $\Delta\theta/\tau$ has expectation $m = 0.594$ and variance $v = 1.037$, based on $E(U) = 0.9515$, see equation (4.3). Assuming that this distribution is normal, the borderline for which the two verdicts have identical expected losses is $R^* = 2.57$, obtained by applying equation (4.4). The corresponding equilibrium is $E = \sqrt{v}\,\Phi^{-1}\{R^*/(1+R^*)\} = 0.594$ and the corresponding expected loss is $Q_1(R^*) = Q_2(R^*) = 0.7195$. If we observed $\hat{\theta}$ equal to E, the expected losses, approximated by simulation based on 500 000 replications, would be $Q_1 = 0.7203$ and $Q_2 = 0.7196$, indicating that the approximation is not exact, because $Q_1 \neq Q_2$, but the error is only slight. We can find by a simple iterative scheme, or by trial and error, that the equilibrium is at 0.5945, to be compared with the normal approximation 0.5939. Thus, the normal approximation is quite precise even for very few degrees of freedom, but the exact evaluation is also quite simple.

4.2   Verdicts for variances

In many studies, the target of inference is the contrast (difference) of the expectations of a continuous variable defined in two disjoint subgroups of a population. In some other cases, the variance, as a characteristic of the dispersion (heterogeneity), is of interest. When the studied variable is normally distributed, the established estimator of the variance based on a random sample of size n has a scaled χ^2 distribution with $n-1$ degrees of freedom (scaled χ^2_{n-1} distribution). In this section, we develop decision rules for such estimators.

Let $\mathbf{y} = (y_1, \ldots, y_n)^\top$ be a random sample from a normal distribution $\mathcal{N}(\mu, \sigma^2)$ with both the expectation μ and the (positive) variance σ^2 unknown. The established unbiased estimator of σ^2 is

$$
\hat{\sigma}^2 \;=\; \frac{1}{n-1} \sum_{i=1}^{n} (y_i - \bar{y})^2
$$

$$
\;=\; \frac{1}{n-1} \, \mathbf{y}^\top \mathbf{J}_n \mathbf{y}, \tag{4.5}
$$

using the notation introduced following equation (4.1). The sampling distribution of $\hat{\sigma}^2$ is such that $(n-1)\hat{\sigma}^2/\sigma^2$ has χ^2_{n-1} distribution. The density of this distribution, f_k, is given by (2.3) in Section 2.5.1, with $n-1$ substituted for k. Denote by F_k the corresponding distribution function. Exercise 4.7 explores some alternative estimators of σ^2.

We address the problem of deciding whether σ^2 is smaller or greater than a given positive value σ_0^2. No generality is lost by setting $\sigma_0^2 = 1$, because we can work with $X_1/\sigma_0, \ldots, X_n/\sigma_0$, which is a random sample from $\mathcal{N}(\mu/\sigma_0, \sigma^2/\sigma_0^2)$.

The fiducial distribution of $\hat{\sigma}^2$ is derived by inverting the sampling-distribution identity

$$
k \, \frac{\hat{\sigma}^2}{\sigma^2} \;\sim\; \chi^2_k \,.
$$

We have the fiducial identity

$$
\sigma^2 \;=\; \frac{k\hat{\sigma}^2}{Y} \,,
$$

in which Y is a random variable with χ^2_k distribution. That is, the fiducial distribution of σ^2 is scaled inverse χ^2_k. The density of this distribution is

$$
\frac{k\hat{\sigma}^2}{x^2} \, f_k\!\left(\frac{k\hat{\sigma}^2}{x} \right) . \tag{4.6}
$$

We refer to the state $\Theta_S = (0, \sigma_0^2)$ for variance σ^2 as 'small' (S) and to

$\Theta_L = (\sigma_0^2, +\infty)$ as 'large' (L). Suppose one lossile is incurred for inappropriate verdict 'small' (V_s), and the loss for inappropriate verdict 'large' (V_l) is $R > 0$.

Denote $\rho_0 = (n-1)\hat\sigma^2/\sigma_0^2$. The expected loss for verdict V_s

$$Q_s = \int_{\sigma_0^2}^{+\infty} \frac{(n-1)\hat\sigma^2}{x^2} f_{n-1}\left\{ \frac{(n-1)\hat\sigma^2}{x} \right\} dx$$

$$= \int_0^{\rho_0} f_{n-1}(y)\, dy = F_{n-1}(\rho_0)\,.$$

The expected loss for verdict V_l is derived similarly;

$$Q_l = R \int_0^{\sigma_0^2} \frac{(n-1)\hat\sigma^2}{x^2} f_{n-1}\left\{ \frac{(n-1)\hat\sigma^2}{x} \right\} dx$$

$$= R\{1 - F_{n-1}(\rho_0)\}\,.$$

The equilibrium, where $Q_s(\hat\sigma^2, R) = Q_l(\hat\sigma^2; R)$, is at

$$E = \frac{\sigma_0^2}{n-1} F_{n-1}^{-1}\left(\frac{R}{R+1} \right)\,.$$

The equilibrium function $E(R)$ is plotted in Figure 4.3 for a selection of degrees of freedom. The functions are increasing—for greater R we are more averse to calling verdict V_l, and so the equilibrium is higher. On the log scale for R, the functions have only slight curvature for $R > 1$. The curves converge to zero as $R \to 0$ for all n, so they are bound to be nonlinear.

We can reuse the argument in equation (3.4), Example 2, to explain why the equilibrium function is nearly linear in $\log(R)$, especially for many degrees of freedom $k = n - 1$. The χ_k^2 distribution with large k is well approximated by the normal distribution $\mathcal{N}(k, 2k)$, so E can be approximated by a scalar multiple of the composition of the probit and inverse logit of R. The probit and logit functions are nearly linearly related in a wide range of values centred on zero, so this composition is close to a linear function.

The equilibrium functions in Figure 4.3 intersect at a point R close to 1.0 and at a value E close to 1.0. That is, the equilibrium for $R = 1$ is close to σ_0^2 for any degree of freedom k. The deviation from unity is due to the asymmetry of the χ_k^2 distribution, especially for small k. The χ_{100}^2 distribution, for $n = 101$, is very close to symmetry (and normality), and the equilibrium for it at $R = 1.0$ is 0.9933, very close to unity. In contrast, $E = 0.9342$ for $n = 11$.

In the evaluations with linear and quadratic loss functions, we encounter integrands of the forms $x\, f_k(x)$ and $x^2\, f_k(x)$. It is advantageous to relate them to other χ^2 densities. First, we have

$$x f_k(x) = \frac{1}{\Gamma_2(k)} \left(\frac{x}{2}\right)^{\frac{k}{2}} \exp\left(-\frac{x}{2}\right)$$

$$= \frac{2\Gamma_2(k+2)}{\Gamma_2(k)} f_{k+2}(x) = k f_{k+2}(x)\,. \tag{4.7}$$

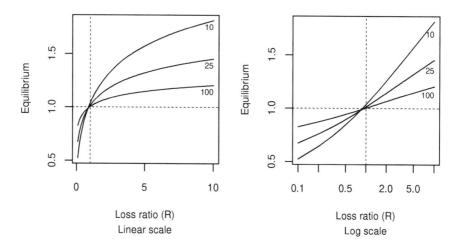

FIGURE 4.3
Equilibrium functions for verdict on the variance σ^2 as to whether it is smaller or greater than $\sigma_0^2 = 1$, on the linear and log scales of the loss ratio R; piecewise constant loss. The numbers of degrees of freedom, $k = 10$, 25, and 100, are marked at the right-hand margin of each panel.

(Recall that $\Gamma_2(x)$ is defined as $\Gamma(\frac{1}{2}x)$.) By reusing these steps, we obtain the identity
$$x^2 f_k(x) = k(k+2)f_{k+4}(x).$$
These two identities imply that
$$\frac{1}{x} f_k(x) = \frac{1}{k-2} f_{k-2}(x)$$
$$\frac{1}{x^2} f_k(x) = \frac{1}{(k-2)(k-4)} f_{k-4}(x), \qquad (4.8)$$
so long as $k > 2$ and $k > 4$, respectively.
    More generally, we have the identity
$$x^m f_k(x) = 2^m \frac{\Gamma_2(k+2m)}{\Gamma_2(k)} f_{k+2m}(x) \qquad (4.9)$$
for $m > -k/2$. It is derived by absorbing the factor x^m in the expression for the χ^2 density, and compensating for it by the appropriate Γ_2 factor.

4.2.1   Linear loss for variances

Suppose the loss with the verdict of 'small' variance, with the threshold σ_0^2, is proportional to the magnitude of the error; $L_{\mathrm{sL}} = S_\mathrm{L} e$, where $e = \sigma^2 - \sigma_0^2$

if $\sigma^2 > \sigma_0^2$ and $e = 0$ otherwise. Suppose also that the number of observations is $n \geq 4$. The corresponding expected loss is

$$
\begin{aligned}
Q_s &= S_L \int_{\sigma_0^2}^{+\infty} (x - \sigma_0^2) \frac{(n-1)\hat{\sigma}^2}{x^2} f_{n-1}\left\{ \frac{(n-1)\hat{\sigma}^2}{x} \right\} dx \\
&= S_L \int_0^{\rho_0} \left\{ \frac{(n-1)\hat{\sigma}^2}{y} - \sigma_0^2 \right\} f_{n-1}(y)\, dy \\
&= S_L \sigma_0^2 \left\{ \frac{\rho_0}{n-3} F_{n-3}(\rho_0) - F_{n-1}(\rho_0) \right\}.
\end{aligned}
\tag{4.10}
$$

The second line in this string of identities is obtained by the transformation $y = (n-1)\hat{\sigma}^2/x$, which simplifies the argument of the density f_{n-1}, and the third line is obtained by applying the first identity in equation (4.8).

Suppose the loss with the verdict of 'large' variance is also proportional to the magnitude of the error; $L_{IS} = -S_S\, e$ if $\sigma^2 < \sigma_0^2$ and $L_{IS} = 0$ otherwise. The expected loss is

$$
\begin{aligned}
Q_l &= S_S \int_0^{\sigma_0^2} (\sigma_0^2 - x) \frac{(n-1)\hat{\sigma}^2}{x^2} f_{n-1}\left\{ \frac{(n-1)\hat{\sigma}^2}{x} \right\} dx \\
&= S_S \int_{\rho_0}^{+\infty} \left\{ \sigma_0^2 - \frac{(n-1)\hat{\sigma}^2}{y} \right\} f_{n-1}(y)\, dy \\
&= S_S \sigma_0^2 \left\{ 1 - F_{n-1}(\rho_0) - \frac{\rho_0}{n-3} + \frac{\rho_0}{n-3} F_{n-3}(\rho_0) \right\} \\
&= S_S \sigma_0^2 \left(1 - \frac{\rho_0}{n-3} \right) + \frac{S_S}{S_L} Q_s.
\end{aligned}
$$

The second and third expressions are obtained by the same steps as their counterparts for Q_s in equation (4.10), and the concluding expression exploits the similarity of the loss structures L_{sL} and L_{IS}. In fact, the scalar multiple of Q_l could have been extracted from the defining identity for Q_s, since

$$
\begin{aligned}
\frac{Q_s}{S_L} - \frac{Q_l}{S_S} &= \int_0^{+\infty} \left\{ \frac{(n-1)\hat{\sigma}^2}{y} - \sigma_0^2 \right\} f_{n-1}(y) dy \\
&= \sigma_0^2 \int_0^{+\infty} \left\{ \frac{\rho_0}{n-3} f_{n-3}(y) - f_{n-1}(y) \right\} dy \\
&= \sigma_0^2 \left(\frac{\rho_0}{n-3} - 1 \right).
\end{aligned}
$$

The expected losses with linear loss functions, such as $L_{sL} = R_L + S_L(\sigma^2 - \sigma_0^2)$ when $\sigma^2 > \sigma_0^2$, are obtained by combining the expressions for Q_L based on constant loss and proportional loss, exploiting the additive nature of the loss and expectation.

4.2.2   Verdicts for standard deviations

Some calculations are easier to conduct for variances, especially with the assumption of normality of the original data. The connection with the χ^2 distribution is particularly convenient. Also, the calculus of variances is simple: the operations of variance and summation can be interchanged for a set of mutually independent random variables—the variance is additive for such variables. However, the variance is on a different scale than the original data. For example, if observations are recorded in grams, then their variance is in grams-squared. This provides an incentive to work with standard deviation. In this section, we discuss making decisions about standard deviations. We use the same estimator $\hat{\sigma}^2$ which has a scaled inverse χ^2 distribution with $n-1$ degrees of freedom.

We consider the decision problem for an unknown standard deviation σ, whether it is small, $\sigma < \sigma_0$, or large, $\sigma > \sigma_0$. With piecewise constant loss, this problem is equivalent to deciding whether σ^2 is smaller or greater than σ_0^2. In contrast, piecewise linear loss for σ differs substantially from piecewise linear loss for σ^2. We confirm this analytically.

Suppose the loss with the inappropriate verdict V_s, when $\sigma^2 > \sigma_0^2$, is proportional to the error $\hat{\sigma} - \sigma_0$ on the scale of σ; $L_{sL} = S_L (\sigma - \sigma_0)$ when $\sigma > \sigma_0$. Then the expected loss is

$$
\begin{aligned}
Q_s &= S_L \int_{\sigma_0^2}^{+\infty} \left(\sqrt{x} - \sigma_0 \right) \frac{(n-1)\hat{\sigma}^2}{x^2} f_{n-1}\left\{ \frac{(n-1)\hat{\sigma}^2}{x} \right\} dx \\
&= S_L \int_0^{\rho_0} \left(\hat{\sigma}\sqrt{\frac{n-1}{y}} - \sigma_0 \right) f_{n-1}(y)\, dy .
\end{aligned}
$$

By relating $f_{n-1}(y)/\sqrt{y}$ to the density of χ^2_{n-2} distribution, according to the identity in equation (4.9), we obtain the expected loss

$$
Q_s = S_L \sigma_0 \left\{ \sqrt{\frac{\rho_0}{2}} \frac{\Gamma_2(n-2)}{\Gamma_2(n-1)} F_{n-2}(\rho_0) - F_{n-1}(\rho_0) \right\} .
$$

With the loss $L_{lS} = S_S(\sigma_0 - \sigma)$, when $\sigma < \sigma_0$ but we issue verdict V_1, the expected loss is

$$
Q_l = S_S \sigma_0 \left\{ 1 - \sqrt{\frac{\rho_0}{2}} \frac{\Gamma_2(n-2)}{\Gamma_2(n-1)} \right\} + \frac{S_S}{S_L} Q_s .
$$

No generality is lost by setting $S_L = 1$.

A variance is already on a squared scale, so it would seem not to be appropriate to use quadratic loss for a variance. However, quadratic loss for a standard deviation may be meaningful. In the derivations that follow we show that this loss is different from the linear loss for an error related to a variance.

Suppose the loss functions are $L_{\mathrm{sL}} = (\sigma - \sigma_0)^2$ when we issue verdict V_{s} but $\sigma > \sigma_0$ and $L_{\mathrm{IS}} = S_{\mathrm{S}}(\sigma_0 - \sigma)^2$ when we issue verdict V_{I} even though $\sigma < \sigma_0$. The expected loss with verdict V_{s} is

$$
\begin{aligned}
Q_{\mathrm{s}} &= \int_{\sigma_0^2}^{+\infty} \left(\sqrt{x} - \sigma_0\right)^2 \frac{(n-1)\hat{\sigma}^2}{x^2} f_{n-1}\left\{\frac{(n-1)\hat{\sigma}^2}{x}\right\} dx \\
&= \sigma_0^2 \int_0^{\rho_0} \left(\sqrt{\frac{\rho_0}{y}} - 1\right)^2 f_{n-1}(y)\, dy \\
&= \sigma_0^2 \left\{\rho_0 \int_0^{\rho_0} \frac{1}{y} f_{n-1}(y)\, dy - 2\sqrt{\rho_0} \int_0^{\rho_0} \frac{1}{\sqrt{y}} f_{n-1}(y)\, dy \right. \\
&\qquad \left. + \int_0^{\rho_0} f_{n-1}(y)\, dy\right\} \\
&= \sigma_0^2 \left\{\frac{\rho_0}{n-3} F_{n-3}(\rho_0) - \sqrt{2\rho_0}\, \frac{\Gamma_2(n-2)}{\Gamma_2(n-1)} F_{n-2}(\rho_0) + F_{n-1}(\rho_0)\right\},
\end{aligned}
$$

obtained by the transformation that simplifies the argument of the density f_{n-1}, and applying equation (4.9) with $m = \frac{1}{2}$ and $m = 1$.

For the expected loss with verdict V_{I} with the squared proportional loss $L_{\mathrm{IS}} = S_{\mathrm{S}}(\sigma - \sigma_0)^2 I_{\sigma < \sigma_0}$, we obtain the expression

$$
Q_{\mathrm{I}} = S_{\mathrm{S}} \sigma_0^2 \left\{1 + \frac{\rho_0}{n-3} - \sqrt{2\rho_0}\, \frac{\Gamma_2(n-2)}{\Gamma_2(n-1)}\right\} - S_{\mathrm{S}} Q_{\mathrm{I}}.
$$

4.3 Comparing two variances

Another common task is to compare two variances. It is more natural to study their ratio σ_1^2/σ_2^2 than the difference $\sigma_2^2 - \sigma_1^2$. Suppose we have unbiased estimators $\hat{\sigma}_1^2$ and $\hat{\sigma}_2^2$ of respective variances σ_1^2 and σ_2^2, based on independent normally distributed random samples of respective sizes n_1 and n_2, and the two estimators have scaled χ^2 sampling distributions with respective degrees of freedom $k_1 = n_1 - 1$ and $k_2 = n_2 - 1$. Note that the ratio $\hat{\sigma}_1^2/\hat{\sigma}_2^2$ is not an unbiased estimator of σ_1^2/σ_2^2. The sampling distribution of the ratio $\hat{\sigma}_1^2/\hat{\sigma}_2^2$ is related to the F distribution with k_1 and k_2 degrees of freedom:

$$
\frac{\hat{\sigma}_1^2}{\sigma_1^2} \frac{\sigma_2^2}{\hat{\sigma}_2^2} \sim F_{n_1 - 1,\, n_2 - 1}.
$$

By inverting this sampling-distribution identity we obtain the fiducial identity for the ratio $r = \sigma_1^2/\sigma_2^2$:

$$
\frac{\sigma_1^2}{\sigma_2^2} = \frac{\hat{\sigma}_1^2}{\hat{\sigma}_2^2} Z,
$$

where Z is a random variable with F distribution with n_2-1 and n_1-1 degrees of freedom ($F_{n_2-1,\,n_1-1}$). Note that the degrees of freedom are interchanged from the sampling distribution of $\hat{\sigma}_1^2/\hat{\sigma}_2^2$. In the derivation of the fiducial distribution we used the following property. If X has F distribution with k_1 and k_2 degrees of freedom, then $1/X$ has F distribution with k_2 and k_1 degrees of freedom. This is obvious from the definition of the F distribution by the ratio of two independent scaled χ^2-distributed random variables.

The density of the $F_{k_2,\,k_1}$ distribution is

$$f(x; k_2, k_1) = \frac{1}{B_2(k_1, k_2)} \left(\frac{k_2}{k_1}\right)^{\frac{1}{2}k_2} x^{\frac{1}{2}k_2-1} \left(1 + \frac{k_2}{k_1}x\right)^{-\frac{1}{2}(k_1+k_2)}$$

($x > 0$), where B_2 is the half-beta function, defined as

$$B_2(k_1, k_2) = \frac{\Gamma_2(k_1)\,\Gamma_2(k_2)}{\Gamma_2(k_1+k_2)},$$

that is, the beta function, with both its arguments halved. The expectation of this distribution is $k_1/(k_1 - 2)$, so long as $k_1 > 2$. Its unique mode is at

$$\kappa = \frac{k_1}{k_2}\frac{k_2 - 2}{k_1 + 2}$$

when $k_2 > 2$. For $k_2 \leq 2$, the density is a decreasing function throughout $(0, +\infty)$, so the mode is at $\kappa = 0$. These results are derived by differentiating $\log\{f(x; k_1, k_2)\}$.

As with χ^2 distributions, it is useful to relate the product $xf(x; k_1, k_2)$ to another F density. Assume that $k_2 > 2$, as it would be in any realistic setting. Since the exponent of x in the expression for $xf(x; k_1, k_2)$ is raised from $\frac{1}{2}k_1-1$ to $\frac{1}{2}k_1$, corresponding to increasing k_1 to k_1+2, we compensate for it in the exponent of $(1+k_1x/k_2)$ by reducing k_2 to k_2-2. That precipitates further changes that enable us to extract the function $f(\kappa x; k_1 + 2, k_2 - 2)$:

$$
\begin{aligned}
x\,f(x; k_1, k_2) &= \frac{1}{B_2(k_1+2, k_2-2)} \left(\frac{k_1+2}{k_2-2}\right)^{\frac{1}{2}k_1+1} (\kappa x)^{\frac{1}{2}k_1} \\
&\quad \times \left(1 + \frac{k_1+2}{k_2-2}\kappa x\right)^{\frac{1}{2}(k_1+k_2)} \frac{\Gamma_2(k_1+2)\,\Gamma_2(k_2-2)}{\Gamma_2(k_1)\,\Gamma_2(k_2)} \frac{k_2-2}{k_1+2} \\
&= \frac{k_1}{k_1+2}\,f(\kappa x; k_1 + 2, k_2 - 2)\,.
\end{aligned}
$$

In the first two lines we recognise the density of an F distribution evaluated at κx (the first four factors). The last two (constant) factors reduce to $k_1/(k_1+2)$.

By re-using this result, assuming that $k_2 > 4$, we obtain the identity

$$
\begin{aligned}
x^2\,f(x; k_1, k_2) &= \frac{k_1}{k_1+2}\frac{k_1+2}{k_1+4}\frac{1}{\kappa}\,f(\kappa\,\kappa_2\,x; k_1 + 4, k_2 - 4) \\
&= \frac{(k_1+2)\,k_2}{(k_1+4)\,(k_2-2)}\,f\left(\frac{k_1}{k_2}\frac{k_2-4}{k_1+4}x; k_1 + 4, k_2 - 4\right),
\end{aligned}
$$

where

$$\kappa_2 = \frac{k_2 - 4}{k_1 + 4} \frac{k_1 + 2}{k_2 - 2}.$$

Example 7

Suppose our task is to decide whether two variances, σ_1^2 and σ_2^2, are similar. Denote their ratio by r; $r = \sigma_1^2/\sigma_2^2$. We define similarity as $T < r < 1/T$, so that, in the notation introduced earlier, the thresholds are $T_1 = T$ and $T_2 = 1/T$. We set T to 0.90.

Suppose the losses are piecewise constant. If we issue verdict 'different' (V_d) inappropriately, the loss is one lossile; $L_{dS} = 1$. If we issue verdict 'similar' (V_s) inappropriately, the loss is $L_{sD} = R$. Then the expected losses are

$$
\begin{aligned}
Q_d &= \frac{1}{\hat{r}} \int_T^{1/T} f\left(\frac{x}{\hat{r}}; k_2, k_1\right) dx \\
&= F_{k_2, k_1}(z_2) - F_{k_2, k_1}(z_1) \\
Q_s &= \frac{R}{\hat{r}} \left\{ \int_0^T f\left(\frac{x}{\hat{r}}; k_2, k_1\right) dx + \int_{1/T}^{+\infty} f\left(\frac{x}{\hat{r}}; k_2, k_1\right) dx \right\} \\
&= R(1 - Q_d),
\end{aligned}
$$

where $z_1 = T/\hat{r}$ and $z_2 = 1/(\hat{r}T)$.

The balance equation, $Q_d - Q_s = 0$, is

$$B = (R+1)Q_d - R.$$

We issue verdict V_d if $B < 0$, and verdict V_s otherwise. The expected loss Q_d is equal to the probability that an F distribution falls in the range (z_1, z_2). If this range is sufficiently narrow and R is not very small, then $Q_d < R/(R+1)$ for all values of $\hat{r}$. In that case, verdict V_d is issued for all values of $\hat{r}$; we do not have to evaluate $\hat{r}$. This setting can be interpreted as a very high standard for similarity (T close to unity), combined with being averse to false similarity (R not very small). Both are incentives for verdict V_d. In particular, if state S degenerates to $r = 1$, then D is a safe bet because the fiducial distribution is continuous, and so the fiducial probability of $r = 1$ vanishes. This is a replay of the argument against testing the corresponding null hypothesis in Section 2.8.

The balance function $B(\hat{r})$ increases up to its unique maximum and then it decreases; this follows from unimodality of the F_{k_2, k_1} distribution. Therefore, $B(\hat{r})$ has two roots when the maximum it attains exceeds $R/(R+1)$, one root in the anomalous case when the maximum is equal to $R/(R+1)$, and no root otherwise. The roots, if there are any, can be found by the Newton-Raphson algorithm, taking advantage of a simple expression for the differential,

$$\frac{\partial Q_d}{\partial \hat{r}} = \frac{1}{\hat{r}^2} \left\{ Tf(z_1; k_2, k_1) - \frac{1}{T} f(z_2; k_2, k_1) \right\}.$$

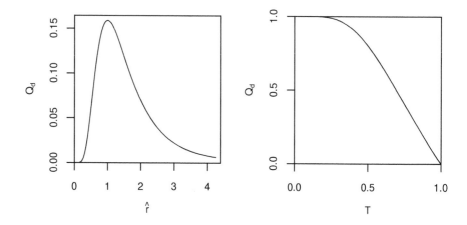

FIGURE 4.4
Expected loss Q_d as a function of $\hat{r}$ for $T = 0.9$ (left-hand panel) and of T for $\hat{r} = 1.127$ (right-hand panel); $k_1 = 12$ and $k_2 = 20$.

An example is given in Figure 4.4 based on $k_1 = 12$ and $k_2 = 20$. In the left-hand panel, the expected loss Q_d is plotted as a function of $\hat{r}$ for $T = 0.90$. In the right-hand panel, Q_d is plotted as a function of T for $\hat{r} = 1.127$, an arbitrarily selected value. The maximum of $Q_\mathrm{d}(\hat{r})$ is 0.159, attained at $\hat{r}^* \doteq 1$. For $R > Q_\mathrm{d}(\hat{r}^*)/\{1 - Q_\mathrm{d}(\hat{r}^*)\} = 0.189$, the balance B would be negative even if $\hat{r} = \hat{r}^*$, so we would conclude with verdict V_d for any value of $\hat{r}$.

The right-hand panel of Figure 4.4 confirms that $Q_\mathrm{d}(T)$ is indeed a decreasing function of T, with its respective limits of unity and zero at $T = 0$ and $T = 1$. Thus, values of T close to unity, when $(T, 1/T)$ is a narrow interval, make V_d an attractive verdict, whereas V_s is attractive for T close to zero, when $(T, 1/T)$ is very wide. Note that $B(T)$ is very close to linearity for $T \in (0.6, 1.0)$.

The borderline function $R^*(T)$ is defined as the loss ratio R for which the balance function $B(T)$ has a single root. With T fixed, there are two equilibria for $R < R^*(T)$ and none for $R > R^*(T)$. We construct the borderline function from the maxima of the expected losses Q_d regarded as a function of $\hat{r}$. These maxima are bound to be close to $\hat{r} = 1$, the value that might be expected to support the hypothesis that $r = 1$ most strongly. In fact, owing to asymmetry of the F distribution, the maxima occur for $\hat{r}$ in the left-hand vicinity of 1.0. For example, for $T = 0.80$, the largest value of Q_d, equal to 0.3287, occurs for $\hat{r} = 0.9979$. The maxima of Q_d are found by the Newton method.

The maxima as a function of T, denoted as $Q_\mathrm{d,max}(T)$, are plotted in the left-hand panel of Figure 4.5 for $k_1 = 12$ and $k_2 = 20$ degrees of freedom. The function is very close to linearity, with $Q_\mathrm{d,max} \to 0$ as $T \to 1$ and $Q_\mathrm{d,max}(0.7) \doteq 0.5$.

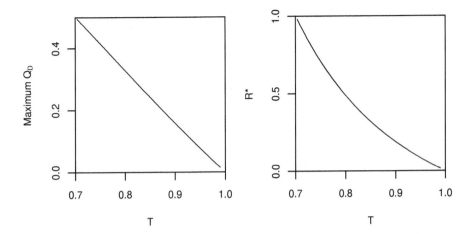

FIGURE 4.5
The largest expected loss $Q_{d,max}(T)$ and the borderline function $R^*(T)$ for scaled F distribution with $k_2 = 20$ and $k_1 = 12$ degrees of freedom.

The right-hand panel displays the function $R^*(T)$ where $R^* = Q_{d,max}/(1 - Q_{d,max})$ is the lower bound of the loss ratios for which verdict V_d is issued for all values of $\hat{r}$. We have $R^* \doteq 1$ for $T = 0.7$, where $Q_{d,max} \doteq 0.5$. The function $R^*(T)$ is decreasing and converges to zero as $T \to 1$. That is, for T close to unity, verdict Q_s is plausible only for very small values of R, which correspond to extreme aversion to inappropriate verdict of dissimilarity. □

Proportional, linear or quadratic loss functions are not well suited for decisions about a variance ratio because such losses are limited for small ratios r and not limited for large ratios. Linear loss for log-ratio is better suited because it introduces a symmetry; r is in the vicinity of $-\infty$ for small log-ratios $\log(r)$, and in the vicinity of $+\infty$ for large log-ratios. However, there are no closed-form expressions for the corresponding expected losses. Such an expression is not essential for solving a single problem because the requisite integrals can be evaluated numerically. For a detailed exploration of the properties of balance function it may introduce some computational complexity.

A transformation with a tractable solution is $h(x) = \max(x, 1/x)$. See Appendix for details.

4.4 Statistics with binomial and Poisson distributions

The binomial distribution is derived as the number of successes in a sequence of a set number of replications of an experiment with a binary outcome (success/failure, 0/1, or Y/N). The random variable is denoted by X, the number of replications (trials) by n, and the probability of success by p. The probabilities of the distribution are given by the binomial formula

$$P(X = k) = \binom{n}{k} p^k (1-p)^{n-k}.$$

Estimation of p is a common statistical task. The estimator $\hat{p} = X/n$, with expectation p (i.e., with no bias) and variance $p(1-p)/n$, has no worthy competitor in the frequentist paradigm. In a Bayesian analysis, a prior is specified for p. When this specification entails no profound enterprise, a beta distributed prior, $\mathcal{B}(\alpha, \beta)$, is selected, to draw on some analytical advantage discussed below. The uniform distribution, with $\alpha = \beta = 1$, has an appeal as being noninformative, although we contradict it below.

The density of the beta distribution is

$$g(p; \alpha, \beta) = \frac{1}{B(\alpha, \beta)} p^{\alpha-1} (1-p)^{\beta-1},$$

defined for $p \in (0, 1)$; the beta function B is defined as

$$B(\alpha, \beta) = \frac{\Gamma(\alpha)\,\Gamma(\beta)}{\Gamma(\alpha + \beta)}.$$

The distribution function of the beta distribution does not have a closed form, but in most statistical packages, including R, it is approximated with very high precision by the first few terms of an expansion.

The beta distribution is conjugate for the binomial. Conjugacy is defined as the condition that the prior and posterior distributions belong to the same class, in this case the class of beta distributions. Indeed, the posterior density of p based on the prior distribution $\mathcal{B}(\alpha, \beta)$ is

$$\frac{1}{C} p^k (1-p)^{n-k} p^{\alpha-1} (1-p)^{\beta-1} = \frac{1}{C} p^{k+\alpha-1} (1-p)^{n-k+\beta-1},$$

where the constant C collects all the factors that do not depend on p. The result has to be a density, so $C = B(k + \alpha, n - k + \beta)$, and this density is of the beta distribution, $\mathcal{B}(X + \alpha, n - X + \beta)$.

So, the prior parameters α and β can be interpreted as additional (prior, or extra-data) numbers of successes and failures. In other words, the effect of the prior is to boost the numbers of successes and failures, by α and β, respectively. Therefore, we do not have to elicit from a client or expert a beta distribution that would adequately describe the information the client has prior to conducting the study. A typical client (or any non-statistician) would be more comfortable with the request:

> Please express what you know about this probability in terms
> of numbers of successes and failures in a real or imagined study.

We refer to such information as the *basis* of the prior. The analysis that follows proceeds by adding these counts to the numbers of successes and failures in the study proper. The uniform distribution is $\mathcal{B}(1,1)$. Its density is constant in the range $(0,1)$, so it would seem to be an obvious default when we have no prior information about p, or do not want to make use of it. Is it appropriate to represent such ignorance by adding one success and one failure to the data? Well, maybe, sometimes.

We do not have to insist of the prior counts, α and β, to be integers. For example, $\alpha = 0.1$ and $\beta = 0.1$ would clearly amount to less prior information than $\alpha = 1$ and $\beta = 1$, and the absence of any prior information is best captured by setting $\alpha = 0$ and $\beta = 0$. It does not correspond to a proper (prior) distribution, but we can work with the 'posterior' distribution $\mathcal{B}(X, n - X)$. This distribution or, more generally, $\mathcal{B}(X + \alpha, n - X + \beta)$ is adopted as the fiducial distribution for p. The corresponding density is $g(p; X + \alpha, n - X + \beta)$ and the distribution function is denoted as $G(p; X + \alpha, n - X + \beta)$.

Suppose we have a threshold probability T_1, and of interest is whether $p < T_1$ or $p > T_1$, because for 'small' p one course of action, denoted by S, and for 'large' p another course of action, L, would be appropriate to take. Suppose the losses are piecewise constant, not dependent on the magnitude of the error, and the loss ratio, $R = L_{sL}/L_{lS}$, is a constant. The expected losses are

$$
\begin{aligned}
Q_s &= R \int_{T_1}^1 g(p; X + \alpha, n - X + \beta)\, dp \\
&= R\{1 - G(T_1; X + \alpha, n - X + \beta)\} \\
Q_l &= G(T_1; X + \alpha, n - X + \beta).
\end{aligned}
$$

Hence the balance equation

$$
G(T_1; X + \alpha, n - X + \beta) = \frac{R}{R+1}, \tag{4.11}
$$

to be solved for X. When $\alpha = \beta = 0$, a solution exists because the left-hand side converges to unity as $X \to 0$ and to zero as $X \to n$. An iterative method could be applied to find the equilibrium. A practical alternative is to plot the left-hand side of equation (4.11), as done in Figure 4.6 for a set of thresholds T_1 indicated in the diagram, and $R = 9$, $\alpha = 5$, $\beta = 12$ and $n = 50$. The equilibria are $E = 0.151, 0.212, 0.273, 0.337, 0.401$ and 0.466 for the respective thresholds $T_1 = 0.25, 0.30, \ldots, 0.50$. The equilibria are smaller than the corresponding thresholds T_1 because $R = 9$ represents considerable aversion to false negatives. However, the equilibria do not match T_1 for $R = 1$ because the beta distributions involved are not symmetric. For example, for $T_1 = 0.35$ the equilibrium is at 0.371.

The beta distribution $\mathcal{B}(\alpha, \beta)$ converges to a normal as $\alpha \to +\infty$ and $\beta \to +\infty$, and α and β diverge in such a way that α/β converges to a finite

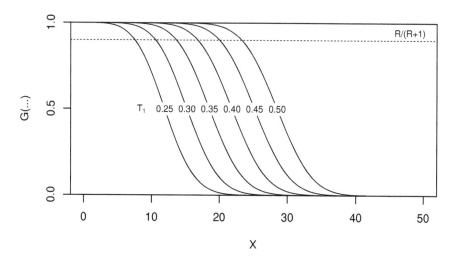

FIGURE 4.6
Illustration of the equilibrium equation (4.11).

positive value. Therefore the evaluations with beta distribution are relevant only for relatively small counts X and small numbers of trials n, especially when the probability p is likely to be distant from both zero and unity.

As with other fiducial or posterior distributions, we find an expression for $p\,g(p; \alpha, \beta)$ in terms of another beta density.

$$
\begin{aligned}
p\,g(p; \alpha, \beta) &= \frac{1}{B(\alpha+1, \beta)}\, p^\alpha\,(1-p)^{\beta-1}\frac{\Gamma(\alpha+1)}{\Gamma(\alpha+\beta+1)}\frac{\Gamma(\alpha+\beta)}{\Gamma(\alpha)} \\
&= \frac{\alpha}{\alpha+\beta}\, g(p; \alpha+1, \beta)\,. \tag{4.12}
\end{aligned}
$$

Further,

$$
\begin{aligned}
p^2\,g(p; \alpha, \beta) &= \frac{\alpha}{\alpha+\beta}\, p\,g(p; \alpha+1, \beta) \\
&= \frac{\alpha(\alpha+1)}{(\alpha+\beta)(\alpha+\beta+1)}\, g(p; \alpha+2, \beta)\,.
\end{aligned}
$$

In some problems, the probability of one-half, or 50%, is the focal value, and the purpose of the study is to arbitrate as to whether a probability is or is not close to $\frac{1}{2}$. We address this problem by defining a range around $\frac{1}{2}$ that corresponds to similarity, or symmetry, such as $(\frac{1}{2}-c, \frac{1}{2}+c)$ for a small positive c. A range symmetric around zero on the log-odds scale is equivalent to a range symmetric around 0.5 on the probability scale. For linear and quadratic losses such equivalence holds only approximately when the probabilities concerned are not too distant from $\frac{1}{2}$, where the logit function has very little curvature.

Suppose proximity to $\frac{1}{2}$ is defined by the interval $(\frac{1}{2} - c, \frac{1}{2} + c)$ and the loss for verdict V_a of asymmetry, $p \notin (\frac{1}{2} - c, \frac{1}{2} + c)$, when it is not appropriate, is one lossile, $L_{aS} = 1$, while the loss for the inappropriate verdict V_s of symmetry is $L_{sA} = R|p - \frac{1}{2}|$. Then the expected loss with V_a is

$$Q_a = G\left(\frac{1}{2} + c; X + \alpha, n - X + \beta\right) - G\left(\frac{1}{2} - c; X + \alpha, n - X + \beta\right),$$

that is, the fiducial probability of $p \in \Theta_S$.

The expected loss with V_s has two components, one for $p < \frac{1}{2} - c$ and the other for $p > \frac{1}{2} + c$. The former is

$$Q_{s-} = R\int_0^{\frac{1}{2} - c} \left(\frac{1}{2} - p\right) \frac{p^{X + \alpha - 1}(1 - p)^{n - X + \beta - 1}}{B(X + \alpha, n - X + \beta)}\, dp$$

$$= \frac{R}{2} G\left(\frac{1}{2} - c; X + \alpha, n - X + \beta\right)$$

$$- R\frac{X + \alpha}{n + \alpha + \beta} G\left(\frac{1}{2} - c; X + \alpha + 1, n - X + \beta\right),$$

obtained by applying equation (4.12). The second component is obtained by similar steps;

$$Q_{s+} = R\int_{\frac{1}{2} + c}^1 \left(p - \frac{1}{2}\right) \frac{p^{X + \alpha - 1}(1 - p)^{n - X + \beta - 1}}{B(X + \alpha, n - X + \beta)}\, dp$$

$$= R\frac{X + \alpha}{n + \alpha + \beta}\left\{1 - G\left(\frac{1}{2} + c; X + \alpha + 1; n - X + \beta\right)\right\}$$

$$- \frac{R}{2}\left\{1 - G\left(\frac{1}{2} + c; X + \alpha; n - X + \beta\right)\right\}.$$

The balance function, $Q_a - Q_{s+} - Q_{s-}$, is

$$\left(1 - \frac{R}{2}\right) G\left(\frac{1}{2} + c; X + \alpha, n - X + \beta\right)$$

$$- \left(1 + \frac{R}{2}\right) G\left(\frac{1}{2} - c; X + \alpha, n - X + \beta\right)$$

$$+ \frac{R(X + \alpha)}{n + \alpha + \beta}\left\{G\left(\frac{1}{2} + c; X + \alpha + 1, n - X + \beta\right)\right.$$

$$\left. + G\left(\frac{1}{2} - c; X + \alpha + 1, n - X + \beta\right)\right\} + \frac{R}{2}\frac{n - 2X - \alpha + \beta}{n + \alpha + \beta}.$$

The binomial distribution, suitably scaled, converges to the normal with increasing n quite rapidly when the probability of success, p, is distant from both zero and unity. Therefore the methods described in this section are relevant only for problems with relatively small samples or extreme probabilities. For the latter condition, the Poisson distribution is applicable when the number of trials, n, is sufficiently large.

4.4.1 Poisson distribution

The Poisson distribution is defined by the probabilities

$$P(X = k) = \frac{e^{-\lambda} \lambda^k}{k!},$$

$k = 0, 1, \ldots$. The conjugate prior for the Poisson is the gamma distribution, given by the density

$$h(x; \gamma, \delta) = \frac{\delta^\gamma}{\Gamma(\gamma)} x^{\gamma-1} \exp(-\delta x), \qquad (4.13)$$

where $\gamma > 0$ and $\delta > 0$ are parameters. This distribution has expectation γ/δ and variance γ/δ^2. With a gamma prior, the posterior distribution of λ is also gamma, with parameters $k + \gamma$ and $1 + \delta$. Thus, γ can be interpreted as additional (extra-data) events and δ as an increase in the relative concentration, defined as the ratio of the mean and variance. This motivates the fiducial distribution gamma with parameters k and 1, obtained by letting γ and δ converge to zero. Note that this is different from setting γ and δ to zero, for which the gamma distribution is not defined. For piecewise constant loss, we obtain a balance equation similar to equation (4.11).

Example 8

Let $k_1, k_2, \ldots, k_m$ be a random sample from a Poisson distribution with unknown parameter λ. We would like to establish whether λ is smaller or greater than a given positive constant λ_0. The loss with inappropriate verdict V_s (small) is $L_{sL} = R + S\lambda/\lambda_0$, and the loss with inappropriate verdict V_l (large) is $L_{lS} = 1.0$.

Let $k_+ = k_1 + \cdots + k_m$. The sampling distribution of k_+ is Poisson with expectation and variance both equal to $m\lambda$. As the fiducial distribution of λ we use the gamma distribution with parameters k_+ and unity.

The expected loss for verdict V_s is

$$
\begin{aligned}
Q_s &= \int_{m\lambda_0}^{+\infty} \left(R + S \frac{x}{m\lambda_0} \right) \frac{1}{\Gamma(k_+)} x^{k_+ - 1} e^{-x} dx \\
&= R\{1 - G(m\lambda_0; k_+, 1)\} + \frac{Sk_+}{m\lambda_0} \{1 - G(m\lambda_0; k_+ + 1, 1)\},
\end{aligned}
$$

where G is the distribution function of the gamma distribution. The expected loss for V_1 is

$$
\begin{aligned}
Q_1 &= \int_0^{m\lambda_0} \frac{1}{\Gamma(k_+)} x^{k_+ - 1} e^{-x} \mathrm{d}x \\
&= G(m\lambda_0 ; k_+ , 1) \ .
\end{aligned}
$$

Now we proceed by solving the balance equation $Q_s = Q_1$,

$$
R - (R + 1)\, G(m\lambda_0 ; k_+ , 1) + \frac{Sk_+}{m\lambda_0} \{1 - G(m\lambda_0 ; k_+ + 1, 1)\} = 0 \,,
$$

and verifying that verdict V_s is appropriate when $k_+ < E$ and verdict V_1 is appropriate otherwise. The details are left for the reader to complete. There is no neat analytical solution.

4.5 Further reading

An alternative to the fiducial distribution for the Student t distribution is developed in Longford (2012a). An esoteric problem arises when no successes are recorded in a sequence of independent binary trials. Longford (2009) addresses this problem using the method of Section 4.4. See Tuyl, Gerlach and Mengersen (2008) for an exploration of the impact of prior information in this problem. An application of the method in Section 4.4.1 with Poisson distribution is presented in Longford (2010).

4.6 Exercises

4.1. Is the statement 'A convex combination of loss functions is a loss function' equivalent to 'The total of a set of loss functions is a loss function?' Is the (pointwise) limit of a sequence of loss functions, when it exists, also a loss function?

4.2. Draw in a single plot the densities of the standard normal distribution and a selection of t distributions standardised to have unit variance. Make a judgement about the fewest degrees of freedom for which the t distribution is indistinguishable from the standard normal distribution. Re-draw the plot with the t densities not standardised. Would you have to alter your judgement about the fewest degrees of freedom for which the t distribution cannot be distinguished from the standard normal? Plot as functions of the degrees of freedom a few extreme quantiles, such as 0.95 and 0.99, of the t

distribution, and compare them with the corresponding quantiles of the standard normal, or the quantiles of the centred normal with the matched variances. Review your conclusion about the convergence of the t distributions to the normal.

4.3. Derive (analytically) the expression in equation (4.9) for integer m. Verify graphically the relationship of the chi-squared densities $f_k(x)$ and $f_{k+2}(x)$ in equation (4.7) and of $f_k(x)$ and $f_{k+4}(x)$ in the following equation.

4.4. Use the identities in equations (4.7) and (4.8) to derive the expectation and variance of the χ^2 distribution and of its reciprocal.

4.5. Derive the density of $\sqrt{Y/k}$ in equation (4.2), where $Y \sim \chi_k^2$, and its moments given by equation (4.3). Verify by simulations the expressions for the expectation and variance in equation (4.3).

4.6. Rerun Example 5 with different settings, for example, $\Delta\hat{\theta} = 1.42$, $s^2 = 30.63$ with 35 degrees of freedom and $R \in (0.04, 0.10)$.

4.7. The estimator of the variance of a normally distributed sample, $\hat{\sigma}^2$, is given by equation (4.5). Find the efficient estimator in the class of the estimators $c\hat{\sigma}^2$, where $c > 0$ is a constant. Find the bias of $\sqrt{\hat{\sigma}^2}$ as an estimator of σ, correct for it by estimating σ by $d_1\sqrt{\hat{\sigma}^2}$ for an appropriate constant d_1. Find the constant factor d_2 for which $d_2\sqrt{\hat{\sigma}^2}$ is the efficient estimator of σ in the class of estimators $d\sqrt{\hat{\sigma}^2}$.

4.8. Verify from basic principles the formula for the scaled inverse chi-squared distribution in equation (4.6). Derive the expectations and variances of these distributions, as functions of the degrees of freedom. Explain why they differ from the reciprocals of their counterparts for the scaled chi-squared distributions.

4.9. Check at your own pace the expressions for the expected loss in Section 4.2.1. Verify that they are in accord with the expressions for the expectations of the χ^2 distributions.
Hint: Consider the limits of Q_s and Q_l for $\sigma_0^2 \to 0$ and $\sigma_0^2 \to +\infty$.

4.10. Follow up on the suggestion at the end of Section 4.2.1 to derive the expected loss with a linear loss function. Are balance equations similarly additive? What about equilibria?

4.11. Derive the balance equation for the problem of deciding whether the variance of a normally distributed random sample is smaller or greater than $\sigma_0^2 = 1$ when the losses are $L_{sL} = 1$ and $L_{lS} = R + S(\sigma_0^2 - \sigma^2)$. Compile a programme to find the equilibrium and evaluate it for $\hat{\sigma}^2 = 1.24$, $R = 0.4$ and $S = 0.6$.

4.12. Conduct a sensitivity analysis for the dependence of the equilibrium on the parameters R and S in Exercise 4.11 in the respective ranges $(0.3, 0.5)$ and $(0.4, 0.75)$.

4.13. Verify the expressions for the expected losses Q_s and Q_l for the linear loss for a standard deviation in Section 4.2.2.

4.14. Construct an example on which you would demonstrate that, in expectation, the linear loss for a variance differs from the quadratic loss for the corresponding standard deviation.

4.15. Compare the expected losses in Section 4.2.2, or with a setting of your choice, with the expectations based on an incorrect number of degrees of freedom for $\hat{\sigma}^2$. Formulate a rule for the magnitude of the error that can be forgiven.

4.16. Verify the expressions for $x\, f(x; k_1, k_2)$ and $x^2\, f(x; k_1, k_2)$, where f is the density of the F distribution, by plotting these functions and the expressions derived for them in Section 4.3.

4.17. An established bioassay instrument, used for measurement of the concentration of a particular antigen in a (human) blood sample, is subject to measurement error but the values it yields are not biased. Unbiasedness is ensured by a calibration procedure. The measurement error is well described by a lognormal distribution. A new instrument is tested with the intention to replace the established instrument. Suppose its measurements are also unbiased (arranged by calibration). Discuss how you would go about collecting evidence that the proposed instrument has smaller measurement error (i.e., smaller variance of the measurement error) than the established instrument.

4.18. Present arguments for and against using linear loss functions for log-variance and logarithm of the ratio of two variances of normal random samples. Consider the likely views of both the client and the analyst.

4.19. A procedure that yields a binary outcome (Y/N) is fair if the probability of both outcomes Y and N is equal to $\frac{1}{2}$. Discuss the parameters that would be involved in a study and its analysis to support the claim that the procedure is fair. Describe the hypothesis test approach and the method outlined in Section 4.4.

4.20. For the binomial distribution as the model, compare the limiting posterior distribution of the probability p as the beta-prior parameters α and β converge to zero with the posterior distribution for the limit of the priors. Can the operations of limit and evaluating the posterior distribution be interchanged?
Note: This exercise is equally appropriate for the Poisson distribution with gamma prior.

4.21. Derive the posterior distribution for the Poisson mean λ using the parametrisation with $\nu = 1/\delta$ for the gamma-distributed prior. Comment on which parametrisation you prefer, δ or ν, or is there some other that you would prefer to use?

4.22. Work out the intermediate steps in the derivations of the expected loss in Example 8. Work out the details of solving the balance equation.

4.23. Check in a few instances whether the results for binomial distribution converge to the corresponding results for the Poisson when the rate p and number of trials n converge/diverge appropriately.

Appendix

This appendix derives an expression for the expected loss for a verdict about the variance ratio (Section 4.3) when the losses are $L_{\mathrm{dS}}(r) = h(T) - h(r) = 1/T - h(r)$ when $r \in (T, 1/T)$, and $L_{\mathrm{sD}}(r) = R\{h(r) - h(T)\} = R\{h(r) - 1/T\}$ when $r \notin (T, 1/T)$; $h(r) = \max(r, 1/r)$. In the identities that follow, we write $f_{k_2,k_1}(x)$ instead of $f(x; k_2, k_1)$. Assuming that both $k_1 > 2$ and $k_2 > 2$, the expected losses are

$$
\begin{aligned}
Q_{\mathrm{d}} &= \frac{1}{\hat{r}} \int_T^1 \left(\frac{1}{T} - \frac{1}{x}\right) f_{k_2,k_1}\left(\frac{x}{\hat{r}}\right) \mathrm{d}x + \frac{1}{\hat{r}} \int_1^{1/T} \left(\frac{1}{T} - x\right) f_{k_2,k_1}\left(\frac{x}{\hat{r}}\right) \mathrm{d}x \\
&= \int_{z_1}^{1/\hat{r}} \left(\frac{1}{T} - \frac{1}{\hat{r}y}\right) f_{k_2,k_1}(y)\,\mathrm{d}y + \int_{1/\hat{r}}^{z_2} \left(\frac{1}{T} - \hat{r}y\right) f_{k_2,k_1}(y)\,\mathrm{d}y \\
&= \frac{1}{T} \{F_{k_2,k_1}(z_2) - F_{k_2,k_1}(z_1)\} - \frac{1}{\hat{r}} \int_{z_1}^{1/\hat{r}} \frac{1}{y} f_{k_2,k_1}(y)\,\mathrm{d}y \\
&\quad - \hat{r} \int_{1/\hat{r}}^{z_2} y f_{k_2,k_1}(y)\,\mathrm{d}y \\
&= \frac{1}{T} \{F_{k_2,k_1}(z_2) - F_{k_2,k_1}(z_1)\} - \frac{k_2 - 2}{k_2 \hat{r}} \int_{z_1}^{1/\hat{r}} f_{k_2-2,k_1+2}(\kappa_1 y)\,\mathrm{d}y \\
&\quad - \frac{k_2 \hat{r}}{k_2 + 2} \int_{1/\hat{r}}^{z_2} f_{k_2+2,k_1-2}(\kappa_2 y)\,\mathrm{d}y \\
&= \frac{1}{T} \{F_{k_2,k_1}(z_2) - F_{k_2,k_1}(z_1)\} \\
&\quad - \frac{k_1 \hat{r}}{k_1 - 2} \left\{ F_{k_2-2,k_1+2}(\kappa_2 z_2) - F_{k_2-2,k_1+2}\left(\frac{\kappa_2}{\hat{r}}\right) \right\} \\
&\quad - \frac{k_1 + 2}{k_1 \hat{r}} \left\{ F_{k_2-2,k_1+2}\left(\frac{\kappa_1}{\hat{r}}\right) - F_{k_2-2,k_1+2}(\kappa_1 z_1) \right\},
\end{aligned}
$$

where $z_1 = T/\hat{r}$, $z_2 = 1/(\hat{r}T)$, $\kappa_1 = k_1(k_2 + 2)/\{k_2(k_1 - 2)\}$ and $\kappa_2 = 1/\kappa_1$. The expression for Q_{d} may at first seem quite formidable, but it is derived by elementary methods, and its evaluation is easy. An expression for Q_{s} is derived by similar steps.

5

Small or large?

In this chapter, we deal with a class of problems in which a parameter θ may have a 'special' value, such as zero. One course of action is appropriate for $\theta = 0$ and another course is advisable if θ is distant from zero. For example, when θ is a regression coefficient associated with a covariate X, it would be wise discard it from the model when $\theta = 0$. For a sufficiently large value of $|\theta|$ we would certainly retain X in the model.

This reflects our preference for parsimonious models, which contain no unimportant covariates, but also models that are adequate, which contain all the important covariates. This rule is not formulated clinically because it does not state with clarity what action to take (exclude covariate X or retain it in the model) when θ is small. More to the point, the threshold between values of $|\theta|$ that should be regarded as small and large is not specified. It is preferable to discard a covariate not only when its coefficient θ vanishes, $\theta = 0$, but also when $|\theta|$ is sufficiently small. With the covariate discarded, the model is not valid, but the bias of the estimator based on it is small in relation to the variance inflation that would result from including X in the model.

In an established approach, the hypothesis that $\theta = 0$ is tested against the two-sided alternative that $\theta \neq 0$. By such a test, we control the probability of inappropriate inclusion (false positive) by setting on it an upper limit of 0.05 (the test size), but leave ourselves exposed to a possibly large probability of inappropriate exclusion (false negative). Better than such a 'one-sided' control may be an arrangement in which the two probabilities are held in a balance. Information about the relative magnitudes of the squared bias and variance inflation, as functions of θ, would help us define this balance.

In our perspective, the null hypothesis ($\theta = 0$) should be rejected unconditionally, unless there is some prior information, external to the data, that supports the singular value of zero over all other values, even those extremely small, such as 10^{-14}. In the absence of such a support, zero is but one of uncountably many values, uncountably many of them in arbitrarily close distance from zero. Betting on any single value of θ is a losing proposition. But such a proposition is advanced by the null hypothesis and is applied in most methods of model selection.

The two-stage procedure, in which we first select a model and then apply the estimator that we would have applied if we knew that the selected model is parsimonious and valid, has two drawbacks. First, it ignores the model uncertainty and pretends that model selection is unerring. That is, the appropriate model is selected only with a limited probability; with the complementary probability inappropriate models are selected, and their influence on the selected-model based estimator is difficult to assess. Second, we rule out the possibility that an invalid model might yield an efficient estimator. Related to this is the possibility that some models may be useful for certain targets, and not for some others. This suggests that the two-stage procedure is a poor strategy. Instead of asking 'Which model?' we should ask 'Which estimator?' or, by adhering to the original goal, 'How to estimate?'. And with the latter two questions we might ask whether selection is how to get the best out of a collection of estimators. Chapter 11 takes this idea further.

In the next three sections, in which we develop our perspective, we assume that there are two thresholds, $T_1 < 0$ and $T_2 > 0$. One course of action (e.g., model reduction) is appropriate if $\theta \in (T_1, T_2)$, and the other (model retention) if $\theta \notin (T_1, T_2)$. That is, the two states are $\Theta_S = (T_1, T_2)$ and $\Theta_L = (-\infty, T_1) \cup (T_2, +\infty)$, referred to as 'small' (S) and 'large' (L), respectively. Denote $\bar{T} = \frac{1}{2}(T_1 + T_2)$. In most settings that we consider, $T_1 = -T_2$, and so $\bar{T} = 0$, but in the derivations that follow this is never assumed and they apply more generally. In any case, such symmetry can be arranged by the linear transformation $\theta' = \theta - \bar{T}$. Results are derived for an unbiased normally distributed estimator $\hat{\theta}$, but they are easy to adapt to estimators with other distributions, and symmetric distributions in particular.

5.1 Piecewise constant loss

We assume first that the loss for declaring S (verdict V_s) when $\theta \in \Theta_L$ is one lossile, and for declaring L (verdict V_l) when $\theta \in \Theta_S$ is R lossiles. See Figure 5.1 for an illustration with $R = 3$, $T_1 = -0.3$ and $T_2 = 0.3$. Black horizontal segments indicate the loss with the inappropriate verdict and grey colour is used for their 'appropriate' counterparts, when no loss is incurred. We refer to this loss structure as symmetric piecewise constant. Asymmetry is introduced later in this section. Some alternative loss structures, which lead to generalisations, are introduced in Sections 5.2 and 5.3. The model selection problem corresponds to $R = 1$—the errors of the two kinds are equally grave.

Suppose we have for θ a normally distributed unbiased estimator $\hat{\theta}$, with variance τ^2. No generality is lost by assuming that $\tau^2 = 1$ because we could consider the parameter $\theta' = \theta/\tau$ and its estimator $\hat{\theta}' = \hat{\theta}/\tau$. The simplification that setting $\tau^2 = 1$ would introduce is only slight, so we do not assume this.

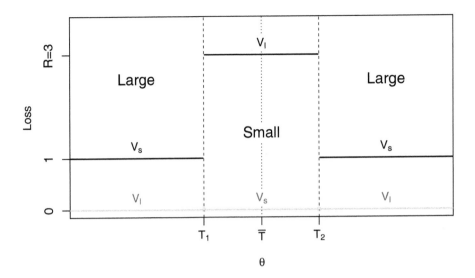

FIGURE 5.1
Losses with verdicts V_s and V_l; symmetric piecewise constant loss.

The expected losses for the two kinds of error are

$$Q_l = R \int_{T_1}^{T_2} \varphi\left(x; \hat{\theta}, \tau\right) dx$$

$$= R\left\{\Phi(z_2) - \Phi(z_1)\right\}$$

$$Q_s = \int_{-\infty}^{T_1} \varphi\left(x; \hat{\theta}, \tau\right) dx + \int_{T_2}^{+\infty} \varphi\left(x; \hat{\theta}, \tau\right) dx$$

$$= 1 - \Phi(z_2) + \Phi(z_1),$$

where $z_h = (T_h - \hat{\theta})/\tau$, $h = 1, 2$. We also introduce $\bar{z} = (\bar{T} - \hat{\theta})/\tau = \frac{1}{2}(z_1 + z_2)$. As a function of $\hat{\theta}$, Q_l is symmetric around $\bar{T}$ and has zero limits at $\pm\infty$. It increases for $\hat{\theta} \in (-\infty, \bar{T})$ and decreases for $\hat{\theta} \in (\bar{T}, +\infty)$. In contrast, Q_s decreases in $(-\infty, \bar{T})$ and increases in $(\bar{T}, +\infty)$. Its limits at $+\infty$ and $-\infty$ are 1.0. Thus $Q_l < Q_s$ for sufficiently large $|\hat{\theta}|$, warranting the verdict V_l. But V_s is not always appropriate for sufficiently small $|\hat{\theta}|$, sometimes not even for $\hat{\theta} = 0$. Two contrasting examples of pairs of expected losses for the scenario in Figure 5.1, with values of τ set to 0.4 and 1.0, are displayed in Figure 5.2.

As functions of $\hat{\theta}$, the two expected losses attain their extremes at $\hat{\theta} = \bar{T}$. This point is the maximum of Q_l and the minimum of Q_s. This confirms that, as a value of $\hat{\theta}$, $\bar{T}$ is the most favourable for S and the least favourable for L. If $Q_l(\bar{T}) < Q_s(\bar{T})$, then $Q_l(\hat{\theta}) < Q_s(\hat{\theta})$ for all $\hat{\theta}$, and so verdict V_l can be issued without evaluating $\hat{\theta}$. This scenario, illustrated in the right-hand panel

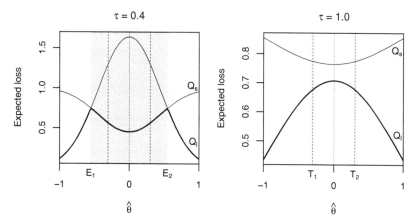

FIGURE 5.2
Expected loss functions with piecewise constant loss; scenario of Figure 5.1,
with $\tau = 0.4$ in the left-hand panel and $\tau = 1.0$ in the right-hand panel. The
expected loss with the appropriate verdict is highlighted. The range in which
verdict V_s is issued, where $Q_s < Q_1$, is indicated by shading.

of Figure 5.2, arises when $T_2 - T_1$ is sufficiently small, that is, when S is too
exclusive (Θ_S too narrow), when R is sufficiently small, when we are averse
to the inappropriate choice of S, or when τ is sufficiently large. In these cases,
the support for S is insufficient to overcome the aversion to the inappropriate
choice of S even when $\hat{\theta} = \bar{T}$.

In the other scenario, Q_1 and Q_s intersect at two points, the equilibria,
denoted by E_1 and E_2. They are located symmetrically around $\bar{T}$; they satisfy
the identity $E_1 + E_2 = T_1 + T_2 \, (= 2\bar{T})$. In the left-hand panel of Figure 5.2, the
equilibria are at $E_1 = -0.55$ and $E_2 = 0.55$. In the range (E_1, E_2), marked
by shading, verdict V_s is issued.

The borderline between these two scenarios is given by the solution of the
equation $Q_1(\hat{\theta}) = Q_s(\hat{\theta})$ for $\hat{\theta} = \bar{T}$. This is equivalent to

$$\Phi\left(\frac{T_2 - \bar{T}}{\tau}\right) - \Phi\left(\frac{T_1 - \bar{T}}{\tau}\right) = \frac{1}{R+1} \tag{5.1}$$

and, owing to symmetry of the normal distribution, to

$$2\Phi\left(\frac{T_2 - \bar{T}}{\tau}\right) - 1 = \frac{1}{R+1}.$$

The respective solutions for T_2 and τ are $T_2^{(0)} = \bar{T} + U_0\tau$ and $\tau^{(0)} = (T_2 - \bar{T})/U_0$, where

$$U_0 = \Phi^{-1}\left(\frac{R+2}{2R+2}\right).$$

For τ and $\bar{T}$ given, and $T_2 < T_2^{(0)}$, verdict V_1 is issued with no regard for $\hat{\theta}$. This reinforces the argument presented earlier that a (two-sided) null hypothesis, which corresponds to $T_2 = 0$, should always be rejected. When T_2 is given and $\tau > \tau^{(0)}$, V_1 is issued with no regard for $\hat{\theta}$. In the two panels of Figure 5.2, $T_2^{(0)} = 0.13$ for $\tau = 0.4$ and $T_2^{(0)} = 0.32$ for $\tau = 1.0$. The solution for R, with T_2 and τ fixed, is also straightforward to derive from equation (5.1).

There are two equilibria when V_s would be the verdict if $\hat{\theta} = \bar{T}$. They are the solutions of the balance function

$$B\left(\hat{\theta}\right) = \Phi(z_2) - \Phi(z_1) - \frac{1}{R+1}.$$

They can be found by the Newton-Raphson algorithm, exploiting the simple expression for the differential

$$\frac{\partial B}{\partial \hat{\theta}} = \frac{\varphi(z_1) - \varphi(z_2)}{\tau}.$$

Thus, starting with an initial guess $E^{(0)} \neq \bar{T}$, iteration $t = 1, 2, \ldots$ evaluates the next (provisional) solution

$$E^{(t)} = E^{(t-1)} + \frac{\tau}{R+1} \frac{(R+1)\{\Phi(z_2) - \Phi(z_1)\} - 1}{\varphi(z_2) - \varphi(z_1)},$$

where $E^{(t-1)}$ is substituted for $\hat{\theta}$ in the expressions for z_1 and z_2, $z_h = (T_h - E^{(t-1)})/\tau$. One solution, E, is obtained at convergence; by symmetry, the other solution is $2\bar{T} - E$.

The dependence of the equilibria on R is confirmed by applying the implicit function theorem:

$$\frac{\partial E}{\partial R} = -\frac{\partial B}{\partial R} \bigg/ \frac{\partial B}{\partial E} = \frac{\tau}{(R+1)^2} \frac{1}{\varphi(z_2) - \varphi(z_1)}, \qquad (5.2)$$

where E is used for the argument $\hat{\theta}$. The left-hand equilibrium, for which $\bar{z} > 0$ and therefore $\varphi(z_1) > \varphi(z_2)$, decreases with R and, by symmetry, the right-hand equilibrium increases. Thus, the domain of V_s widens with increasing R. Similarly, we can show that this domain widens with increasing T_2 (with T_1 held constant or changing so that $\bar{T}$ is held constant), and with decreasing τ.

Note the role of τ as a factor in the differential. The standard deviation can be absorbed in the density φ, since $\varphi(z_h)/\tau = \varphi(T_h; \hat{\theta}, \tau)$, $h = 1, 2$. With greater uncertainty about θ, the difference $\varphi(T_1; \hat{\theta}, \tau) - \varphi(T_2; \hat{\theta}, \tau)$ diminishes in absolute value, and so the differential in (5.2) increases in absolute value. Therefore, for greater τ the equilibria diverge with increasing R to $\pm\infty$ faster.

5.1.1 Asymmetric loss

The losses may be different for positive and negative values of θ, with one or both verdicts. Figure 5.3 illustrates such a scenario, with losses $L_{sL-} = 1$,

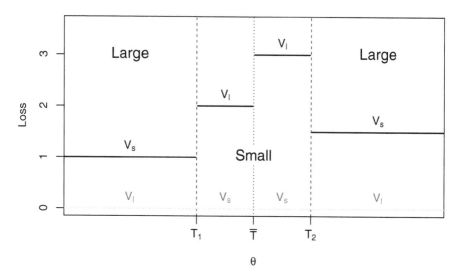

FIGURE 5.3
Asymmetric piecewise constant loss. The losses are $L_{\mathrm{sL}-} = 1.0$, $L_{\mathrm{lS}-} = 2.0$, $L_{\mathrm{lS}+} = 3.0$ and $L_{\mathrm{sL}+} = 1.5$; $T_1 = -0.3$ and $T_2 = 0.3$.

$L_{\mathrm{lS}-} = 2$, $L_{\mathrm{lS}+} = 3$ and $L_{\mathrm{sL}+} = 1.5$ in the respective intervals $(-\infty, T_1)$, $(T_1, \bar{T})$, $(\bar{T}, T_2)$ and $(T_2, +\infty)$. In this example, $L_{\mathrm{lS}+}/L_{\mathrm{lS}-} = L_{\mathrm{sL}+}/L_{\mathrm{sL}-} = 1.5$; errors of a given type, sL or lS, have 50% graver consequences when θ is positive than when it is negative.

The losses need not be proportional; for instance, the ratios of the losses may be 1.5 for V_{s} and 1.2 for V_{l}. Further, the boundary of the segments of constant loss may be moved from $\bar{T}$ in either direction, and the ranges $(-\infty, T_1)$ and $(T_2, +\infty)$ may be partitioned to several intervals, not necessarily in a symmetric fashion, and a different loss may apply in each of them. Evaluation of the expected losses and a description of the behaviour of these functions of $\hat{\theta}$ for these settings is left for an exercise. It suffices to say that a vast variety of loss structures can be generated in this way. In any particular application, the temptation to define a loss function with some 'interesting' features for the sake of demonstrating analytical flexibility should be resisted, unless it reflects the client's perspective and preferences.

Figure 5.4 displays the expected losses Q_{s} and Q_{l} for the 'asymmetric' scenario in Figure 5.3 for two values of the standard error τ. For $\tau = 0.3$, there are two equilibria, -0.59 and 0.51, whereas for $\tau = 1.3$ the expected loss functions do not intersect and V_{l} is issued for any value of $\hat{\theta}$. In this case, estimation of θ is superfluous.

Combinations of symmetric losses (as well as asymmetric losses) and standard deviations τ can be found for which the two equilibria are on the same side of $\bar{T}$ (e.g., $E_1 < E_2 < \bar{T}$), or even both of them are smaller than T_1

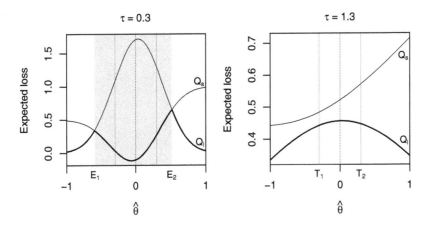

FIGURE 5.4
Expected loss functions with asymmetric piecewise constant loss; scenario of
Figure 5.3, with $\tau = 0.3$ in the left-hand panel and $\tau = 1.3$ in the right-hand
panel. The expected loss with the appropriate verdict is highlighted.

(or greater than T_2). While such a set of losses may appear to be unusual
or extreme, the equilibria for them are not counterintuitive. They reflect the
(extreme) disparity of the losses across the ranges in which the loss is constant.

A drawback of the piecewise constant loss is that a small error (θ being
close to a threshold) is treated on par with a large error (θ distant from
both thresholds). This can be addressed to a limited extent by partitioning
the states to several intervals and defining losses within them according to
a particular trend. A 'smooth' version of such a scheme is preferable and its
analytical treatment is less tedious.

5.2 Piecewise linear loss

A linear loss function may be more appropriate for verdict V_s. Such loss is
greater the further away θ is from the nearer of the thresholds T_1 (when
$\theta < T_1$) and T_2 (when $\theta > T_2$). In this case, it is more practical to define
separate loss functions for the two intervals that comprise Θ_L. These functions
are $L_{sL+} = \theta - T_2$ for $\theta > T_2$ and $L_{sL-} = T_1 - \theta$ for $\theta < T_1$, or their positive
scalar multiples.

A linear loss with verdict V_l is defined similarly. The magnitude of the
error is the smaller of the distances of θ from T_1 and T_2, so the loss is $L_{lS+} =
R(T_2 - \theta)$ when $\theta \in (\bar{T}, T_2)$ and $L_{lS-} = R(\theta - T_1)$ when $\theta \in (T_1, \bar{T})$. An

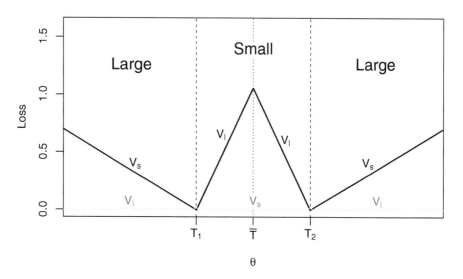

FIGURE 5.5
Symmetric piecewise linear loss.

illustration is presented in Figure 5.5 with $T_1 = -0.3$, $T_2 = 0.3$ and $R = 3.5$, so that $L_{\text{lS}+} = L_{\text{lS}-} = 1.05$ at $\theta = 0$. The loss structure is referred to as symmetric linear, to distinguish it from asymmetric loss functions introduced later in this section.

Recall the notation $z_h = (T_h - \hat{\theta})/\tau$ for $h = 1, 2$, introduced in Section 5.1. The expected loss with verdict V_s is $Q_s = Q_{s-} + Q_{s+}$, where

$$
\begin{aligned}
Q_{\text{s}-} &= \int_{-\infty}^{T_1} (T_1 - x)\, \varphi\!\left(x; \hat{\theta}, \tau\right) \mathrm{d}x \\
&= \tau \int_{-\infty}^{z_1} (z_1 - y)\, \varphi(y)\, \mathrm{d}y \\
&= \tau \Phi_1(z_1)\,.
\end{aligned}
$$

This is obtained by the linear transformation $y = (x - \hat{\theta})/\tau$ followed by integration by parts. The function $\Phi_1(x) = x\Phi(x) + \varphi(x)$, the primitive function of $\Phi(x)$, was introduced in Section 3.2.

Similar operations yield the complementary identity

$$
Q_{\text{s}+} = \tau \Phi_1\left(-z_2\right)\,.
$$

Asymmetry of the loss can be arranged by introducing a factor $M_{\text{L}+}$ in $L_{\text{sL}+}$, altering it to $M_{\text{L}+}(\theta - T_2)$. No generality is lost by keeping $M_{\text{L}-} = 1$. Linear loss may be combined with constant loss. With such a combination, the loss may be substantial even in the vicinity of T_1 and T_2. The two parts of the loss,

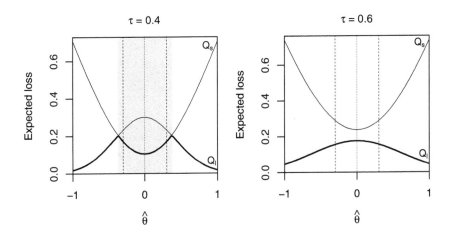

FIGURE 5.6
Expected loss functions with symmetric piecewise linear loss; scenario of Figure 5.5, with $\tau = 0.4$ in the left-hand panel and $\tau = 0.6$ in the right-hand panel. The expected loss with the appropriate verdict is highlighted.

constant and linear, can be regarded as separate because the overall expected loss is the total of the expected losses that correspond to the two parts.

Figure 5.6 displays two examples of pairs of expected loss functions based on the scenario of Figure 5.5, with the two values of the standard error τ indicated in the panel titles. For $\tau = 0.6$, in the right-hand panel, verdict V_1 is appropriate for all values of $\hat{\theta}$. The borderline between the balance equation having two solutions and having none is at $\tau = 0.54$. This borderline can be found by an iterative method, although trial and error may be just as fast and convenient.

Although the expected losses with (symmetric) constant and linear loss functions have distinct functional forms, it is difficult to distinguish them based on their plots. A 'piecewise linear' expected loss can be matched quite closely by a 'piecewise constant' expected loss, although the corresponding parameters usually differ substantially. This suggests that elicitation of the loss ratios and other parameters associated with the loss by inspecting the shapes of expected loss functions is unlikely to be constructive.

Asymmetry can be introduced by replacing the loss ratio R with factors $R^{(-)}$ and $R^{(+)}$ specific to the two subintervals of Θ_S separated by $\bar{T}$. The discontinuity at $\bar{T}$ is of no concern, just as it is not with piecewise constant loss. A piecewise linear loss function may have breakpoints (or even discontinuities) additional to T_1, $\bar{T}$ and T_2, with a distinct linear loss function within each interval delimited by the breakpoints. However, meaningful loss functions are decreasing throughout $(-\infty, T_1)$ and increasing throughout $(T_2, +\infty)$, so the

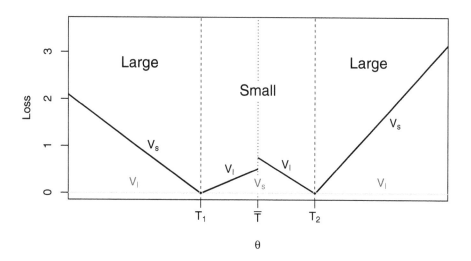

FIGURE 5.7
Asymmetric piecewise linear loss.

loss functions have to be continuous or have jumps with the same signs as $\theta - \bar{T}$.

An example of an asymmetric linear loss, with breakpoints only at T_1, $\bar{T}$ and T_2, is given in Figure 5.7. Here $T_1 = -0.3$, $T_2 = 0.3$, the loss functions are $3(T_1 - \theta)$ for $\theta < T_1$ and $4(\theta - T_2)$ for $\theta > T_2$ with verdict V_s, and $1.7(\theta - T_1)$ for $\theta \in (T_1, \bar{T})$ and $2.5(T_2 - \theta)$ for $\theta \in (\bar{T}, T_2)$ with verdict V_1.

The expected loss with verdict V_1 is the sum of two terms,

$$
\begin{aligned}
Q_{1-} \;&=\; R^{(-)} \int_{T_1}^{\bar{T}} (x - T_1)\, \varphi\!\left(x; \hat{\theta}, \tau\right) dx \\
&=\; \tau R^{(-)} \int_{z_1}^{\bar{z}} (y - z_1)\, \varphi(y)\, dy \\
&=\; \tau R^{(-)} \left\{ (\bar{z} - z_1)\, \Phi(\bar{z}) - \Phi_1(\bar{z}) + \Phi_1(z_1) \right\}
\end{aligned}
$$

and

$$
\begin{aligned}
Q_{1+} \;&=\; R^{(+)} \int_{\bar{T}}^{T_2} (T_2 - x)\, \varphi\!\left(x; \hat{\theta}, \tau\right) dx \\
&=\; \tau R^{(+)} \left\{ -(z_2 - \bar{z})\, \Phi(\bar{z}) + \Phi_1(z_2) - \Phi_1(\bar{z}) \right\},
\end{aligned}
$$

where $\bar{z} = (\bar{T} - \hat{\theta})/\tau$. In the symmetric case, when $R^{(-)} = R^{(+)}$ $(= R)$,

$$
Q_1 = Q_{1-} + Q_{1+} = R\tau \left\{ \Phi_1(z_1) + \Phi_1(z_2) - 2\Phi_1(\bar{z}) \right\}. \tag{5.3}
$$

In Appendix A, we show that $Q_1(\hat{\theta})$ is increasing for $\hat{\theta} < \bar{T}$ and decreasing for $\hat{\theta} > \bar{T}$, so it attains its maximum at $\hat{\theta} = \bar{T}$.

Example 9

In this example, we explore the properties of the balance function, which are related to the existence of equilibria. Suppose the loss function is piecewise linear, with symmetry in both kinds of error. The expected losses with the respective verdicts V_s and V_l are

$$Q_s = \tau\{\Phi_1(z_1) + \Phi_1(-z_2)\}$$
$$Q_l = R\tau\{\Phi_1(z_1) + \Phi_1(z_2) - 2\Phi_1(\bar{z})\} .$$

We explore first whether there is an equilibrium. The function Φ_1 has the property

$$\Phi_1(z) - \Phi_1(-z) = z ,$$

proved by substituting the identity $\Phi_1(z) = z\Phi(z) + \varphi(z)$ and exploiting the symmetry of the normal distribution. Hence $Q_s = \tau\{\Phi_1(z_1) + \Phi_1(z_2) - z_2\}$, and after elementary reorganisation, the balance function, $B(\hat{\theta}) = Q_l - Q_s$, becomes

$$\tau[(R-1)\{\Phi_1(z_1) + \Phi_1(z_2)\} - 2R\Phi_1(\bar{z}) + z_2] .$$

The function diverges to $-\infty$ as $|\hat{\theta}| \to +\infty$, but its value at $\hat{\theta} = 0$,

$$B(0; T_2) = \tau\left[(R-1)\left\{\Phi_1\left(\frac{T_2}{\tau}\right) + \Phi_1\left(-\frac{T_2}{\tau}\right)\right\} - 2R\Phi_1(0)\right] + T_2, \quad (5.4)$$

can have either sign. Its value for $T_2 = 0$ is $-2\tau\varphi(0) < 0$, whereas for $T_2 \to +\infty$ it diverges to $+\infty$. Therefore, the balance function is negative throughout for sufficiently small $T_2 > 0$, and verdict V_l is then issued for all values of $\hat{\theta}$. In contrast, for large T_2 there is a strong preference for V_s when $\hat{\theta}$ is close to zero. Appendix B contains further exploration of the equilibria.

5.3 Piecewise quadratic loss

Similarly to a linear loss, a piecewise quadratic loss function is defined by its elements, or segments, each of them a quadratic function given by the threshold and a coefficient. For example, the symmetric quadratic loss with verdict V_s is $L_{sL+} = (\theta - T_2)^2$ when $\theta > T_2$ and $L_{sL-} = (T_1 - \theta)^2$ when $\theta < T_1$. That is, $L_{sL} = \min\{(T_1 - \theta)^2, (\theta - T_2)^2\}$ for $\theta \notin (T_1, T_2)$. The loss with verdict V_l is $L_{lS+} = R(T_2 - \theta)^2$ when $\theta \in (\bar{T}, T_2)$ and $L_{lS-} = R(\theta - T_1)^2$ when $\theta \in (T_1, \bar{T})$. There are the same options for injecting asymmetry in the loss as they are for piecewise linear loss. Figure 5.8 presents an example with symmetric quadratic loss, with $R = 12$ and $T_1 = -0.3$ and $T_1 = 0.3$.

The two examples of expected loss displayed in Figure 5.9, both based on the loss function in Figure 5.8, have features similar to the examples in

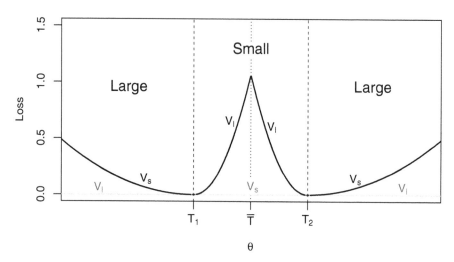

FIGURE 5.8
Symmetric piecewise quadratic loss; $T_1 = -0.3$, $T_2 = 0.3$ and $R = 12$.

Figure 5.6 for symmetric linear loss. There are two equilibria for $\tau < 0.593$
(see the left-hand panel, where $\tau = 0.3$), and no equilibrium for $\tau > 0.593$
(right-hand panel, where $\tau = 0.7$). The borderline ($\tau = 0.593$) can be found
by trial and error or a line search. As τ increases towards the borderline, the
domain of V_s gets narrower. It is wider than (T_1, T_2) for $\tau = 0.3$, it coincides
with (T_1, T_2) for $\tau = 0.50$ and is only half as wide for $\tau = 0.57$.

Further flexibility is introduced by convex combinations of loss functions.
Of particular importance are combinations with piecewise constant loss functions. Both linear and quadratic functions attain arbitrarily small values in
the vicinity of a threshold. In some settings, the loss is non-trivial even in this
case. This can be taken care of by combining the linear or quadratic loss with
constant loss.

A related example is the loss function $(\theta - \bar{T})^2$ for verdict V_s when $\theta \notin$
(T_1, T_2). It is a convex combination of piecewise quadratic, linear and constant
loss functions. This follows from the decomposition

$$\left(\theta - \bar{T}\right)^2 = \left(\theta - T_2\right)^2 + 2\left(T_2 - \bar{T}\right)\left(\theta - T_2\right) + \left(T_2 - \bar{T}\right)^2$$

for $\theta > T_2$ and the analogous decomposition for $\theta < T_1$.

The quadratic loss has some affinity with its namesake in established forms
of estimation, as applied in the mean squared error (MSE). The quadratic loss
in MSE refers to the squared estimation error $(\hat{\theta} - \theta)^2$, whereas the quadratic
loss in the context of a decision refers to $(\theta - T_h)^2$, where T_h is a threshold.
There is no loss when the verdict based on $\hat{\theta}$ coincides with the verdict that
would have been issued with $\hat{\theta} = \theta$; an estimation error is 'forgiven' when the
appropriate verdict is issued.

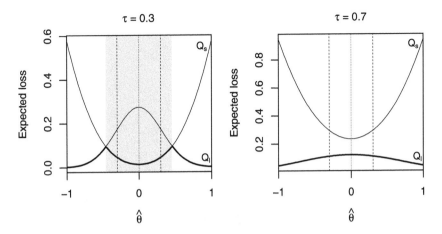

FIGURE 5.9
Expected loss functions with symmetric piecewise quadratic loss; scenario of Figure 5.8, with $\tau = 0.3$ in the left-hand panel and $\tau = 0.7$ in the right-hand panel. The expected loss with the appropriate verdict is highlighted.

Example 10

We explore the quadratic counterpart of the symmetric loss structure studied in Example 9. The four components of loss are $L_{\mathrm{sL}+} = (\theta - T_2)^2$ for $\theta > T_2$ and $L_{\mathrm{sL}-} = (T_1 - \theta)^2$ for $\theta < T_1$ with verdict V_{s}, and $L_{\mathrm{lS}+} = R(T_2 - \theta)^2$ for $\theta \in (\bar{T}, T_2)$ and $L_{\mathrm{lS}-} = R(\theta - T_1)^2$ for $\theta \in (T_1, \bar{T})$ with verdict V_{l}. The corresponding contributions to the expected loss are derived by the same steps as for the linear loss, except for applying integration by parts twice. For verdict V_{s}, we have

$$
\begin{aligned}
Q_{\mathrm{s}-} &= \int_{-\infty}^{T_1} (T_1 - x)^2\, \varphi\!\left(x; \hat{\theta}, \tau\right) \mathrm{d}x \\
&= \tau^2 \int_{-\infty}^{z_1} (z_1 - y)^2\, \varphi(y)\, \mathrm{d}y \\
&= 2\tau^2 \int_{-\infty}^{z_1} (z_1 - y)\, \Phi(y)\, \mathrm{d}y \\
&= 2\tau^2 \int_{-\infty}^{z_1} \Phi_1(y)\, \mathrm{d}y = 2\tau^2\, \Phi_2(z_1)\,,
\end{aligned}
$$

and

$$Q_{s+} = \int_{T_2}^{+\infty} (x - T_2)^2 \, \varphi\left(x; \hat{\theta}, \tau\right) dx$$

$$= \tau^2 \int_{-\infty}^{-z_2} (y + z_2)^2 \, \varphi(y) \, dy$$

$$= -2\tau^2 \int_{-\infty}^{-z_2} (y + z_2) \, \Phi(y) \, dy$$

$$= 2\tau^2 \, \Phi_2 \, (-z_2) = \tau^2 \left\{ 1 + z_2^2 - 2\Phi_2 \, (z_2) \right\} .$$

The concluding identity follows from the expression $2\Phi_2(z) = (1 + z^2)\Phi(z) + z\varphi(z)$ and symmetry of the normal distribution. The expected loss with verdict V_s is

$$Q_s = Q_{s-} + Q_{s+} = \tau^2 \left\{ 1 + z_2^2 - 2\Phi_2 \, (z_2) + 2\Phi_2 \, (z_1) \right\} .$$

Since Φ_2 is increasing and $z_2 > z_1$, $Q_s < \tau^2 (1 + z_2^2)$. As the domain of S is narrowed around a fixed value of $\bar{T}$, when $T_2 - T_1 \to 0$ and therefore $z_2 - z_1 \to 0$, Q_s converges to $\tau^2(1 + \bar{T}^2)$. The expected loss diverges to $+\infty$ as $z_1 \to -\infty$ and $z_2 \to +\infty$.

The first- and second-order partial differentials of Q_s are

$$\frac{\partial Q_s}{\partial \hat{\theta}} = 2\tau \left\{ -z_2 + \Phi_1(z_2) - \Phi_1(z_1) \right\}$$

$$\frac{\partial^2 Q_s}{\partial \hat{\theta}^2} = 2 \left\{ 1 - \Phi(z_2) + \Phi(z_1) \right\} .$$

The latter is positive, so the former is increasing, and therefore Q_s is convex. Further, the first-order differential vanishes at $\hat{\theta} = \bar{T}$, where $z_1 = -z_2$, so Q_s has its minimum at $\hat{\theta} = \bar{T}$, which is also implied by the symmetry of Q_s.

The contributions to the expected loss with V_l are evaluated similarly. First,

$$Q_{l-} = R \int_{T_1}^{\bar{T}} (x - T_1)^2 \, \varphi\left(x; \hat{\theta}, \tau\right) dx$$

$$= R\tau^2 \int_{z_1}^{\bar{z}} (y - z_1)^2 \, \varphi(y) \, dy$$

$$= R\tau^2 \left\{ (\bar{z} - z_1)^2 \, \Phi(\bar{z}) - 2 \, (\bar{z} - z_1) \, \Phi_1(\bar{z}) + 2\Phi_2(\bar{z}) - 2\Phi_2(z_1) \right\} .$$

Compared to the evaluations for Q_s, additional terms arise owing to both limits of integration being finite. However, one of these terms cancels out in

$Q_1 = Q_{1-} + Q_{1+}$. We have

$$
\begin{aligned}
Q_{1+} &= R \int_{\bar{T}}^{T_2} (T_2 - x)^2 \, \varphi\left(x; \hat{\theta}, \tau\right) dx \\
&= R\tau^2 \int_{\bar{z}}^{z_2} (z_2 - y)^2 \, \varphi(y) \, dy \\
&= R\tau^2 \left\{ -(z_2 - \bar{z})^2 \, \Phi(\bar{z}) - 2\,(z_2 - \bar{z})\,\Phi_1(\bar{z}) + 2\Phi_2(z_2) - 2\Phi_2(\bar{z}) \right\},
\end{aligned}
$$

and so

$$
Q_1 = 2R\tau^2 \left\{ \Phi_2(z_2) - \Phi_2(z_1) - (z_2 - z_1)\,\Phi_1(\bar{z}) \right\}.
$$

The balance function is

$$
\begin{aligned}
B\!\left(\hat{\theta}\right) &= Q_1 - Q_s \\
&= \tau^2 \left[2(R+1)\left\{\Phi_2(z_2) - \Phi_2(z_1)\right\} - 2R\,(z_2 - z_1)\,\Phi_1(\bar{z}) - 1 - z_2^2 \right].
\end{aligned}
\tag{5.5}
$$

It is symmetric around $\hat{\theta} = \bar{T}$ and its limits are $-\infty$ as $\hat{\theta}$ diverges to $+\infty$ or $-\infty$. These properties are obvious from the general setting. Appendix C presents an analytical proof that can be adapted for a variety of other settings.

Example 11

Consider the problem of deciding whether a quantity θ, estimated without bias with standard error $\tau = 0.15$, is small, meaning $|\theta| < T_2 = 0.2$, or large ($|\theta| > 0.2$). If we declare V_l inappropriately, when $|\theta| < 0.2$, then the magnitude of the error is immaterial and one lossile is incurred; $L_{lS} = 1$. In contrast, if we declare V_s inappropriately, then the magnitude of the error matters, and the loss function is asymmetric. If $\theta < T_1 = -0.2$, then the loss is

$$
L_{sL-} = 2 + \frac{(T_1 - \theta)^2}{2},
$$

whereas if $\theta > T_2 = 0.2$, the loss is

$$
L_{sL+} = \frac{1}{2} + \frac{(\theta - T_2)^2}{8}.
$$

The loss function is plotted in the left-hand panel of Figure 5.10. It shows that we are averse to V_s when $\theta < T_1$, but much less so when $\theta > T_2$.

The expected losses are

$$
\begin{aligned}
Q_1 &= \Phi(z_2) - \Phi(z_1) \\
Q_s &= \frac{1}{2}\left\{1 - \Phi(z_2)\right\} + 2\Phi(z_1) + \tau^2 \left\{ \frac{1 + z_2^2 - 2\Phi_2(z_2)}{8} + \Phi_2(z_1) \right\},
\end{aligned}
$$

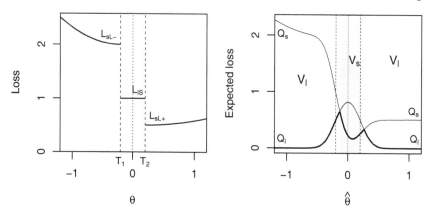

FIGURE 5.10
Loss and expected loss functions in Example 11; an asymmetric combination
of constant and quadratic loss functions.

where $z_1 = (-0.2-\hat{\theta})/0.15$ and $z_2 = (0.2-\hat{\theta})/0.15$. These functions are plotted
in the right-hand panel of Figure 5.10. The domain of V_s is $(-0.138, 0.258)$,
marked by shading. The domain is slightly narrower than Θ_S and is shifted
to the right, in response to our aversion to V_s when $\theta < T_1$. This is tempered
by the relatively large loss L_{1S} compared to L_{sL+} when θ exceeds T_2 only
moderately. Contrary to the impression one might have about Q_s from how
flat it is for $\hat{\theta} \in (0.5, 1.0)$, it diverges to $+\infty$ as $\hat{\theta} \to +\infty$, although the
divergence is rather slow.

5.4   Ordinal categories

Suppose we have solved a problem of deciding whether a parameter is small
(close to zero) or large by issuing the verdict V_1. We may respond to the
client's (or our own) ambition to make a stronger, more specific statement and
conclude that this parameter has the same sign as its estimate: $\theta > 0$ if $\hat{\theta} > 0$
and $\theta < 0$ otherwise. This amounts to a poorly justified improvisation because,
even conditionally on verdict V_1 being appropriate, such a statement ignores
the consequences of getting the sign of θ wrong. We should have foreseen this
in the plan of analysis by declaring interest in three states: being close to zero,
distinctly positive and distinctly negative.

A similar issue is commonly ignored in hypothesis testing. If we reject the null hypothesis that $\theta = 0$, it would seem a waste not to strengthen the well-founded statement of evidence for the alternative $\theta \neq 0$ by claiming to have evidence that θ has a particular sign, the same as $\hat{\theta}$. Only a purist would object to this because the error in the sign, conditional on rejecting the null hypothesis would seem to be negligible. Integrity is absent when the protocol for the analysis does not state how the evidence of non-zero value is converted to the evidence of a particular sign of θ. Can we ignore the uncertainty about $\text{sign}(\theta)$ and substitute $\text{sign}(\hat{\theta})$ for $\text{sign}(\theta)$?

In this section, we address this problem and its generalisations with greater integrity. We have outlined the approach in Section 1.2, where we briefly considered a set of $K \geq 2$ states as a partition of $(-\infty, +\infty)$ defined by $K - 1$ cutpoints $T_1, \ldots, T_{K-1}$. In the example we just introduced, the states are $(-\infty, T_1)$, (T_1, T_2) and $(T_2, +\infty)$, and $-T_1 = T_2 > 0$. In an application, we may have the task of classifying a unit in production or in provision of a particular service as unsatisfactory (U or $-$), satisfactory (S or 0) or excellent (X or $+$). Thus, we contemplate three options, and corresponding verdicts V_-, V_0 and V_+.

Each verdict is associated with two possible errors, so we have to specify six loss functions. They can be presented in a verdict $\times$ state (loss) matrix

$$
\mathbf{L} = \begin{pmatrix} L_{--} = 0 & L_{-0} & L_{-+} \\ L_{0-} & L_{00} = 0 & L_{0+} \\ L_{+-} & L_{+0} & L_{++} = 0 \end{pmatrix},
$$

with the indication that the diagonal, corresponding to the appropriate verdicts, is associated with no loss. Symmetry of $\mathbf{L}$, that is, $\mathbf{L} = \mathbf{L}^{\top}$, corresponds to symmetry of the consequences: if we issue verdict V_0, the loss does not depend on the sign of θ, even though it may depend on $|\theta|$; if we issue verdict V_+ or V_- inappropriately and θ is in state 0, the loss does not depend of the sign either. Another element of symmetry is that if we confuse the two signs, or states $+$ and $-$, we incur the same loss in both scenarios; $L_{-+} = L_{+-}$. In general, symmetry should not be taken for granted. For $K = 2$ it corresponds to the loss ratio $R = 1$, a special case that is often dismissed outright as implausible.

The six types of error form three pairs. For instance, (V_+, S_-) is paired with (V_-, S_+). The former may be referred to as (gross) overrating, and the latter as (gross) underrating. We qualify them as 'gross' to distinguish them from the remaining two pairs of 'minor' errors, (V_+, S_0) paired with (V_0, S_+) and (V_-, S_0) with (V_0, S_-). In these cases, the verdict (selected state) is a neighbour of the state. In a special case of asymmetry, overrating (gross or minor) is associated with R times greater loss than the paired underrating for each of the three pairs.

A gross error can be regarded as a compendium of two minor errors; with verdict V_+ when in state S_-, we commit the errors (V_+, S_0) and (V_0, S_-).

That suggests the loss L_{+-} equal to $L_{+0}+L_{0-}$. We do not have to subscribe to this rule. An extreme deviation from it regards every error as equally grave, or having one loss common to the three types of error of underrating, and another loss common to the three types of error of overrating. At the other extreme, $L_{+-} \gg L_{+0} + L_{0-}$ and $L_{-+} \gg L_{-0} + L_{0+}$. Losses such that $L_{+-} < \max(L_{+0}, L_{0-})$ or $L_{-+} < \max(L_{-0}, L_{0+})$ would seem not to be sensible in any setting, unless the ordering of the options, $-$, 0 and $+$, is not meaningful.

We adhere to the principle of minimising the expected loss. Piecewise constant loss corresponds to (not necessarily identical but) constant entries of $\mathbf{L}$. If we issue verdict V_+, then the expected loss is

$$Q_+ = L_{+-} \, \mathrm{P}\left(\theta < T_1 \mid \hat{\theta}\right) + L_{+0} \, \mathrm{P}\left(T_1 < \theta < T_2 \mid \hat{\theta}\right),$$

where the two probabilities are fiducial or posterior. Denote by $\mathbf{L}_v$. the row v of $\mathbf{L}$, that is, the row-vector of losses associated with verdict v, and by $\mathbf{p} = (p_-, p_0, p_+)^\top$ the column-vector of fiducial probabilities of the respective states $-$, 0 and $+$. Then $Q_+ = \mathbf{L}_v.\mathbf{p}$ and the vector of expected losses with the three verdicts is $\mathbf{Q} = \mathbf{Lp}$.

In the framework of hypothesis testing, three options are addressed by testing two related hypotheses, with complexity arising in the control of the error of the first kind because the two test statistics are correlated, and the outcomes of the two tests are dependent.

In contrast, dealing with $K = 3$ options is conceptually no more difficult than dealing with a dilemma ($K = 2$). We could split the selection into two subselection tasks, first comparing the expected losses associated with verdict V_- and its complement, the union of 0 and $+$, and if the latter option is selected, then comparing the expected losses associated with $+$ and its complement. But comparing the three option-specific expected losses can be accomplished also without such a diversion.

We reap the benefits of additivity, adding up the expected losses for each type of error. This advantage comes at the cost of more extensive elicitation leading to the specification of the loss matrix $\mathbf{L}$. The various plausible constraints on the entries of $\mathbf{L}$ (the structure of $\mathbf{L}$) leave us with a lot of freedom to represent a great variety of perspectives, value judgements or remits. Defining the cutpoints T_1 and T_2 is an additional responsibility.

5.4.1 Piecewise linear and quadratic losses

Dependence of the loss on the magnitude of the error is introduced by loss functions, setting the off-diagonal elements of $\mathbf{L}$ to functions. Linear and quadratic functions are the obvious candidates for these losses, as they are in problems with $K = 2$ options. The expected loss is evaluated for each type of error (verdict-state pair), so we require no calculus additional to that for $K = 2$, although we have to apply it several times, once for each of the $K(K-1)$ types of error. Concern for the number of evaluations is usually misplaced, especially when the losses have similar functional forms. In the elicitation of the loss functions, it is good practice to discuss with the client the various inequality constraints, and seek agreement about them in the specific setting. Such constraints may assist in the elicitation by anchoring the process – reducing the variety of loss functions and their coefficients (parameters) that have to be contemplated. Illustrating these functions graphically may be effective.

5.5 Multitude of options

If there are only a few options it is feasible to explore each of them and establish the matrix **L**, element by element, or even specify plausible ranges for (a few of) its parameters. When there are many options, we have to look for some shortcuts. We explore them in this section. We distinguish between discrete options and continua of options. The former include finite numbers of options and settings in which every option has only a few neighbours. The latter can be regarded as a problem of estimation, although not necessarily by minimising the mean squared error (MSE).

5.5.1 Discrete options

Suppose the options are S_k, $k = 1, 2, \ldots$. There may be a finite number K, or infinitely (but countably) many options. We assume that these options are in a space in which a distance $\mathcal{D}$ is defined. The distance is nonnegative, $\mathcal{D}(s_1, s_2) = 0$ only when $s_1 = s_2$, and it satisfies the triangular inequality,

$$\mathcal{D}(s_1, s_3) \geq \mathcal{D}(s_1, s_2) + \mathcal{D}(s_2, s_3)$$

for any three elements s_1, s_2 and s_3. We assume that for every option S_k there is a positive number ε such that $\mathcal{D}(s_k, s_{k'}) > \varepsilon$ for all $s_k \in S_k$ and $s_{k'} \in S_{k'}$ for $k' \neq k$. This is the property of discreteness, that every option is isolated from the rest, and the options have no limiting point that is also an option.

To distinguish from the problem addressed in the previous section, we assume that it is beyond our resources or capacity to establish the entire loss matrix **L**. Nevertheless, we will seek ways of converting this problem to one or several problems, each with a manageable number of options. In the first variant, we form a small number of groups of options and solve first the problem of selecting one of these groups. Then we focus on the selected group of options. If it is still too large for specifying every loss function, then we partition it further to subgroups and select one of them. We continue with these cycles of selection until we reach a verdict of a single option. The cycles (or a hierarchy) of decisions make the problem manageable, but at the same time it is open-ended—the number of cycles cannot be fixed, especially when the groups are of unequal sizes. As an extreme example, a single (important) state may form a group on its own. No cycles are needed if this state is selected; otherwise further cycles are necessary. Each cycle represents a distinct problem, so the cycles have to combine elicitation of the losses and evaluation (analysis), and the client has to be contacted at each cycle. The initial grouping as well as the grouping at each cycle requires input from the client.

When uncertainty is represented by plausible ranges (or regions), impasse may arise in every cycle, and so there are many types of impasse. Another weakness of this approach is that the losses within a group may be very uneven. Thus, in the first cycle we may select a group that contains several strong candidate options, whilst another group may contain mostly poor options except for one that is superior to all. So, care has to be exercised in forming the groups of options at each cycle.

In another approach, we eliminate first all the options that have very small probabilities and their selection is likely to involve appreciable losses. If only a few options are retained, we find among them the one with the smallest expected loss. If this expected loss is so large that one of the discarded options may have a smaller expected loss, then we revise the first step and retain a larger set of options. This is equivalent to defining a partition of the options to sets. The first set contains the prime candidates for selection, the second and subsequent sets contain less and less likely candidates. The search starts with the first set and the winner of the contest is either selected as the overall winner, or is included in the contest with the second set. This involves some improvisation and errors in the overall selection may be committed when the partitioning (division) of the options is inappropriate or there are some 'surprises'—options that have very small probabilities and yet also very small expected losses.

The probabilities used for partitioning may be based on an estimate $\hat{\theta}$ or solely on prior information. Using $\hat{\theta}$ would seem to be preferable. However, this may be a laborious exercise that cannot be conducted when the client wants to avoid delays due to data collection and evaluation of $\hat{\theta}$. Much more time may be available before data collection and it is more convenient to have a single long session (meeting) than many short sessions of elicitation. The data may be collected in stages. In the first stage, a small dataset is collected that is sufficient for eliminating a large fraction of the options. The second stage can then focus on selection of one of the retained options by exact evaluations.

Instead of the probabilities, the options may be grouped according to their expected losses. These need not be calculated (with precision) but may be based on guesses. Plausible ranges (ball-park figures) may suffice to form a suitable partitioning of the options. The decision may be so involved, and thus expensive, that a second-rate solution arrived at with a limited effort and low cost may be preferable to the option that has the smallest expected loss. The cost of the analysis should be regarded as a component of the (expected) loss.

5.5.2 Continuum of options

When the options form a continuum, such as a real interval, selecting the option that corresponds to the correct state, even with a small probability, is a tall order. Getting to it as close as possible in expectation is a more realistic proposition. This corresponds to estimation in the established form of inference. However, we reject the ubiquitous choice of the mean squared error (MSE). Here the qualifier 'mean' is the same as 'expected'. The MSE has the property that d times greater error in estimation, $e = \hat{\theta} - \theta$, is associated with d^2 times greater loss. Further, it is symmetric; only the absolute value of the error matters. An alternative for the first property is the mean absolute error (MAE), in which the loss is equal (or proportional) to the absolute value of the error.

Suppose $\hat{\theta}$ is an unbiased normally distributed estimator of θ, with sampling variance τ^2. Then the fiducial distribution of θ is $\mathcal{N}(\hat{\theta}, \tau^2)$. We consider the class of estimators $\hat{\theta} + c$, where c is a constant, and show that the smallest MAE is attained for $c = 0$. Let $z = c/\tau$. The MAE of $\hat{\theta} + c$ for θ is

$$\int_{-\infty}^{+\infty} |\hat{\theta} + c - x| \, \varphi\left(x; \hat{\theta}, \tau\right) dx$$

$$= \tau \int_{-\infty}^{z} (z - y)\varphi(y) \, dy + \tau \int_{z}^{+\infty} (y - z)\varphi(y) \, dy$$

$$= \tau \left[(z - y)\Phi(y)\right]_{-\infty}^{z} + \tau \int_{-\infty}^{z} \Phi(y) \, dy + \tau \int_{-\infty}^{-z} (-y - z)\varphi(y) \, dy$$

$$= \tau \left\{\Phi_1(z) + \Phi_1(-z)\right\}.$$

The differential of this function of c is $\Phi(z) - \Phi(-z) = 2\Phi(z) - 1$. It vanishes at $z = 0$, that is, $c = 0$, where the MAE attains its minimum, with estimator $\hat{\theta}$. The minimum MAE is equal to $2\tau\varphi(0)$. The same result may be deduced by appeal to symmetry, ruling out any $c \neq 0$ as a solution.

The MAE is a special case of piecewise linear loss, in which the estimation error $e = \hat{\theta} - \theta$ is associated with loss e when $e > 0$ and with loss $-Re$ when $e < 0$; R is a positive constant. The MAE corresponds to $R = 1$. For the asymmetric case, $R \neq 1$, we consider the same class of estimators $\hat{\theta} + c$. The expected loss with the piecewise linear loss is evaluated using the by-now familiar calculus.

$$\text{MAE}\left(\hat{\theta} + c; \theta, \tau\right) = R\tau \int_{-\infty}^{z} (z - y)\varphi(y) \, dy + \tau \int_{z}^{+\infty} (y - z)\varphi(y) \, dy$$

$$= \tau \left\{R\Phi_1(z) + \Phi_1(-z)\right\}.$$

We find the minimum of this function of $c = \tau z$ as the root of the differential

$$\frac{\partial \text{MAE}}{\partial c} = (R + 1)\,\Phi(z) - 1,$$

which is

$$c^* = \tau\Phi^{-1}\left(\frac{1}{R+1}\right). \tag{5.6}$$

The minimum MAE is $(R+1)\tau\varphi(c^*/\tau)$. For $R = 1$ we recover the earlier result of symmetry: $c^* = 0$, with expected loss $2\tau\varphi(0)$. The optimal shift c^* is a decreasing function of R. Greater R corresponds to relatively stronger aversion to negative errors, to which we respond by smaller (more negative) adjustment c^* in the estimator $\hat{\theta} + c^*$.

The piecewise quadratic loss is motivated similarly by reflecting the uneven consequences of the positive and negative errors. Derivation of the expected loss involves the same steps as for piecewise linear loss, except for applying integration by parts twice. We obtain the expression

$$Q = 2\tau^2\left\{R\Phi_2\left(z\right) + \Phi_2\left(-z\right)\right\}.$$

Recall that $\Phi_2(z) = \frac{1}{2}(1 + z^2)\Phi(z) + \frac{1}{2}z\varphi(z)$, which implies that $\Phi_2(z) + \Phi_2(-z) = \frac{1}{2}(1 + z^2)$. Therefore

$$Q = \tau^2\left(1 + z^2\right) + 2\tau^2(R - 1)\Phi_2\left(z\right).$$

For symmetry, that is, $R = 1$, this reduces to $Q = \tau^2(1 + z^2)$, which is minimised for $c^* = 0$, confirming that $\hat{\theta}$ is the efficient (minimum-MSE) estimator in the class of estimators $\hat{\theta} + c$. More generally, the expected loss is minimised for the solution of the equation

$$(G(z, R) =)\ \ z + (R - 1)\Phi_1(z) = 0, \tag{5.7}$$

which can be found by the Newton-Raphson method. Details are left for an exercise. The solution is a decreasing function of R. This is shown by applying the implicit function theorem. Both $\partial G(z, R)/\partial z = 1 + (R - 1)\Phi(z)$ and $\partial G(z, R)/\partial R = \Phi_1(z)$ are positive throughout, therefore $\partial z/\partial R < 0$ for all $R > 0$.

We conclude this section with an argument that a continuum of options has to be considered only in some esoteric settings. Suppose a space of options is represented by a parameter θ in the unit interval $(0, 1)$. This implies that there is a material difference between the values $\theta = 0.42$ and, say, $\theta = 0.42 + e^{-72}$; that is, different actions would be taken if θ were known to be equal to one or the other value. A piece of equipment may be fine-tuned according to the value of θ, so that it has a continuum of settings, but in practice this continuum is always reduced to a finite, albeit large, set of options, usually to a regular grid of values.

These settings may not be known at the time of the analysis, and so a solution for a continuum of options might protect the client from a solution based on a set of values of θ that is too coarse, such as $0.0, 0.2, \ldots, 1.0$, where $0.0, 0.05, \ldots, 1.0$ may later turn out to be more appropriate. The client can be protected in this respect by the analysis based on the set of options $0.00, 0.01$,

..., 1.00; the selected option could then be rounded if necessary. An analysis with a continuum of options is an extravagant addition to the compexity of the analysis which could be avoided by postponing the analysis until the list of options, or at least their level of coarseness, is established. The analyst's task is to tailor the analysis to the client's perspective, and one is bound to be handicapped in this endeavour if the client's perspective is not established and disclosed, at least in an outline, upfront to the analyst.

5.6 Further reading

The editorial Longford (2005) questions the general idea of model selection, pointing to problems in which its application would result in failure. There is an extensive statistical literature on classification, and classification is an important task in machine learning, but its main concern is with finding suitable classes (partitions), not with assignment to a priori defined classes. The literature is rarely concerned with the consequences of errors; in some contexts, an error is not well defined.

5.7 Exercises

5.1. Review the definitions of parsimony and validity of models. Can these two properties be proven for any particular dataset and model, leaving no doubt that the model contains all the relevant variables and none of the redundant ones?

5.2. Suppose a modelling exercise, in the context of ordinary regression, involves 14 covariates and concludes with the selection of a model with six of them. Model selection was performed by backwards selection, starting with the model with all the covariates and excluding one covariate at a time if its absolute value of the t ratio was smaller than 1.645 (the 95th percentile of the standard normal). Discuss how likely is the selected model valid and parsimonious. How could the model selection algorithm be adapted to increase the chances that the selected model would be valid?

5.3. Explain why one should reject the singular zero as a value of the difference of the expectations of two normal random samples. Or of two random samples from any other continuous distribution. How about zero and six other values, all of them small in absolute value? Or a sequence of values that converges to zero?

5.4. Define asymmetric loss in general and give examples in which it may be useful or relevant.

5.5. Reproduce a panel of Figure 5.2 for a different value of R, and redraw it for a selection of values of τ. By trial and error, find the borderline, where the two equilibria converge as τ is increased. Apply a more sophisticated method (e.g., Newton method) to find the borderline, and confirm that the two methods yield the same result. Consider the transformation $\omega = 1/\tau$; does it simplify the search for the borderline? Show that the two equilibria converge as the thresholds T_1 and T_2 are moved closer to $\bar{T}$.

5.6. Describe the changes that have to be made in the formulae for the expectation of the piecewise constant loss when the normal distribution is replaced by a different centred symmetric distribution. What about a distribution that is not symmetric (but has zero expectation)?

5.7. Describe the behaviour of the function $\Phi(z + \Delta) - \Phi(z)$, for a positive constant Δ: what are its limits at $\pm\infty$, ranges where it is monotone, and its extremes? How would your conclusions differ if Φ is replaced by the distribution function of the t distribution with a small number of degrees of freedom?

5.8. The piecewise constant loss function is discontinuous at the thresholds T_1 and T_2; see Figure 5.1. Is that a problem; in general, or in any specific settings?

5.9. Derive the expected loss functions and the balance function for the asymmetric piecewise constant loss for the small vs. large dilemma.

5.10. Prove that the extreme of a quadratic function is found by a single iteration of the Newton-Raphson algorithm.

5.11. Draw a set of expected loss functions Q_s, based on piecewise constant loss, for a selection of states separated by symmetric thresholds T_1 and T_2 ($T_1 = -T_2$), to illustrate the effect of the thresholds. Repeat this exercise with piecewise linear function in place of piecewise constant. Draw the balance functions instead of the expected loss functions.

5.12. In an example of asymmetric piecewise linear loss, explore how the domain of V_s, the interval delimited by the two equilibria, moves as R is increased. Can the domain move all the way to the left of $\bar{T}$, when $E_2 < \bar{T}$? Or even to the left of T_1? How about moving it to the right of $\bar{T}$ and to the right of T_2?

5.13. Justify by intuition that with increasing standard error τ the verdict gravitates towards option L when the loss is piecewise constant. What about piecewise linear or quadratic loss?

5.14. Approximate the (asymmetric) piecewise linear loss function by a piecewise quadratic function, with one segment for $\theta < \bar{T}$ and another for $\theta > \bar{T}$, so that the segments have their minima at the respective thresholds T_1 and T_2.

5.15. Describe the entire variety of loss functions introduced in Chapter 5. Does the product of two loss functions have all the properties of a loss function? What about the square root, logarithm and exponential of a loss function?

5.16. Derive the balance function in Example 11 and explore its properties. Plot the equilibrium function(s).

5.17. List the properties of a loss matrix for the setting in which the states are a set of ordinal categories. Discuss the difficulties in defining a loss matrix when some or all of its elements are functions.

5.18. A community is planning to build a new elementary school. The key question is: how many classrooms should the school have? Describe this problem in the context of a decision with a multitude of options. Which factors should be taken into account in defining the losses? Consider the perspectives of the various stakeholders: parents, the local (government/educational) authority, the professional organisation (union) of teachers, the building contractors, and the like.

5.19. Collect the arguments for and against using MAE and MSE in estimation. Which of them can or should be informed by the perspective of the client? How are MAE and MSE adapted to be asymmetric?

5.20. How can (or should) a continuum of options be reduced to a finite set of (ordinal) options?

5.21. Derive the minimum MAE given in text following equation (5.6).

5.22. Implement a method for solving equation (5.7). Plot the solution as a function of R. Confirm the results by simulations. That is, generate a large number of replicates of $\hat{\theta}$ and average the replicate losses for a set of values of c. The average should be smallest for c^* given by (5.6).

5.23. Compile a list of conditions that the loss functions in a 3×3 loss matrix have to satisfy. If you find the task too onerous, assume that one of the loss functions is constant. If the task is too easy, solve it also for a 4×4 loss matrix.

5.24. Compile a function in R that draws the panels of Figures 5.2, 5.4, 5.6 and 5.10.

Appendix

This Appendix contains some technical material that is not essential for the continuity of the text in the chapter.

A. Expected loss Q_1 in equation (5.3)

We show that $Q_1(\hat\theta)$ is increasing for $\hat\theta < \bar{T}$ and decreasing for $\hat\theta > \bar{T}$. The partial differential of this function with respect to $\hat\theta$ is RF_1, where

$$F_1\left(\hat\theta\right) = 2\Phi(\bar{z}) - \Phi(z_1) - \Phi(z_2) . \tag{5.8}$$

Consider the function $G(\Delta; z) = 2\Phi(z) - \Phi(z+\Delta) - \Phi(z-\Delta)$, which is equal to $F_1(\hat\theta)$ for $\Delta = z_2 - \bar{z}$ and $z = \bar{z}$. Its partial differential with respect to Δ is $g(\Delta; z) = \varphi(z - \Delta) - \varphi(z + \Delta)$. So, for a fixed z, $G(\Delta; z)$ attains its sole extreme at $\Delta = 0$, and there $G(0, z) = 0$ for all z. For $z > 0$, $g(\Delta; z)$ switches its sign from negative for $\Delta < 0$ to positive for $\Delta > 0$. Therefore G is decreasing for $\Delta < 0$ and increasing for $\Delta > 0$, and so the extreme at $\Delta = 0$ is a minimum of G for any fixed $z > 0$. By symmetry, the extreme is a maximum of G when $z < 0$. In either case, the value at the extreme is $G(0; z) = 0$. In summary, G is non-positive for $z < 0$ and nonnegative for $z > 0$, and attains zero only when $\Delta = 0$. Therefore F_1 is positive when $\hat\theta < \bar{T}$ and negative when $\hat\theta > \bar{T}$, and so Q_1 has the shape of a 'mound' symmetric around zero.

B. Continuation of Example 9

The partial differential of $B(0; T_2)$ in equation (5.4) with respect to T_2 is

$$(R - 1)\left\{2\Phi\left(\frac{T_2}{\tau}\right) - 1\right\} + 1 > 0,$$

so the function is increasing. Therefore, there is a critical value $T_2^{(0)}$ such that for $T_2 < T_2^{(0)}$ verdict V_1 is issued for all $\hat\theta$ (including $\hat\theta = \bar{T} = 0$). For $T_2 > T_2^{(0)}$, V_s is issued for $\hat\theta$ in an interval $(-E, E)$ for $E > 0$ that depends on T_2, and V_1 is issued outside this interval.

The differential of the balance function with respect to $\hat\theta$, involved in $\bar{z}$, z_1 and z_2, is

$$F\left(\hat\theta\right) = (R - 1) F_1\left(\hat\theta\right) + F_2\left(\hat\theta\right) ,$$

where F_1 is given by equation (5.8) and $F_2(\hat\theta) = 2\Phi(\bar{z}) - 1$. Both functions F_1 and F_2 vanish at $\hat\theta = \bar{T}$; $F_1(\bar{T}) = F_2(\bar{T}) = 0$. Further, both functions are positive for $\hat\theta < \bar{T}$ and negative for $\hat\theta > \bar{T}$. We proved this for F_1 in Appendix A and it is obvious for F_2. From this we conclude that the balance function increases for $\hat{\theta} < \bar{T}$, attains its maximum at $\hat{\theta} = \bar{T}$ and then decreases for $\hat{\theta} > \bar{T}$. As a result, either there is no equilibrium and V_1 is the appropriate verdict with no regard for the value of $\hat{\theta}$, or there are two (symmetrically located) equilibria $E > \bar{T}$ and $2\bar{T} - E$.

The behaviour of the balance function with respect to $\Delta = \frac{1}{2}(T_2 - T_1)$ or τ can be explored by the same approach, studying the appropriate partial differentials of the balance function.

C. Continuation of Example 10

To prove symmetry of the balance function in (5.5), consider how z_1 and z_2 are altered when we change $\hat{\theta}$ to $-\hat{\theta}$. If $\hat{\theta}$ corresponds to z_1, then $-\hat{\theta}$ corresponds to $(T_1 + \hat{\theta})/\tau = -z_2$. Similarly, z_2 is paired with $-z_1$ and $\bar{z}$ with $-\bar{z}$. Recall that $\Phi_2(-z) = \frac{1}{2}(1 + z^2) - \Phi_2(z)$ and $\Phi_1(-z) = \Phi_1(z) - z$. These identities imply that

$$\frac{1}{\tau^2} B\left(-\hat{\theta}\right) = (R+1)\left\{2\Phi_2(z_2) - 2\Phi_2(z_1) + z_1^2 - z_2^2\right\}$$
$$+ 2R(z_2 - z_1)\bar{z} - 2R(z_2 - z_1)\Phi_1(\bar{z}) - 1 - z_1^2,$$

and this coincides with $B(\hat{\theta})/\tau^2$ after substituting $z_1 + z_2$ for $2\bar{z}$ and consolidating the factors of z_1^2 and z_2^2.

To study the limiting behaviour of $B(\hat{\theta})$ for $\hat{\theta}$ at $+\infty$, we use the fact that z_1, z_2 and $\bar{z}$ diverge to $-\infty$, while their differences remain constant. As $z \to -\infty$, $\Phi_1(z)$ and $\Phi_2(z)$ converge to zero, so $B(\hat{\theta})/\tau^2$ has the same limit as $-(1 + z_2^2)$. Hence $B(\hat{\theta}) \to -\infty$ as $\hat{\theta} \to +\infty$. By symmetry, B diverges to $-\infty$ also as $\hat{\theta} \to -\infty$.

Denote $\Delta T = (T_2 - \bar{T})/\tau$; we use ΔT^2 as a shorthand for $(\Delta T)^2$. The value of the scaled balance function at $\hat{\theta} = \bar{T}$ is

$$\frac{1}{\tau^2} B(\bar{T}) = 2(R+1)\left\{\Phi_2(\Delta T) - \Phi_2(-\Delta T)\right\} - 4R\Delta T\Phi_1(0) - 1 - \Delta T^2$$
$$= 4(R+1)\Phi_2(\Delta T) - (R+2)\left(1 + \Delta T^2\right) - 4R\Delta T\varphi(0),$$

after using the identity $2\Phi_2(-z) = 2\Phi_2(-z) = 1 + z^2 - 2\Phi_2(z)$. As a function of ΔT, $B(\bar{T})/\tau^2$, with $\bar{T}$ fixed, can attain either sign. For $\Delta T \to 0$, it converges to $2(R+1)\Phi(0) - R - 2 = -1$, and its limit as $\Delta T \to +\infty$ is $+\infty$; more precisely, $B(\bar{T})/\tau^2/\Delta T^2 \to R$. In fact, the function is increasing, although the proof of this is rather intricate, and for a purist, incomplete, as we show next.

The respective first- and second-order differentials of the scaled balance function with respect to ΔT are

$$4(R+1)\Phi_1(\Delta T) - 2(R+2)\Delta T - 4R\varphi(0)$$
$$4(R+1)\Phi(\Delta T) - 2(R+2).$$

The latter is increasing in ΔT and its unique root is

$$\Delta T^{(0)} = \Phi^{-1}\left(\frac{R+2}{2R+2}\right).$$

The first-order differential is decreasing for $\Delta T < \Delta T^{(0)}$ and increasing for $\Delta T > \Delta T^{(0)}$. Its value at $\Delta T^{(0)}$, its minimum, is

$$4(R+1)\,\varphi\left(\Delta T^{(0)}\right) - 4R\varphi(0).$$

We do not have an analytical proof that this function is positive for all $R > 0$, but this is easy to check by plotting the function. Its limit at $R \to 0$ is zero, and it is concave and increasing, with the limit of 1.596 as $R \to +\infty$.

6

Study design

Most studies that are of consequence are expensive undertakings, and experience accumulated over time is conclusive about the value of designing them—investing in their preparation and planning by reviewing the know-how and experience from past studies with similar agenda and context. Playing out plausible scenarios to understand the advantages of alternative approaches and becoming wise to pitfalls that would undermine the purpose and intent of the study are also parts of this process. When a lot is at stake, we consult, inquire, deliberate, replace hope with controlled intent, rehearse the planned procedures, and then commit ourselves to the execution of the study: recruitment and intervention, if applicable, measurement, data collection and construction of the database, followed by analysis. We follow the tailors' dictum of measuring three times before cutting the cloth.

6.1   Design and analysis

The result of planning a study, that is, designing it, is a protocol. It is a document that prescribes all activities in the conduct of the study. An ideal protocol would be entirely self-contained, so that the personnel charged with conducting the study would not have to consult any other document, other than the references cited in it, nor any personnel other than persons and offices listed in it, with the understanding that such personnel would provide competent and constructive responses without any delay.

There is a strong rationale to plan also the analysis but there is often much less at stake. An analysis usually requires much more modest resources than the conduct of a study up to finalising its dataset. This is not to disparage the analysts and their expertise nor to underrate the importance of the analysis. But their work can be improved or corrected, if this turns out to be necessary, at an expense that is a small fraction of the expense on collecting a new dataset by an amended protocol, or by adhering to the original protocol, when an irreversible fault or failure occurs in its implementation.

A different rationale for planning an analysis, and disclosing the plan, for instance by publishing it before the data collection commences, is to be transparent and dismiss any suspicion that the method of analysis might have been selected after data collection to fit an undisclosed agenda. A study should have its inferential goals clearly formulated at the design stage and should not be conducted with the sole purpose of searching for some finding that would have been remarkable and worth reporting had it been declared at the design stage. The value (or remarkability) of such a finding should always be considered in the context of the entire range of inferences that were contemplated. That is, even the search (a data-fishing expedition) should have a protocol, outlining the variety of contemplated inferences—the variety of fish we will try to catch, and the pond(s) where we'll be fishing.

Without disclosing the plan in advance, the analyst may be accused of having tried several analyses, and then admitted to having conducted only the one that yielded the most favourable result, as judged by an agenda that is alien to client's perspective. Note that the agenda may not be alien to a client's representative, who fails to adhere faithfully to his or her remit.

There is a vast variety of designs and each type and circumstance require a different approach, with an emphasis on different aspects. While analysis may appear to be a more attractive and valuable activity for an analyst, the design is more challenging and often far more valuable for the project as a whole. In some settings, the design may appear to deflate the value of the analysis by making it simple. Many complex analyses are brought about by poor design—they are comparable to acrobatic rescue acts that command a great kudos when they are accomplished with success. However, the kudos is even greater when the study is accomplished without having to employ any analytical acrobatics and its analysis entails no contentious caveats.

There is nevertheless ample scope for intricate analysis when the theoretically optimal design cannot be implemented, not even after some adaptation. In a general example of such a study, the data have been collected in the past for another purpose, and so they are now available at a nominal cost. In another example, a good (or optimal) design is readily identified but many of the (human) subjects whom we try to recruit reject it, or ethical and other standards adopted in the relevant field of research bar it outright.

Suppose a study is conducted to compare the prevalence of a disease in subjects with two kinds of diet. In the optimal design, the diet would be assigned to each subject in such a way that the two groups of subjects defined by the diet assigned would be similar in all aspects that are unaffected by the assignment (the background), and therefore a comparison of their outcomes at the end of the study period would be straightforward. Randomisation in a sufficiently large study satisfies this standard. But where are the subjects who would be willing to adhere for several months to a prescribed diet, however benign the prescription?

When we have no control over the assignment of the diet, or we inquire about each subject's diet in the past (in a retrospective study), the comparison is far from straightforward, because the two groups of subjects differ in their background—they are not equivalent, and their comparison is not of like with like. So, keep your analytical and computational powder dry, and use it when

other means fail, in particular when the opportunity to design is limited or
does not exist at all.

6.2   How big a study?

We address the setting in which we are planning to evaluate an estimator $\hat{\mu}$
of a target μ. More precisely, for each contemplated design $\mathcal{D}$ we consider an
estimator $\hat{\mu}_\mathcal{D}$. To avoid some distractions, we assume that the distribution
of each estimator $\hat{\mu}_\mathcal{D}$ is known. This is a reasonable assumption. It would
be singularly unwise to commit the analysis to an estimator with unknown
properties. To deal with a simple problem to start with, we assume that the
designs are fully characterised by their (fixed) sample sizes n, so that the
designs can be denoted by $\mathcal{D}_n$, or by the integer n. Further, we assume that the
estimator associated with $\mathcal{D}_n$, denoted by $\hat{\mu}_n$, has the sampling distribution
$\mathcal{N}(\mu, \tau_n^2)$ with $\tau_n^2 = \sigma^2/n$, and $\sigma^2 > 0$ is known. No generality would be lost
by assuming that $\sigma^2 = 1$.

The client wants to know whether μ is positive or negative, and their loss
is piecewise constant with $L_{+-} = 1$ and $L_{-+} = R = 6$. In the notation
introduced in Section 1.2, $K = 2$ and $T_1 = 0$. How large should n be so that
the interests of the client would be served best?

The starting point is to establish the cost of the study and the expected
losses as functions of the sample size n and, more generally, of the design $\mathcal{D}$.
Suppose the conduct of the study and its analysis cost $C_n = an + b$ lossiles,
where a and b are known positive constants.

In the planned analysis, we would evaluate $\hat{\mu}$ and then issue verdict V_+ if

$$\hat{\mu} > \frac{\sigma}{\sqrt{n}} \Phi^{-1}\left(\frac{1}{R+1}\right),\tag{6.1}$$

and verdict V_- otherwise. The probabilities of these verdicts are $P(V_+; \mu, n)$
$= 1 - P(V_-; \mu, n)$ and

$$
\begin{aligned}
P(V_-; \mu, n) &= P\left\{\hat{\mu} < \frac{\sigma}{\sqrt{n}} \Phi^{-1}\left(\frac{1}{R+1}\right)\right\} \\
&= P\left\{Z < \Phi^{-1}\left(\frac{1}{R+1}\right) - \frac{\mu\sqrt{n}}{\sigma}\right\} \\
&= \Phi\left\{\Phi^{-1}\left(\frac{1}{R+1}\right) - \frac{\mu\sqrt{n}}{\sigma}\right\},
\end{aligned}
$$

where $Z = \sqrt{n}(\hat{\mu} - \mu)/\sigma$; Z has the standard normal distribution. We study
these two probabilities as functions of μ and n; $P(V_-)$ is relevant for $\mu > 0$,
and $P(V_+)$ is relevant for $\mu < 0$. We want to select (a sample size) n for

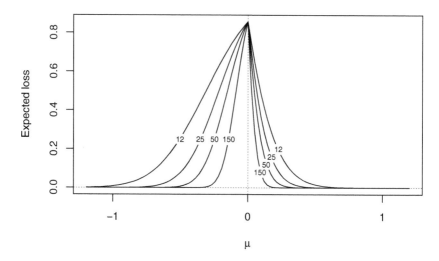

FIGURE 6.1
Expected losses $P(V_+)$ and $RP(V_-)$ in the design of a study of the sign (positive or negative) of the mean of a normal random sample. The sample sizes, $n = 12, 25, 50$ and 150, are indicated in the plot; $R = 6$.

the study. The range of plausible (possible) values of μ, that is, the range of scenarios that we may encounter in the study, is an important factor.

For a given $\mu > 0$, $P(V_-)$ is a decreasing function of n, and for $\mu < 0$, $P(V_-)$ increases in n. For fixed n, $P(V_-)$ increases for $\mu < 0$ and decreases for $\mu > 0$. A pessimist would be concerned with the maxima or, more precisely, the suprema of the functions $P(V_+)$ and $RP(V_-)$. Both functions attain their suprema for $\mu \to 0$, and the suprema are both equal to $R/(R+1)$. They do not depend on n, so the pessimistic view leads to a blind alley: expected loss of up to $R/(R+1)$ cannot be ruled out and this upper bound cannot be reduced by increasing the sample size n. The optimist's proposal to discount the possibility that μ might be close to zero is not credible or constructive either.

The two expected loss functions, $P(V_+)$ and $RP(V_-)$, are plotted in Figure 6.1 for sample sizes $n = 12, 25, 50$ and 150, indicated on the curves. The diagram confirms that all the functions attain their suprema at $\mu = 0$ and that the expected losses are very small for μ distant from $T_1 = 0$. For greater n the expected loss is large, say greater than 0.6, in a narrower range. So, greater sample size pays off by reduced chance of a large expected loss.

We proceed by integrating (averaging) the expected loss over a prior distribution of μ. A noninformative prior, such as $\mathcal{N}(0, s^2)$ for a large positive s^2, is not appropriate because it is too optimistic, assigning a very small probability (or density) to values of μ in the proximity of zero, where the decision about the size of μ is most contentious. A centred normal distribution with a small variance is perhaps too pessimistic because it assigns a large probability of μ being close to zero.

We adopt the uniform distribution on $(-0.5, 0.5)$ as the prior for μ. We emphasise that this is not 'exact' science, especially without delving into the minutiae of the planned study. The only way to address such uncertainty is by applying a few other prior distributions, after sorting out how to deal with a single prior, and then gaining an understanding of the effect a prior has on the integrated expected loss.

Thus, we evaluate the expression

$$E(n, R; \sigma) = \int_{-\infty}^{0} P(V_+ ; \mu, n) f_{\text{pri}}(\mu) \, d\mu + R \int_{0}^{+\infty} P(V_- ; \mu, n) f_{\text{pri}}(\mu) \, d\mu$$

$$= \int_{-0.5}^{0} \left[1 - \Phi \left\{ \Phi^{-1} \left(\frac{1}{R+1} \right) - \frac{\mu\sqrt{n}}{\sigma} \right\} \right] d\mu$$

$$+ R \int_{0}^{0.5} \Phi \left\{ \Phi^{-1} \left(\frac{1}{R+1} \right) - \frac{\mu\sqrt{n}}{\sigma} \right\} d\mu$$

$$= 0.5 + \int_{-0.5}^{0.5} \{-1 + (R+1)I_{\mu>0}\} \Phi \left\{ \Phi^{-1} \left(\frac{1}{R+1} \right) - \frac{\mu\sqrt{n}}{\sigma} \right\} d\mu,$$

where $f_{\text{pri}}(\mu) = 1$ for $\mu \in (-0.5, 0.5)$ is the density of the uniform distribution and I is the indicator function, equal to unity when its logical argument in the subscript ($\mu > 0$ in our case) is true and equal to zero when it is false.

This integral cannot be evaluated analytically, so we have to resort to an approximation. We define a fine regular grid of values of μ in the interval $(-0.5, 0.5)$, evaluate the integrand for these values and approximate the integral by their average. This simple method has a variety of refinements but they are not particularly useful in our case. A reliable diagnostic for the closeness of the approximation is to compare the result after repeating the evaluation with a coarser grid of evaluation points.

Figure 6.2 presents the integrated losses as functions of the sample size for a selection of values of R. The integrated loss decreases with sample size, more steeply for small sample sizes. We might set an upper bound and, for a given value of R, find the smallest sample size for which the integrated loss falls below this bound. A more appropriate approach takes into account the cost of the study and the value of the knowledge acquired by the study, that is, the *potential* of the study. It is denoted by κ.

As an aside, the grey dashes illustrate that the approximation is adequate. The solid lines are based on integration with 1000 evaluation points. The grey dashes mark the results for $R = 15$ with 500 evaluation points. The difference from the results with 1000 points cannot be discerned. For example, the values for the two approximations at sample size $n = 10$ are 0.5019 and 0.5017. For $n \in (10, 200)$, the differences range from 0.00014 to 0.00029, and for $R < 15$ the differences are even smaller.

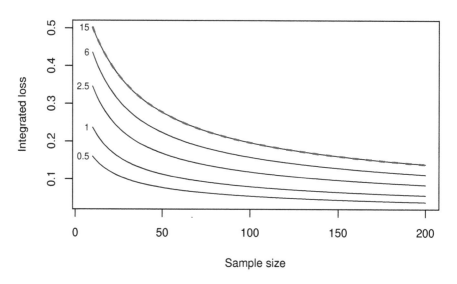

FIGURE 6.2
Integrated loss as a function of sample size n and for values of R indicated
at the left-hand margin. Numerical integration based 1000 evaluation points.
Thick grey dashes mark the result for $R = 15$ with 500 evaluation points.

Suppose knowing whether $\mu < 0$ or $\mu > 0$ is worth $\kappa = 0.9$ lossiles and
the study would cost $C_n = 0.44 + 0.005n$. Then the conduct of the study
is a profitable undertaking for sample sizes n for which $C_n < \kappa - E(n, R)$.
The two sides of this inequality are plotted in Figure 6.3. The linear cost C_n
exceeds the knowledge gained, $\kappa - E(n, R)$, for $n < 14$ and $n > 45$. A small
study, with $n < 14$, is not useful because what we learn from it does not
compensate the outlay on it, dominated by the set-up cost of 0.44 lossiles.
For $n > 45$, the study is too large and expensive, so that what we learn
is a poor return on the outlay. The unit-cost component increases with n
(equal to $44 \times 0.005 = 0.22$ for $n = 44$), but offers diminishing returns as
the 'knowledge' function $\kappa - E(n, R)$ gradually becomes flatter and increases
towards the potential $\kappa = 0.9$ very slowly.

Thus, the range of sample sizes n for which the study is useful is $15 - 44$.
We can opt for learning most ($n = 44$), gain largest profit $\kappa - a - bn$ ($n = 27$),
or hedge bets against something untoward by allowing for a few units (say,
10%, that is, 3) to be lost in the process of establishing their values of the
outcome variable.

This part of the study design would be completed if there were no uncertainty, discord or ambiguity about the settings: the cost parameters $a = 0.44$
and $b = 0.005$, the value of the knowledge, $\kappa = 0.9$, the variance σ^2 and the
loss ratio $R = 6$. We set aside the consideration of the loss ratio, because each

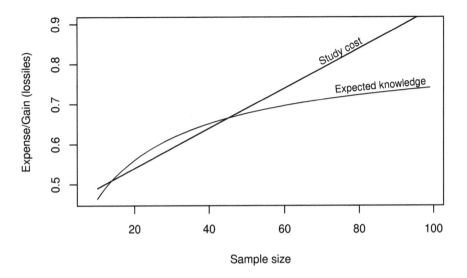

FIGURE 6.3
The knowledge gained and the cost of the study; scenario of Figure 6.2 with
$R = 6$, $\kappa = 0.9$ and $C_n = 0.44 + 0.005n$.

value R corresponds in effect to a different currency, and so its change has to
be reflected by resetting the values of a, b and κ.

First, the knowledge gained (the study 'profit') depends on a and κ only
through the difference $\kappa - a$, so we lose no generality by regarding κ as fixed. If
the set-up cost a is greater by $\Delta > 0$, then the study cost is uniformly greater
by Δ, and the range of sample sizes for which the study is useful is narrower,
centred around $n = 27$, where the margin of knowledge gained over the cost is
largest (0.0266). Therefore, $0.44 + 0.0266 = 0.467$ is the upper bound on the
set-up cost—if the set-up cost is higher the study is not worthwhile. Figure
6.4 presents a more comprehensive view by plotting the study costs based on
coefficients $a = 0.42, 0.44, 0.46$, crossed with $b = 0.004, 0.005$ and 0.006 (nine
settings). The black lines reproduce Figure 6.3; the grey lines represent the
other eight settings recognised by their distinct slopes (for b) and positions
(lowest for $a = 0.42$).

The thin horizontal segments at the bottom margin delineate the ranges of
profitability for the pairs of cost coefficients. There is no segment for the configuration ($a = 0.46; b = 0.006$) which is too expensive; the study would not
be profitable for any sample size. The horizontal segments indicate that the
sample sizes in the range $20 - 26$ are profitable with all the other eight configurations. With the most expensive configuration, $(0.46, 0.006)$, the expected
loss attains its minimum of 0.018 for $n = 23$.

With a fixed, higher unit-cost b (steeper cost function) reduces the range of
sample sizes that are useful, and lower b widens it. The expected loss depends

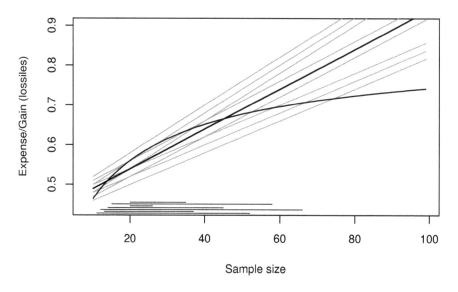

FIGURE 6.4
Sensitivity analysis of the sample size for which the design is profitable; study costs with $a = 0.42$, 0.44 and 0.46 crossed with $b = 0.004$, 0.005 and 0.006.

on the standard deviation σ and sample size n only through $\sqrt{n}/\sigma$, so multiplying σ^2 by a factor $d > 0$ is negated by changing n to dn. However, C_n has then to be replaced by C_{dn}.

Figure 6.5 explores the dependence of the range of sample sizes on the standard deviation σ, with the scenarios of Figure 6.3 and $\sigma = 0.9$, 1.0, 1.1. Greater standard deviation (or variance) implies greater sample size. The ranges of 'profitable' sample sizes are very sensitive to the value of σ. These ranges, indicated by the segments at the bottom margin of the diagram, are $9-28$, $14-45$ and $22-70$ for the respective variances $\sigma^2 = 0.81$, 1.00 and 1.21. It is rather fortuitous, that their overlap, $22-28$, overlaps also with the profitable range of sample sizes $20-26$ found in Figure 6.4.

It may still be rather premature to conclude the sample size calculation with the proposal of $22 \leq n \leq 26$ because we have explored the impact of uncertainty about (a, b) and σ separately. For instance, re-drawing Figure 6.4 with alternative values of σ, and Figure 6.5 with alternative values of (a, b) has considerable merit.

Further, we could explore the sensitivity of the profitable range with respect to the losses. The parameters a, b, κ and R are all expressed in lossiles anchored at the loss $L_{+-} = 1$. We remain faithful to this currency but, with a, b and κ fixed, we have to allow L_{+-} and L_{-+} to differ from their originally considered values of 1 and $R = 6$.

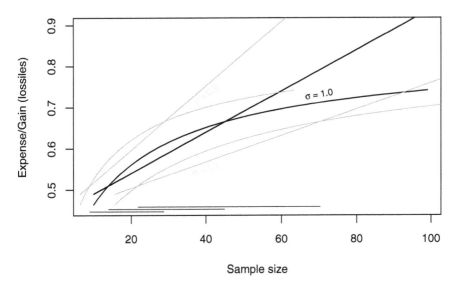

FIGURE 6.5
Sensitivity analysis of the sample size for which the design is profitable; study costs $C_n = 0.44 + 0.005n$ and standard deviations $\sigma = 1.0$ (black), 0.9 and 1.1 (grey).

6.3 Planning for impasse

In the previous section, we established the cost of data collection and the variance of the outcome variable, both functions of the sample size n, as sources of uncertainty in the design of a study. A layer of uncertainty additional to them arises when the loss ratio and the threshold are represented by their plausible ranges. In this case, we have to draw up plans for a possible impasse, when one option would be preferred for some combinations of plausible values and another for other combinations.

It is desirable to avoid impasse; an unequivocal verdict is always preferred. Impasse can be interpreted as not knowing what to do, in effect a waste of the study conducted specifically to elect one of the available options. We may invest in more detailed elicitation, resulting in narrower plausible ranges for the elicited parameters, but this may entail other costs in terms of time spent, contention and inconvenience to the parties involved in the elicitation process, and consequent loss of good disposition towards the analyst. Foremost, the clients must not be under any pressure that might cause them to rig the elicitation by agreeing to a plausible range narrower than what is well supported and justified by the available information and by solid judgement detached from the ubiquitous pressures and false incentives to disregard uncertainty.

A more constructive approach may be to iterate between sampling design and elicitation, identifying the parameters for which further elicitation may be productive, or more productive than for other parameters. This involves a combination of the sample size calculation and discussion with the client (or committee) to balance the threats of impasse with elicitation targeted on particular parameters. For which parameter a relatively small narrowing of the plausible range would be rewarded by substantial reduction of the threat? This may take the form of setting priorities for the committee.

6.3.1   Probability of impasse

In this section, we assume that elicitation has concluded with a plausible range of the loss ratio R and a plausible set of priors for the difference $\Delta = \mu_2 - \mu_1$. The goal is to find designs (sample sizes) for which the posterior probability of impasse is smaller than a set threshold and, more generally, to establish how this probability is related to the sample size and other factors and parameters considered in the design.

Suppose the planned analysis of the data collected by the study compares the means of two random samples, with sample sizes n_1 and n_2. In the design, a given ratio $r = n_2/n_1$ is planned, compromised only by rounding because both n_2 and n_1 have to be integers. We assume that the two random samples have the same variance σ^2 and that the estimators of the means, $\hat{\mu}_1$ and $\hat{\mu}_2$, are independent and normally distributed. Then the difference $\Delta = \mu_2 - \mu_1$ is estimated without bias by $\hat{\Delta} = \hat{\mu}_2 - \hat{\mu}_1$, with sampling variance $m\sigma^2$, where $m = 1/n_1 + 1/n_2$. We assume that σ^2 is known but explore how the sample sizes depend on this variance. No generality would be lost by setting $\sigma^2 = 1$ and adjusting μ and $\hat{\mu}$ accordingly. The planned study has to choose between actions corresponding to $\Delta > 0$ and $\Delta < 0$. For Δ we have a plausible set of prior distributions $\mathcal{N}(\delta, q\sigma^2)$, where $\delta \in (\delta_-, \delta_+)$ and $q \in (q_-, q_+)$.

A prior distribution is called equilibrium for a particular loss function (or loss ratio R), if the balance function evaluated with this prior vanishes. Equilibrium priors, or their parameters (δ, q) are the solutions of the balance equation

$$\Phi(\tilde{z}) = \frac{R}{R+1}, \tag{6.2}$$

where $\tilde{z}$ is the ratio of the posterior mean and posterior standard deviation,

$$\tilde{z} = \frac{m\delta + q\hat{\Delta}}{m+q} \Big/ \sigma\sqrt{\frac{mq}{m+q}}$$

$$= \frac{m\delta + q\hat{\Delta}}{\sigma\sqrt{mq(m+q)}}.$$

For the prior parameter δ we have the class of solutions

$$\delta^*(q) = \sigma z_R^* \sqrt{\frac{q(m+q)}{m}} - \frac{q\hat{\Delta}}{m},$$

where z_R^* is the solution of equation (6.2). In practice, the data contain much more information than the prior, so $m \ll q$. Then the numerator under the square root can be approximated as

$$mq + q^2 \doteq \left(\frac{1}{2}m + q\right)^2,$$

yielding a linear approximation to the equilibrium function,

$$
\begin{aligned}
\delta^*(q) &\doteq \sigma z_R^* \left(\frac{\sqrt{m}}{2} + \frac{q}{\sqrt{m}}\right) - \frac{q\hat{\Delta}}{m} \\
&= \sigma z_R^* \frac{\sqrt{m}}{2} + q\left(\frac{\sigma}{\sqrt{m}} z_R^* - \frac{\hat{\Delta}}{m}\right).
\end{aligned}
\tag{6.3}
$$

For a realised value of $\hat{\Delta}$ we would evaluate this equilibrium line and check whether it intersects the rectangle of plausible prior parameters, $(q_-, q_+) \times (\delta_-, \delta_+)$. At the planning stage, $\hat{\Delta}$ is not realised, so we simulate its values as random draws from its prior-related distribution $\mathcal{N}\{\delta, (m+q)\sigma^2\}$. We obtain replicate equilibrium lines $\delta^*(q; \hat{\Delta})$, and estimate the probability of impasse by the proportion of these lines that intersect the plausible rectangle.

We can forgo simulations by exploiting the approximation in equation (6.3). Suppose first that R, and therefore z_R^*, is given. Otherwise we would consider a plausible range of values of z_R^*. Denote the intercept of $\delta^*(q)$ in equation (6.3) by a and the slope of $\delta^*(q)$ by b, so that $\delta^*(q) = a + bq$. We distinguish three cases for a. If $a > \delta_+$, then impasse occurs when $a + bq_- > \delta_-$ and $a + bq_+ < \delta_+$. If $\delta_- < a < \delta_+$, then impasse occurs when $\delta_- < a + bq_- < \delta_+$. And if $a < \delta_-$, then impasse occurs when $a + bq_+ > \delta_-$ and $a + bq_- < \delta_+$. These three scenarios are illustrated in Figure 6.6. The shaded region (an angle with its tip at $q = 0$) covers the equilibrium lines that intersect the plausible rectangle. For them, the conclusion of the analysis would be impasse.

Next we evaluate the probability of impasse in each scenario. The evaluations are simple because the intercept a does not depend on $\hat{\Delta}$ and b is a linear function of $\hat{\Delta}$. For scenario A, impasse occurs when

$$(A_- =)\ \sigma z_R^* \sqrt{m} - m\frac{\delta_- - a}{q_-} > \hat{\Delta} > \sigma z_R^* \sqrt{m} - m\frac{\delta_+ - a}{q_+}\ (= A_+), \tag{6.4}$$

and the probability of this event, $A_- > \hat{\Delta} > A_+$, denoted by $\mathcal{I}_A$, is

$$P(\mathcal{I}_A) = \Phi\left(\frac{A_- - \delta}{\sigma\sqrt{m+q}}\right) - \Phi\left(\frac{A_+ - \delta}{\sigma\sqrt{m+q}}\right).$$

For δ and q we have only their plausible ranges, so we are interested in the largest plausible probability of impasse. This is evaluated as the largest of the probabilities for pairs (δ, q) at the vertices of the plausible rectangle.

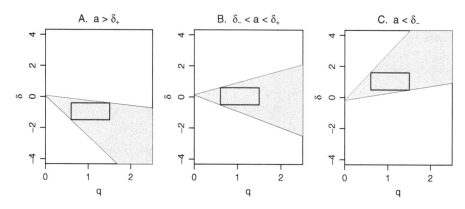

FIGURE 6.6
Scenarios of impasse with plausible priors.

For scenario B, we obtain the identity

$$P(\mathcal{I}_B) = \Phi\left(\frac{B_- - \delta}{\sigma\sqrt{m+q}}\right) - \Phi\left(\frac{B_+ - \delta}{\sigma\sqrt{m+q}}\right),$$

where

$$B_- = \sigma z_R^*\sqrt{m} - m\frac{\delta_- - a}{q_-}$$

$$B_+ = \sigma z_R^*\sqrt{m} - m\frac{\delta_+ - a}{q_-}.$$

For scenario C, we have

$$P(\mathcal{I}_C) = \Phi\left(\frac{C_- - \delta}{\sigma\sqrt{m+q}}\right) - \Phi\left(\frac{C_+ - \delta}{\sigma\sqrt{m+q}}\right),$$

where

$$C_- = \sigma z_R^*\sqrt{m} - m\frac{\delta_- - a}{q_+}$$

$$C_+ = \sigma z_R^*\sqrt{m} - m\frac{\delta_+ - a}{q_-}.$$

In an implementation, we evaluate z_R^* on a grid of values in the plausible range of R, identify the scenario A–C for each R, and evaluate the probability of impasse. Usually the same scenario applies throughout (R_-, R_+) and the probabilities are monotone functions of R. Then the evaluations can be reduced to $R = R_-$ and R_+, but this does not result in any substantial reduction of the programming or computing effort. Instead of identifying the scenario A–C, we can evaluate the quartet of values of $\hat{\Delta}$ for which the equilibrium lines run through the vertices of the plausible rectangle, and find their maximum and minimum. The probability of impasse is then obtained by substituting them for C_- and C_+ in the formula for scenario C, and similarly for the other two scenarios.

The evaluations can be reduced to the limits of the plausible range of R because both the intercept and slope of $\delta^*(q)$ are increasing functions of z_R^*, and z_R^* is an increasing function of R. Therefore, the impasse zone is bounded from above by $\delta^*(q)$ evaluated for R_+ and from below by $\delta^*(q)$ evaluated for R_-. The two bounds do not intersect at $q = 0$, as they do in Figure 6.6; the identity in equation (6.3) implies that the impasse zone at $q = 0$ extends from $\frac{1}{2}\sigma z_{R_-}^* \sqrt{m}$ to $\frac{1}{2}\sigma z_{R_+}^* \sqrt{m}$.

The final product of the analysis is a function that relates the probability of impasse to the sample size. Impasse may be associated with a loss. Then the design should be set by pitting this loss against the cost of the study. The loss decreases from unity at $n_1 = 0$ ($m = +\infty$), to a small positive value as $n_1 \to +\infty$, whereas the cost increases with n_1 from the set-up cost to infinity. The two functions have a single intersection unless the set-up costs are so large as to make the study prohibitive.

The solution is readily generalised from piecewise constant loss to other loss functions by replacing equation (6.2) with its counterpart, such as

$$(R-1)\Phi_1(\tilde{z}) = R\tilde{z}$$

for piecewise linear loss functions, and

$$(R+1)\Phi_2(\tilde{z}) = R\left(1 + \tilde{z}^2\right)$$

for piecewise quadratic functions. These equations have to be solved by iterative methods but, owing to monotonicity, the solutions are required only for R_- and R_+.

Example 12

We describe sample size calculation for the following setting. We want to compare two expected values, μ_1 and μ_2, based on normal random samples of equal size n. The two distributions have the same variance σ^2. The prior distribution for the difference $\Delta = \mu_2 - \mu_1$ is normal with plausible range of the prior mean δ from -0.5 to 0.5, and plausible variance ratio q between 0.6 and 1.5. The estimator $\hat{\Delta} = \hat{\mu}_2 - \hat{\mu}_1$ is unbiased for Δ and has variance $2\sigma^2/n$.

We evaluate first, separately for each value of R, the ranges of $\hat{\Delta}$ for which impasse occurs, and then their probabilities. The results are presented in Figure 6.7 as functions of R for $n = 20, 40, \ldots, 100$. The rate of impasse is defined as the 100-multiple of the corresponding probability; this is for both convenience and typographical reasons. The diagram shows that the rate of impasse is a flat function of R. For example, the rate increases for $n = 40$ from 3.24% at $R = 0.2$ to 3.32% at $R = 1.0$ and then decreases to 3.24% at $R = 5.0$. In fact, the rate is a symmetric function of $\log(R)$ for every n.

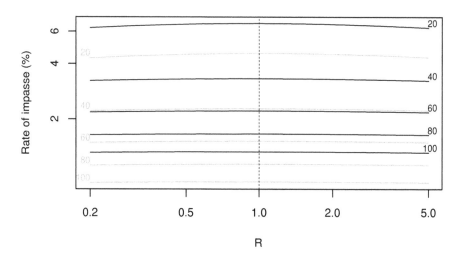

FIGURE 6.7
The rates of impasse as functions of R for sample sizes n printed at the right-hand margin. The plausible ranges of δ and q are $(-0.5, 0.5)$ and $(0.6, 1.5)$, respectively. The rates of impasse for $(\delta, q) \in (-0.3, 0.5) \times (0.7, 1.4)$ are drawn by light grey colour. Both axes are on the log scale.

If for any reason we had to set the sample size without eliciting the value of R, we could simply set the sample size conservatively by assuming that R is equal to 1.0. That would be appropriate only if we were certain that R would later be set to a single value, with no uncertainty.

The rates of impasse for a smaller (tighter, or revised) prior rectangle, $(\delta, q) \in (-0.3, 0.5) \times (0.7, 1.4)$, are drawn by grey colour, with the sample sizes marked at the left-hand margin. The advantage of a narrower plausible range for δ and q is evident. For example, the rate of impasse for $n = 60$ with the original setting is about the same as for $n = 40$ with the revised plausible rectangle. Prior information is invaluable. Pretending that we have none, and resorting to noninformative priors, is always convenient but often scientifically dishonest (or lazy). At the same time, pretending certainty about the prior introduces unjustified optimism that may hurt later, when the client adopts our verdict.

Suppose a plausible range of R, set to the interval $(2, 5)$, is elicited. We adapt the method that leads to Figure 6.7 by evaluating the ranges of $\hat{\Delta}$ that result in an impasse for each plausible R. Instead of evaluating the probability of each interval, we evaluate the probability of the union of these intervals, because only this union covers all outcomes that lead to an impasse. The results are plotted in Figure 6.8 as functions of the sample size n for the same two plausible rectangles for (δ, q) as used in Figure 6.7. Black and grey colours are used for the original (wider) and revised (narrower) plausible rectangles,

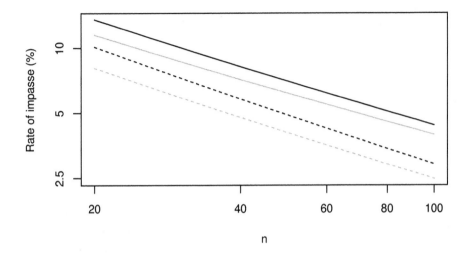

FIGURE 6.8
The rates of impasse as functions of sample size n for plausible range $R \in$
$(2, 5)$, drawn by solid lines and for $R \in (2.5, 4)$ drawn by dashes. The plausible
rectangles for (δ, q) are $(-0.5, 0.5) \times (0.6, 1.5)$, and $(-0.3, 0.5) \times (0.7, 1.4)$, using
black and grey colours, respectively. Both axes are on the log scale.

respectively, and solid line and dashes for the respective plausible ranges $(2, 5)$
and $(2.5, 4)$ for R.

Log scale is used for both axes; on this scale, the rates of impasse have
very little curvature. The linear approximations have higher (less negative)
slopes for wider plausible range of R. The rates of impasse are much higher
than with a single value of R. For example, the rates for $n = 100$ are 4.4%
with the original setting, 4.0% when the plausible priors are narrowed for
(δ, q) and 2.5% when the plausible range for R is also narrowed. In contrast,
when R is set to a single value, the rate of impasse with $n = 100$ is only 1.3%.
In summary, elicitation has a great potential; narrower plausible ranges are
rewarded by smaller sample sizes. Rigging the process, by declaring narrower
ranges than is warranted, results in a false economy by exposing the study to
a greater chance of impasse than planned.

6.4 Further reading

This chapter is based on Longford (2014a), with some simulations replaced by
analytic evaluations. There is a vast literature on sample size calculation for
estimation and hypothesis testing. It is a key aspect of clinical trials and other

studies in which established ethical standards have to be met. See Appendix for a concise derivation for a special case. There is much more to design than sample size calculation. Kish (1987) is an unsurpassed classic on this subject.

6.5   Exercises

6.1. Make inquiries about the budgets of some of the studies that you have read about. Ask about the components of the budget, for the data analysis in particular. You do not have to establish the exact amounts; ball-park figures would suffice.

   If a party does not want to disclose their budget, inquire about the number of workers involved, the length of time contracted and their duties. Include database managers and data analysts in your list.

6.2. Collect materials published about a study involved in Exercise 6.1, identify items and their parts (chapters or sections) that deal with planning and design of the study. Skim the documents for any accounts of deviations from the design, how they were addressed, whether they were anticipated, and if so, whether there were any plans to address them.

6.3. Identify among the materials in Exercise 6.2 the protocol(s) and assess their completeness. In a protocol of your choice, highlight the terms that you are not familiar with and procedures that should be explained (to you) in greater detail. (The ideal way of addressing this exercise would be to have a discussion with a person who was involved in compiling the protocol.)

   Hint: It has become a common practice to publish protocols for studies (both observational and randomised) in medical sciences.

6.4. In a published protocol of your choice, critically assess the leeway afforded by its details; that is, how much improvisation it allows in the analysis. If you can pair the protocol with the publication of the results, assess the correspondence of the plan with the execution of the analysis. Distinguish between details that were necessary to address deviations from the parts of the design up to data collection and details that were not specified by the original protocol.

6.5. Discuss examples of comparisons of two treatments (interventions, exposures, or the like), similar to the example of the effect of diet, in which randomisation is not feasible or not practical, and yet the target is the effect of the treatment, which would be estimated straightforwardly if a randomised study could be conducted.

   Hint: Helenius *et al.* (2019) is a suitable example from neonatal research.

6.6. The local press in a district is all excited about being the first among districts of the country to have had two winners in a weekly lottery in the current year. The two winners live in the same town, only two miles apart. Discuss the veracity of the claim that the district, or the town, is in some way special as regards luck. Consider designing a study that would seek to support this claim.

6.7. Explain why no generality would be lost by assuming that $\sigma^2 = 1$ in the setting of the problem outlined in the introduction to Section 6.2.

6.8. Derive the inequality in equation (6.1) and its probability. Identify the type of each probability involved (sampling, fiducial or some other).

6.9. Check by numerical integration, on a set of examples, that the expected loss $P(V_+) I_{\mu<0} + RP(V_-) I_{\mu>0}$ integrated over a centred normal distribution with a large variance is smaller than integrated over a centred normal distribution with a small variance. Discuss whether (and how) this conforms with intuition.

6.10. Construct a class of alternatives to the uniform distribution on $(-0.5, 0.5)$ as priors for μ in Section 6.2.
Hint: The uniform distribution on $(0, 1)$ is a special case of the beta distribution.

6.11. In small studies, the set-up cost a dominates the per-unit costs which add up to bn. With increasing sample size n, the per-unit cost may decrease. How could the cost function C_n be adapted for such economies of scale? Draw a few alternatives to Figure 6.3 for examples of such cost functions.

6.12. Explain how greater σ^2 can be compensated in sample size calculation by an increase in the sample size n.
Hint: Revisit the expressions for $P(V_+)$ and $P(V_-)$ in Section 6.2.

6.13. Re-draw Figure 6.5 with alternative values of some of the parameters, as suggested in the discussion of the diagram.

6.14. Adapt the method of sample size calculation in the Appendix for settings with unequal variances, when the ratio of the variances, $\rho = \sigma_2^2/\sigma_1^2$, is (assumed to be) known, and/or the within-treatment sample sizes are not equal, but are in a specified ratio π ($= n_2/n_1$). Further, adapt the method for a one-sided test, with the alternative $\Delta > 0$.

6.15. Adapt the method of sample size calculation for hypothesis testing to cater for uncertainty about the parameters involved, the variance σ^2 in particular. Discuss the compatibility of these calculations with the practice of using the t test for comparing the mean outcomes of two groups.

6.16. Review the definitions of unequivocal verdict and impasse and their relationship to plausible values of parameters (Section 3.6).

6.17. Suppose the difference $\Delta = \mu_2 - \mu_1$ of the expectations of two normal distributions with identical variances is estimated by the difference of the sample means $\hat{\Delta} = \hat{\mu}_2 - \hat{\mu}_1$. Show that a small increase in the smaller of the two sample sizes, n_1 or n_2, has a greater impact on the sampling variance of $\hat{\Delta}$ than a similar change of the other sample size. Based on your conclusion, what might be the rationale for designing studies with unequal sample sizes for comparing the expectations of two distributions with identical variances?

6.18. Derive the identity in equation (6.2). Draw the solution z_R^* as a function of R.

6.19. In connection with equation (6.3), explore how good is the approximation of $\sqrt{q(m+q)/m}$ by $\frac{1}{2}\sqrt{m} + q/\sqrt{m}$.
Hint: Show first that the quality of the approximation, assessed on the multiplicative scale, depends only on the ratio q/m.

6.20. Derive the inequalities in equation (6.4) and their counterparts for scenarios B and C. How would the three plots in Figure 6.6 be altered if the linear approximation in equation (6.3) were not applied? What changes would take place if the piecewise constant loss were replaced by another loss structure?

6.21. Develop a scheme in which the three scenarios, A, B and C, do not have to be distinguished. Instead, the slopes b are found for the lines $\delta^*(q)$ that run through the four vertices of the plausible rectangle, and their minimum and maximum delineates the impasse zone.

6.22. Derive the counterparts of Φ_1 and Φ_2 for the Student t distribution with k degrees of freedom, denoted by $\Psi_k^{(1)}$ and $\Psi_k^{(2)}$.
Hint: Express the differential of the t_k density in terms of another t density. Then adjust the function $x\Psi_k(x)$ so that its differential would be equal to the distribution function $\Psi_h(x)$ for suitable h. Proceed similarly with $\Psi_k^{(2)}(x)$.

6.23. Reproduce Example 12 for two random samples of sizes $n_2 = \pi n_1$; $\pi \neq 1$.

6.24. Prove that changing the scale of the outcomes from y to $-y$ corresponds to the switch of the loss ratio R to $1/R$ in linearised sample size calculation. Does this symmetry apply also when $n_1 \neq n_2$ is planned, or when the prior distribution is asymmetric?

Appendix. Sample size calculation for hypothesis testing.

This appendix summarises sample size calculation for testing the null hypothesis that the expectations of two normal distributions with identical variances $(\sigma_1^2 = \sigma_2^2 = \sigma^2)$ coincide; that is, the hypothesis is that $\Delta = \mu_2 - \mu_1 = 0$. The purpose is to set the sample size n, common to the two groups, that would be sufficient for the test to have a prescribed power β for a specified alternative that $\Delta = \Delta^\dagger$. No generality is lost by assuming that $\Delta^\dagger > 0$.

We include this method here because it is of considerable historical importance. It is regarded as the mainstay in designing clinical trials and other experiments. However, we regard it as outdated because it caters for a method of analysis, hypothesis testing, that is inappropriate for the settings we consider, in which the consequences of the two kinds of error are an essential factor.

Suppose the two distributions are represented by mutually independent random samples of size n, so that their sample means, $\hat{\mu}_1$ and $\hat{\mu}_2$, are unbiased for the respective expectations μ_1 and μ_2, and have the common sampling variance σ^2/n. Denote $\hat{\Delta} = \hat{\mu}_2 - \hat{\mu}_1$. The distribution of $\hat{\Delta}$ is $\mathcal{N}(\Delta, 2\sigma^2/n)$.

The probability of the error of the first kind, known also as the size of the test, should be α, usually set to 0.05. This condition,

$$P\left(|\hat{\Delta}| > c;\, \Delta = 0\right) = \alpha,$$

has the solution

$$c_n = \frac{\sigma\sqrt{2}}{\sqrt{n}}\,\Phi^{-1}\left(1 - \tfrac{1}{2}\alpha\right).$$

The hypothesis is rejected when $\hat{\Delta} \in (-\infty, -c_n) \cup (c_n, +\infty)$, that is, for sufficiently large values of $|\hat{\Delta}|$. The sample size n is set as the solution of the inequality for sufficient power at $\Delta^\dagger$:

$$P\left(|\hat{\Delta}| > c_n;\, \Delta = \Delta^\dagger\right) \geq \beta.$$

By relating the argument of P to the standard normal distribution we obtain the inequality

$$\Phi\left(\frac{-c_n - \Delta^\dagger}{\sigma\sqrt{2}}\sqrt{n}\right) + 1 - \Phi\left(\frac{c_n - \Delta^\dagger}{\sigma\sqrt{2}}\sqrt{n}\right) \geq \beta.$$

Since $\Delta^\dagger > 0$, the first term is smaller than the remainder of the left-hand side. It is dropped in further consideration, adding some leeway to the solution and making it simpler. After substituting for c_n we obtain the inequality

$$\Phi^{-1}\left(1 - \tfrac{1}{2}\alpha\right) - \frac{\Delta^\dagger\sqrt{n}}{\sigma\sqrt{2}} \leq \Phi^{-1}(1 - \beta),$$

which is equivalent to

$$n \geq \frac{2\sigma^2}{\Delta^{\dagger 2}} \left\{ \Phi^{-1} \left(1 - \tfrac{1}{2}\alpha \right) + \Phi^{-1} \left(\beta \right) \right\}^2 .$$

In summary, four factors are involved in setting the sample size n: the variance σ^2, a feature of the studied outcome variable, the probabilities α and β, which control the rates of error, and the alternative expectation $\Delta^\dagger$ which is commonly interpreted as the smallest deviation from the null, $\Delta = 0$, that is regarded as important, and for which control of the rate of error of the second kind (failure to discover) is intended.

7

Medical screening

Screening for a particular medical condition, such as a type of cancer, is instituted in a country, or among subscribers to a health-care plan, as a measure to discover the condition in its pre-symptomatic phase. At that phase, it is relatively easy to treat, the treatment is more likely to be successful and it is much cheaper and less invasive than it would be some time later when the disease has progressed to causing discomfort, incapacity and being an imminent threat to life.

By screening, the health-care system provides improved service in the long term and saves resources, by shifting the workload from intervention (chemotherapy, surgery, and the like) to prevention (testing and noninvasive treatment by drugs). For some conditions, these two kinds of treatment are in a fine balance. Intervention is applied to a small fraction of a population, those who contract the condition, but it demands a lot of resources and expertise, and tests the resilience and resolve of the patient. The disease may become symptomatic at a stage when it is no longer curable. In contrast, a typical test used in screening, and the logistics involved (generating public awareness and good disposition towards the programme, arranging appointments, keeping records and communicating the results, and following them up if applicable), require only modest resources per patient, but are applied to a large subpopulation. A test is applied not once, but at regular intervals, such as every five years, and in a specified subpopulation, such as all women aged 55–80 years.

We bypass the discussion as to whether a screening programme should be instituted or not, and for whom. We assume that a programme is in place, or is about to be introduced. In this context, we consider the task of setting the rule for how the patients should be classified. In the simplest classification, there are two categories:

(a) negatives – free of the disease and of all its precursors;

(b) positives – not free, with pre-clinical signs of the disease, and requiring treatment.

Patients in category (a) receive an 'all clear' and will be invited for another test in five years' time (or another specified period). Patients in category (b) will commence a course of treatment that may start with some more involved tests for establishing the details of the condition which are essential for the (clinical) treatment. False positives may be discovered at this stage. The negatives are also referred to as healthy and the positives as diseased. The diseased are the target of screening.

An intermediate category may be introduced for patients who will be retested earlier than others or subjected to another test. The classification has to be simple and unambiguous, so that it would be understood by all clinical staff and patients in the same way and without any confusion.

We consider a test based on the concentration of an antigen in blood. Setting aside the precision of the assay used and some rounding applied, the concentration is a continuous or semicontinuous variable. We denote it by Y. Semicontinuity arises when zero concentration, or concentration recorded as zero (being below the limit of detection), is recorded with positive probability.

7.1   Separating positives and negatives

We address the task of setting the cutpoint T for classification to two categories: a patient is declared free of the disease if $Y < T$ and is declared to be in need of treatment if $Y > T$. We assume that in the process of developing the test, by now concluded, the distribution of Y in the relevant population has been established, or estimated with sufficient precision. This distribution may be bimodal, with modes at typical values for the healthy and diseased patients. It is therefore more practical to work with the conditional distributions within the two categories; these distributions are usually unimodal.

We consider only the case of a continuous variable Y. It is often more convenient to work with the log-concentration, which may attain negative values. In what follows, we assume that a suitable transformation has been applied, and use the notation Y for the transformed variable. The disease status of the patient is denoted by Z; $Z = 0$ for a negative (healthy) and $Z = 1$ for a positive (diseased, or requiring treatment). Denote the conditional densities of Y by f_H and f_D for the respective healthy and diseased subpopulations; that is, $f_H(x) = \partial P(Y \leq x \mid Z = 0)/\partial x$ and $f_D(x) = \partial P(Y \leq x \mid Z = 1)/\partial x$. We assume that the rate of the disease status, $p_D = P(Z = 1)$, is small, $p_D \ll \frac{1}{2}$; that is, the condition is not endemic in the population.

It may at first appear that the probability p_D should have no role in setting the value of the cutpoint T. We disprove this view in our evaluations. Its role can be motivated in the Bayesian paradigm as a prior for the classification of a patient, which is then updated by the data y to form the posterior probability of the patient's status.

If the densities f_D and f_H are well separated, so that there is a constant T such that $P(Y < T \mid H) = 1$ and $P(Y > T \mid D) = 1$, then classifying the patients is simple and error-free; any such separator T is a suitable cutpoint, and all patients with $Y < T$ are classified as healthy and the rest as requiring treatment. In a more realistic scenario, the two densities have no such separator. Any constant T chosen to be the cutpoint entails two kinds of error:

- false discovery: $Y > T$ for a healthy patient;

- failure to discover: $Y < T$ for a diseased patient.

The terms 'false positive' and 'false negative' are also used. Denote by $P_H(T)$ and $P_D(T)$ the conditional probabilities of these two kinds of error:

$$P_H(T) \;=\; \int_T^{+\infty} f_H(x)\,\mathrm{d}x$$

$$P_D(T) \;=\; \int_{-\infty}^{T} f_D(x)\,\mathrm{d}x.$$

In the established literature on screening, the complements of these two functions, $1 - P_H$ and $1 - P_D$, are called sensitivity and specificity, respectively. Rules for setting T include the Youden index, which choses T for which the total $P_H(T) + P_D(T)$ is minimised. Its obvious adaptation is to minimise the convex combination, $cP_H(T) + (1 - c)P_D(T)$, for a given constant $c \in (0, 1)$. There is rarely any clear guidance on how to set c and which factors should be considered in its choice.

We associate the two kinds of error with losses. To motivate the losses, we discuss the consequences of the errors. False discovery leads to the inappropriate application of a treatment; it exposes the patient to unwarranted anxiety and possibly making inappropriate arrangements in anticipation of extensive treatment and associated incapacity or hospitalisation. The risk of side effects is another factor. As for the health-care provider, it introduces a waste of resources and unnecessary application of a treatment, in conflict with commonly adopted ethical and professional standards.

In contrast, failure to discover is an unmitigated disaster, if the noninvasive treatment intended for the pre-symptomatic condition is replaced, some time later, by an invasive treatment with potential side effects and incapacity. The patient loses an opportunity to avoid an unpleasant treatment, hospitalisation, invasive surgery, and the like, and the health-care provider expends substantial resources on the treatment. In this case, screening is a mere distraction.

The two sets of consequences are impossible to quantify in general, to be applied for a wide class of conditions and settings, and they are difficult to quantify even in a specific setting. But it is obvious that failure to discover is, or should be, associated with far greater loss L_{hD} than the loss L_{dH} for a false discovery. Otherwise screening would not be useful.

We set $L_{dH} = 1$ and $L_{hD} = R$. Instead of a single value of the loss ratio R we consider a plausible range (R_-, R_+). This is necessary both to address the difficulties in an (hypothetical) elicitation and to cater for a variety of perspectives and value judgements that the prospective patients and other stakeholders may have. They include the tax payer, the health insurance industry, the community of clinical professionals and the administrative authority, such as the country's Ministry of Health, and the patient's family and friends.

Screening is redundant for those who are undergoing treatment or, having concluded one, are under close clinical surveillance. In a typical screening programme, only the verdict, V_h or V_d, is disclosed to the patient and the clinician involved, and only this verdict is entered in the person's health-care record. If the value of the concentration Y is not disclosed, then the margin of decision $|Y - T|$ is of no consequence. Only the sign, $\text{sign}(Y - T)$, matters. In that case, only piecewise constant loss is relevant, and other loss structures can be ruled out.

Suppose first that the within-category densities are both normal, $f_S = \varphi(x; \mu_S, \sigma_S)$, for $S = H$ and D. In any meaningful setting $\mu_H < \mu_D$. Assuming that $\sigma_H^2 = \sigma_D^2$ may be too restrictive. The expected loss with verdict V_d is

$$
\begin{aligned}
Q_d &= \int_T^{+\infty} \varphi(y; \mu_H, \sigma_H)\, dy \\
&= \int_{z_H}^{+\infty} \varphi(x)\, dx = 1 - \Phi(z_H) ,
\end{aligned}
$$

where $z_H = (T - \mu_H)/\sigma_H$. By similar steps, we obtain the expected loss

$$
Q_h = R\,\Phi(z_D) ,
$$

where $z_D = (T - \mu_D)/\sigma_D$. The expected loss incurred in screening the entire population of N patients is $NQ(T)$, where

$$
Q(T) = p_D\, Q_h(T) + (1 - p_D)\, Q_d(T); \tag{7.1}
$$

$Q(T)$ is the expected loss per patient. We search for the minimum of this function of T. There is no closed-form solution, so we apply the Newton-Raphson algorithm, exploiting the relatively simple expressions for the derivatives of Q of first and second order:

$$
\begin{aligned}
Q'(T) &= \frac{R\,p_D}{\sigma_D} \varphi(z_D) - \frac{1 - p_D}{\sigma_H} \varphi(z_H) \\
Q''(T) &= -\frac{R\,p_D\, z_D}{\sigma_D^2} \varphi(z_D) + \frac{(1 - p_D)\, z_H}{\sigma_H^2} \varphi(z_H) .
\end{aligned}
$$

One might expect that the solution T^* is within the interval (μ_H, μ_D). This need not be the case. As R diverges to $+\infty$, it becomes rational to issue the verdict V_d to more and more patients — T^* diverges to $-\infty$, so T^* is smaller than μ_H for sufficiently large R. In practice, such an outcome would be unsatisfactory because the health care system would be overwhelmed.

Example 13

Suppose the two conditional distributions are $\mathcal{N}(\mu_H = 0.25, \sigma_H^2 = 0.05)$ and $\mathcal{N}(\mu_D = 0.95, \sigma_D^2 = 0.12)$ and the prevalence of positives in the population

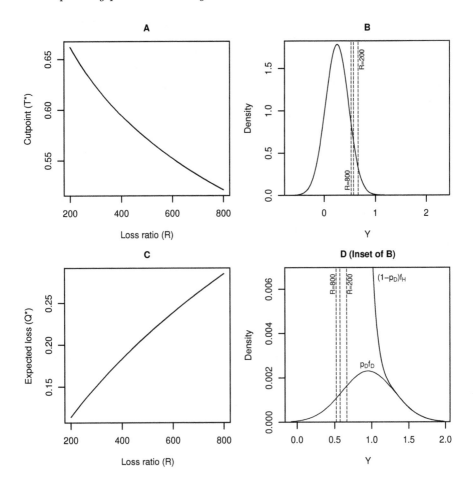

FIGURE 7.1
Screening with conditional distributions $\mathcal{N}(0.25, 0.05)$ and $\mathcal{N}(0.95, 0.12)$ and prevalence 0.2% of diseased patients; the plausible range for R is $(200, 800)$. A – the cutpoint function $T^*(R)$; B – the relative densities $(1 - p_D)f_H$ and $p_D f_D$, the latter completely obscured; C – the expected loss function $Q^*(R)$; D – inset of B in the vicinity of $\mu_D = 0.95$.

is 0.2%. We set the limits of the plausible range for R to $R_- = 200$ and $R_+ = 800$. Figure 7.1 relates to this setting. Its panel A presents the cutpoint function $T^*(R)$ for R in the plausible range (R_-, R_+). Each value T^* of this function is obtained as the minimum of the expected loss function $Q(T; R)$. The cutpoint is a decreasing function of R — the verdict V_d becomes more attractive as the aversion to the false negative is increased. The cutpoint drops to $\mu_D = 0.25$ for $R = 5955$, way beyond R_+.

Panel C displays the expected loss with the optimal choice of the cutpoint, the function $Q\{T^*(R), R\}$, denoted by $Q^*(R)$. It is increasing, from 0.114 at $R = 200$ to 0.285 at $R = 800$. The increase is at a slower pace than R; $0.285/0.114 = 2.5 < 800/200 = 4.0$. Constructing the function T^* is computationally not demanding because the Newton-Raphson algorithm converges rapidly; with the initial solution set to $\frac{1}{2}(\mu_H + \mu_D)$, four iterations suffice to reach the precision of 10^{-6}; that is, the solutions at the third and fourth iterations differ by less than 10^{-6} for both T^* and Q^*. The convergence is so rapid because the objective function Q is 'well behaved'—it does not deviate substantially from a quadratic function.

Panel B presents the density $f = (1 - p_D)f_H + p_D f_D$, with the cutpoints $T^*(200) = 0.662$, $T^*(500) = 0.571$ and $T^*(800) = 0.521$ indicated by vertical dashes. In fact, $f \doteq f_H$ because $p_D \ll \frac{1}{2}$. There is a tiny bulge in the density at around $y = \mu_D = 0.95$, but it could be discerned only after a substantial magnification. Panel D zooms in on this region by plotting the two relative densities, $p_D f_D$ and $(1 - p_D)f_H$. Note that $p_D f_D \ll (1 - p_D)f_H$ even around $\mu_D = 0.95$.

Uncertainty about p_D can be addressed similarly to the exploration in Figure 7.1. By interpreting p_D as a prior probability of being diseased, we conclude that the function $T^*(p_D; R)$ is decreasing in p_D. With an elevated prior probability the posterior probability of being diseased is also higher, and this makes the verdict V_d more attractive—the cutpoint T is reduced. Therefore we can address jointly the uncertainty about R and p_D by evaluating $T^*(R_-, p_{D-})$ and $T^*(R_+, p_{D+})$, where (p_{D-}, p_{D+}) is a plausible range of p_D.

Suppose instead of $p_D = 0.002$ we only have a plausible range $p_D \in (0.001, 0.003)$. Then we have the optimal cutpoints $T^*(R = 200; p_D = 0.001) = 0.725$ and $T^*(R = 800; p_D = 0.003) = 0.474$, widening the plausible range from $(0.521, 0.662)$, derived under the assumption that $p_D = 0.002$. Our intuition, or theoretical derivation, that T^* decreases with p can be partly confirmed by evaluating T^* at the other two vertices of the plausible rectangle. We have $T^*(200, 0.003) = 0.623$ and $T^*(800, 0.001) = 0.594$. □

Further insights can be gained by studying the expression for $Q(T)$ in equation (7.1),

$$Q(T) = R\, p_D\, \Phi(z_D) + (1 - p_D)\left\{1 - \Phi(z_H)\right\},$$

as a function of p_D and R. Now, $\Phi(z_H)$ and $\Phi(z_D)$ involve neither p_D nor R, so we can regard them as constants. Further, the expected loss functions $Q(T)$ and $cQ(T)$ are effectively identical for any constant $c > 0$; certainly, while p_D and R are held constant, they do not alter the value of T^*. A change of R to $R^\dagger = dR$ for a constant $d > 0$ can be compensated by altering p_D to

$$p_D^\dagger = \frac{p_D}{p_D + d(1 - p_D)} \tag{7.2}$$

and altering Q to

$$Q^\dagger = \frac{dQ}{p_D + d(1 - p_D)}.$$

The factor $p_D/\{p_D + d(1 - p_D)\} < 1$ is a decreasing function of d, equal to unity when $d = 0$. Therefore, an increase of R is equivalent to a reduction of p_D, leaving the optimal cutpoint T^* unchanged. Conversely, an increase of R accompanied by an increase of p_D compounds their 'individual' reductions of T^*. This confirms that the extreme plausible values of T^* arise with plausible pairs (R, p_D) in the south-west and north-east vertices of the plausible rectangle.

Uncertainty about the expectations and variances of Y can be addressed similarly. Each element of uncertainty inflates the plausible range of T^*, so we have to be sparing with these uncertainties. By this statement we mean that the assessment of uncertainties should be as rigorous as possible, to keep the uncertainty in T^* in check. But the uncertainties about the parameters involved should be reduced as much as possible, by a combined effort of data collection and elicitation. Also, at the planning stage, an exploration of the extent to which the uncertainty about a parameter converts to wider plausible range of T^* may help us to distribute the available (research) resources among the competing goals of reducing the uncertainties about R, p_D, μ_S and σ_S^2, S = D and H, and, indeed, in establishing the nature and shape of the within-status distributions of Y.

7.2 Cutpoints specific to subpopulations

The rate of the disease status, p_D, is one of the parameters in the process of setting the cutpoint T^*. If in the population targeted by the screening programme we can identify subpopulations with distinctly different rates, there may be a good case for setting a different cutpoint for each subpopulation. If this route is followed, only a few (large) subpopulations can be defined; otherwise the planning process is overwhelmed with the demand for information about the subpopulations, or the plausible ranges of T^* for some of them would be too wide.

The subpopulations have to have simple definitions based on no more than a few commonly recorded variables. The uncertainty about the parameters associated with the subpopulations is usually greater than the uncertainty about the parameters associated with the entire population. The advantage of using different cutpoints in a set of subpopulations has to be weighed against the drawback of wider plausible ranges of the cutpoints.

Regional or country-specific cutpoints T^* are an example of this general idea. The cutpoints may reflect differential rates of the disease in the countries (or regions of a large country), screening different (sub-)populations and different priorities in health care.

7.3   Distributions other than normal

We assumed normality of (transformed) concentration Y only to make the problem formulation specific. Other continuous distributions can be adopted. In the derivations, we used no properties of the distribution function or density that are specific to the normal. One possible exception is the expression for the differential of the density, $\varphi'(y) = -y\varphi(y)$. However, the differentials for most other densities that one might contemplate are also easy to derive, and one can always resort to numerical differentiation. For example, for the gamma density $h(x; \gamma, \delta)$ defined by equation (4.13), we have the expression

$$\frac{\partial h}{\partial x} = \left(\frac{\gamma - 1}{x} - \delta\right) h(x).$$

We do not have to consider a very wide variety of distributions, in particular those with extreme asymmetry of their tails, because such features can be ameliorated by transformations. Even among the transformations, one rarely has to resort to other than the most common functions, such as the logarithm and square-root.

In general, the distribution of the outcome Y is a mixture of the within-status distributions. A mixture of two distributions, with densities f_D and f_H, is defined by a convex combination of these densities, with probabilities p_D and $p_H = 1 - p_D$, that is,

$$f(x) = p_D f_D + p_H f_H. \tag{7.3}$$

In this definition, the densities f_D and f_H are well motivated as the conditional densities given the health status, D or H. A mixture distribution is generated by the following process. First we draw, for n subjects, a random sample of size n from the binary (Bernoulli) distribution, assigning each subject to either status D or H. Then for each subject we draw a value y from the distribution defined by the assigned status. The two draws associated with a subject, the status and the value (concentration) are independent.

A mixture of any finite number M of continuous distributions is defined by a convex combination of the M component densities. In many applications of such mixtures the integer M is not known, and neither is the component for any of the subjects. For $M = 1$ we have a trivial case; it is convenient to regard it as a special case of a mixture.

A generalisation of the normal distribution is a mixture of normals. Such a mixture may be assumed even for a within-status distribution when the normal distribution does not fit well and there is no obvious transformation for which normality would be palatable. There may be no interpretation for the components of such a mixture but the fit may be superior. It is easy to explore by simulations that mixtures of even a small number of normal distributions generate a wide range of distributions, with all manner of asymmetry and several modes.

7.3.1 Normal and t distributions

Our approach to addressing uncertainty by plausible ranges for the parameters cannot be adapted for the uncertainty about the shape of the conditional distributions. The literature is replete with tests for normality and features associated with it, symmetry in particular, but they are merely devices that seek a contradiction with normality or another assumption; they cannot confirm the assumption. Neither can they indicate that the departure from normality is small enough to be ignored because the qualifier 'small' is a highly contextual.

We explore how much the assumption of normality matters by comparing the cutpoints for the normal conditional distributions with their counterparts based on the t distribution with a small number of degrees of freedom. Denote the t distribution with k degrees of freedom by t_k, its density by ψ_k and the distribution function by Ψ_k. In the code or a computational routine for the normal distribution, we only have to change the functions that refer to the normal distribution, φ, Φ and Φ^{-1}, with the corresponding functions for t_k, ψ_k, Ψ_k and Ψ_k^{-1}. The identity $\partial\varphi/\partial x = -x\,\varphi(x)$ for the standard normal density has its counterpart

$$\frac{\partial\psi_k}{\partial x} = -\frac{(k+1)x}{k+x^2}\,\psi_k(x)$$

for the t_k density. As $k \to +\infty$, $\partial\psi_k/\partial x$ converges to $\partial\varphi/\partial x$ for every x.

The t_k distribution has zero expectation and variance $k/(k-2)$, so long as $k > 2$. Therefore, it would seem to be more appropriate to match the standard normal distribution with the $\sqrt{(k-2)/k}$-multiple of the t_k distribution. Figure 7.2 presents the cutpoint and expected loss functions for the three settings, with normal (marked N), t_5 distribution (dashes, t) and the scaled t_5 (long dashes, t-sc), which matches the variance of the normal. The normal setting is based on the parameters used in Figure 7.1.

The vertical axes extend over a range slightly wider than the plausible ranges of the cutpoint and expected loss. The 'error' caused by using the t_5 or scaled t_5 distribution is much smaller for the cutpoint. The largest discrepancy occurs for $R = 200$, equal to $0.684 - 0.662 = 0.022$ for t_5 and $0.637 - 0.662 = -0.025$ for scaled t_5, about 5.6 times smaller in absolute value than the width of the plausible range for T^*, equal to $0.662 - 0.521 = 0.141$. We have no explanation as to why the discrepancy is so small at around $R = 800$, where the cutpoint functions intersect.

The expected loss is affected by the misspecification (normal vs. t_5) more strongly. The largest discrepancy occurs for $R = 800$, reaching 0.071 for t_5 and -0.059 for scaled t_5. These figures should be compared with the width of the plausible range of expected losses, equal to 0.171.

With increasing numbers of degrees of freedom, the discrepancies diminish rapidly. Even for 10 degrees of freedom, the discrepancies in T^* are 0.012 for t_{10} and -0.013 for scaled t_{10}, and the discrepancies in Q^* are 0.027 and -0.025.

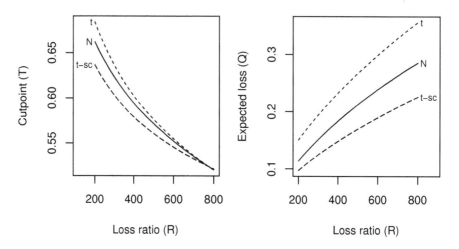

FIGURE 7.2
The cutpoint and expected loss functions in screening with the setting of
Figure 7.1, with the assumptions of normality (N), t_5 distributions, and t_5
distributions scaled to match the variance of the normals.

Scaling, that is, matching the variances of the normal distributions, is
not useful. The tails of a distribution influence the variance strongly. The
verdict for outcomes Y in one tail, left-hand tail for V_h, is correct with high
probability, but the verdict in the other tail is more contentious. However,
with a large loss ratio R, the cutpoint is not in the tail of the conditional
distribution $(Y \mid H)$.

7.4 A nearly perfect but expensive test

Suppose there is a test with a negligible error but it is too expensive for administration on the entire population. The following scheme may save substantial
resources. We set cutpoints $T_1 < T_2$ and issue the verdict V_h to all patients
with $Y < T_1$, verdict V_d to all those with $Y > T_2$, and apply the 'perfect'
test to all patients in the 'inconclusive' category defined by $T_1 < Y < T_2$.

Denote by G the cost of one application of this test, expressed in lossiles.
The verdicts V_h and V_d are associated with respective per-patient expected
losses

$$Q_h = p_D \, R \, \Phi(z_{D1})$$
$$Q_d = (1 - p_D) \, \{1 - \Phi(z_{H2})\} ,$$

where $z_{D1} = (T_1 - \mu_D)/\sigma_D$ and $z_{H2} = (T_2 - \mu_H)/\sigma_H$; z_{D2} and z_{H1}, used below, are defined similarly. Further, the cost of the perfect test, pro-rated for a patient, is

$$
\begin{aligned}
Q_2 &= G \left\{ (1 - p_D) \int_{T_1}^{T_2} \varphi(x; \mu_H, \sigma_H) + p_D \int_{T_1}^{T_2} \varphi(x; \mu_D, \sigma_D) \right\} \\
&= G \left[(1 - p_D) \{ \Phi(z_{H2}) - \Phi(z_{H1}) \} + p_D \{ \Phi(z_{D2}) - \Phi(z_{D1}) \} \right].
\end{aligned}
$$

We set the cutpoints T_1 and T_2 to the values for which $Q = Q_h + Q_d + Q_2$ attains its minimum. The values are found by the Newton-Raphson algorithm. It uses the differentials

$$
\frac{\partial Q}{\partial T_1} = \frac{R p_D}{\sigma_D} \varphi(z_{D1}) - \frac{G(1 - p_D)}{\sigma_H} \varphi(z_{H1}) - \frac{G p_D}{\sigma_D} \varphi(z_{D1})
$$

$$
\frac{\partial Q}{\partial T_2} = -\frac{1 - p_D}{\sigma_H} \varphi(z_{H2}) + \frac{G(1 - p_D)}{\sigma_H} \varphi(z_{H2}) + \frac{G p_D}{\sigma_D} \varphi(z_{D2})
$$

and

$$
\frac{\partial^2 Q}{\partial T_1^2} = -\frac{R p_D z_{D1}}{\sigma_D^2} \varphi(z_{D1}) + \frac{G(1 - p_D) z_{H1}}{\sigma_H^2} \varphi(z_{H1}) + \frac{G p_D z_{D1}}{\sigma_D^2} \varphi(z_{D1})
$$

$$
\frac{\partial^2 Q}{\partial T_2^2} = \frac{(1 - p_D) z_{H2}}{\sigma_H^2} \varphi(z_{H2}) - \frac{G(1 - p_D) z_{H2}}{\sigma_H^2} \varphi(z_{H2}) - \frac{G p_D z_{D2}}{\sigma_D^2} \varphi(z_{D2})
$$

and $\partial^2 Q/(\partial T_1 \partial T_2) = 0$. The latter identity means that the Hessian is a diagonal (2×2) matrix, so the Newton-Raphson algorithm for the two cutpoints can be split into separate univariate procedures in each iteration.

Example 14

We have applied the Newton-Raphson algorithm on several occasions, suggesting that it is a method for all problems involving extremes of smooth functions. This example presents a cautionary tale. We borrow the setting of Example 13, with the respective distributions of the healthy and diseased subjects $\mathcal{N}(\mu_H = 0.25, \sigma_H^2 = 0.05)$ and $\mathcal{N}(\mu_D = 0.95, \sigma_D^2 = 0.12)$ and set $R = 200$ and the cost of the perfect test to $G = 80$. For some initial solutions T_1, the algorithm converges to a local maximum. We explore why this happens by plotting the expected loss Q as a function of T_1 and T_2.

As an alternative to the two cutpoints T_1 and T_2, we consider the lower cutpoint T_1 and the width of the range of the outcomes, $\Delta = T_2 - T_1$, in which the expensive test would be applied. Figure 7.3 displays the functions $Q(T_1; \Delta)$ for a grid of values of Δ. The diagram shows that for $\Delta \geq 0.005$, $Q(T_1; \Delta)$ has a maximum at around 0.2 and is increasing for $T_1 > 0.7$.

The minimum for each value of Δ is marked by a light-grey disc. The absolute minimum occurs at 0.3 for $\Delta = 0.0$. That is, the expensive test should not be applied at all. The value of T_1 at which the minimum is attained

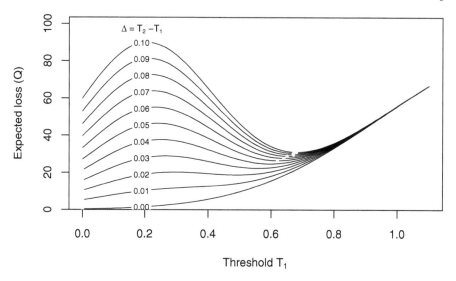

FIGURE 7.3
Expected loss Q as a function of the lower cutpoint T_1 and the width of the range, $\Delta = T_2 - T_1$, in which the expensive test in applied; the value of Δ is indicated on the curves.

increases with Δ, reaching 0.68 for $\Delta = 0.10$. For large T_1, the functions converge to a slowly increasing function. Applying the expensive test for large values of Y is wasteful, but such values are exceedingly rare.

The Newton-Raphson algorithm fails for a wide range of apparently 'good' initial solutions because the expected loss is a shallow function of T_1 and Δ (and therefore also of T_2). The algorithm may converge to the maximum or diverge towards large values of T_1 and T_2.

7.5   Further reading

This chapter is based on Longford (2013b), with a sequel in Longford (2015b). The foundational idea for the established methods for screening in due to Youden (1950). More recent literature proposes various alternatives for how to balance the probabilities of the two kinds of error. Its focus is on the receiver operating characteristic curve (e.g., Baker, 2003), which relates the conditional probabilities of the correct decision given the state (healthy or diseased). See Fluss *et al.* (2015) and Molanes-López and Letón (2011) for examples. A method for analysing outcomes subject to a limit of detection is developed by Longford (2012c).

Mixtures are fitted by the EM algorithm (Dempster, Laird and Rubin, 1977; and McLachlan and Peel, 2000).

7.6   Exercises

7.1. Find information about a national screening programme: its target population, frequency of screening, prevalence of the disease in the (screened or entire) population, claims as to how many lives are saved, information about the rates of false positives and negatives, and other (statistical) information that supports the classification used in screening.

7.2. Compile a list of consequences of a false positive and a false negative in medical screening, in general or in a specific programme, and discuss the factors that should inform how the loss ratio is set.

7.3. In what circumstances can we focus on piecewise constant loss in screening? Why might countries or even their regions have different loss ratios? Why might the loss ratio be periodically revised (e.g., every 10 years)?

7.4. Is it the case in general that the cutpoint T is a decreasing function of the loss ratio R, as in panel A of Figure 7.1? Is the minimum expected loss Q always an increasing function of R, as in panel C?

7.5. Make an honest assessment of the Newton-Raphson algorithm for finding the minimum of the expected loss given by equation (7.1) and the alternative based on plotting the function $Q(T)$, or the plausible functions $Q(T; R, p_D, \ldots)$.

7.6. Explain the role of the prior (or marginal) probability of the disease in the screened population in setting the threshold T.

7.7. Derive the differentials of the densities of some common distributions, such as beta, gamma and t. Express them in terms of densities of the same type. For example, for the t_k density with $k > 0$ we have the identity

$$\psi_k'(x) = -x\sqrt{\frac{k+2}{k}}\,\psi_{k+2}\left(x\sqrt{\frac{k+2}{k}}\right);$$

see Exercise 6.22.

Hint: Use the identity $\partial f/\partial x = f(x)\,\partial \log(x)/\partial x$, which holds for any positive differentiable function f.

Check this result, as well as your own results, by numerical differentiation. Describe the advantages of such expressions for the Newton-Raphson algorithm for finding T^*.

7.8. Derive the expectation and variance of the t_k distribution for $k > 2$ and check the results by simulation.

7.9. Study the mixtures of two normal distributions by evaluating (and plotting) their densities $f(x) = p\,f_1(x) + (1-p)\,f_2(x)$, where f_1 and f_2 are two normal densities and $0 < p < 1$. In particular, explore when the density f has two modes and distinctly different tails. Compare your findings with their counterparts for linearly scaled t distributions with small numbers of degrees of freedom.

7.10. How are the mean and variance of a mixture related to the means and variances of its (two) components?

7.11. A verdict in screening (classification of a patient) is said to be equivocal if it is positive for some plausible values (e.g., R_- in Example 13), and negative for some other plausible values of the parameters. In the setting of Example 13, find the proportion of patients who would be classified equivocally. Represent them graphically, using (or adapting) one of the panels of Figure 7.1.

7.12. The odds ratio for a disease is defined as $r_D = p_D/(1-p_D)$. Express equation (7.2) in terms of r_D and $r_D^\dagger = p_D^\dagger/(1 - p_D^\dagger)$, and describe (interpret) the result, bearing in mind that p_D is small.

7.13. Collect the arguments for and against defining separate cutpoints T^* for two well identified subpopulations in a screening programme. This exercise can be made more realistic by discussing a specific programme in your country.

7.14. Suppose a mixture of a small number of normal distributions fits well to the outcomes (e.g., log-concentrations of a key antigen) for patients in the group of negatives, and another mixture for the group of positives. Discuss the argument that the mixture components may identify subgroups of clinical importance, such as types or stages of the disease.
Hint: Consider a continuous distribution, such as a gamma or t, and assess by trial and error how well it could be approximated by a mixture of normals.

7.15. Explore graphical methods (plotting the expected loss function) as an alternative to finding the optimal cutpoint for screening. Use a setting of your choice, or borrow a setting from one of the examples.

7.16. Construct, by trial and error, an example of distributions of positives and negatives in which an expensive test for screening may be useful.

8

Many decisions

Having time to focus on a single decision in isolation from all other pressing problems is a luxury afforded mostly by armchair advisors, theoreticians and those who face a single momentous problem that dwarfs all the others in its opportunity to gain or its threat to lose. In this chapter, we turn to the other extreme, in which there are many decisions to be made, all of them with consequences of comparable magnitude, so that they are best tackled simultaneously. We address the simplest of such settings in which no decision affects the losses, or loss functions, of any others and the decisions have to be made at the same time but in no particular order.

This general problem has its time-honoured version in hypothesis testing, commonly referred to as multiplicity of testing. There the focus is on controlling the error of one type (false discovery) at the expense of the other type (failure to discover). We dismiss this approach for several reasons. First, the argument that a hypothesis test is oblivious to the consequences of the two kinds of error carries over from a single test to any collection of tests. Next, these consequences may differ across the tests—some tests, or decisions, are more important than others. Further, by controlling the probability of one kind of error we expose ourselves to a high probability of the other kind.

In addition to these reasons, regarding each test as a decision has an important attractive feature. The problem of multiplicity of testing has no optimal solution when the dependence structure of the tests is not known, or is too difficult to elaborate. In contrast, the calculus of expected losses is simple and transparent: the expected losses are additive. However, this feature does not come for free—the more we rely on additivity, the greater the onus on the definition of the loss function that has to be equally applicable (universal) to all the decisions, just like a monetary currency is for all our financial transactions.

Prominent examples or applications of making many decisions include gene expression studies, in which we want to sift through a large number of genes and identify the few that are associated with a particular condition, such as an (inherited) disease, (bank) account transactions, among which we want to identify fraudulent activity, and standardised educational testing, in which we want to select a few papers (response sheets) and examine them as possible cases of cheating.

8.1   Ordinary and exceptional units

In the general case, we have K elementary experiments, studies, items or units (of analysis). Each unit $k = 1, \ldots, K$ has an underlying quantity θ_k estimated by $\hat{\theta}_k$. We assume that these estimators are unbiased, $\mathrm{E}(\hat{\theta}_k) = \theta_k$, and normally distributed with respective sampling variances τ_k^2. The estimators $\hat{\theta}_k$ may be correlated. Further, there is a reference value θ. We want to identify the few units k that have values θ_k distant from θ. We refer to them as *exceptional* units and to their complement as *ordinary* units. As implied by the labels 'exceptional' and 'ordinary', we assume that the ordinary units form a substantial majority.

Exceptional units should not be confused with outliers. A small group of units are outliers if their removal from the sample substantially alters the distribution of the units. An example of such outliers arises when a small set of units (or a single unit) is separated from the rest of the units by a wide gap. We have to draw a distinction between outliers among the targets θ_k and among their estimates $\hat{\theta}_k$. In contrast, exceptionality is a property of the targets θ_k only, and there is a hard threshold that separates exceptional units from the ordinary units.

Following the case of a single decision, we start by specifying the losses (or loss functions) and the threshold, the value that separates large and small distance from θ. Thus, we set a positive constant $\Delta\theta$ and regard all the units k with $\theta_k \notin (\theta - \Delta\theta, \theta + \Delta\theta)$ as exceptional. When there is some uncertainty, ambiguity or discord about the value of $\Delta\theta$, we consider a plausible range of its values, $(\Delta\theta_-, \Delta\theta_+)$, where $\Delta\theta_+ > \Delta\theta_- > 0$.

We use the respective indices O (o) and X (x) for ordinary and exceptional units, values, and the like; capitals are for states and lowercases for verdicts. We set the loss for a failure to discover, with verdict V_o, to $L_{k,oX} = 1$ and the loss for a false discovery, with verdict V_x, to $L_{k,xO} = R$ for all k. The verdicts specific to the units k are denoted as $V_{k,v}$ for v = o or x. The appropriate verdicts are associated with no loss. A plausible range may be declared for R. In principle, each unit k may have its specific loss ratio R_k. If these ratios differ, then the assumption of constant loss $L_{k,oX} = L_{oX} = 1$ is questionable. If we retain the unit loss for each failure to discover, we can address this problem by introducing a factor m_k, called *importance*, and consider the scaled expected loss $Q_k = m_k \mathrm{E}(L_k)$. It is easy to see how one can get into a bind when trying to allow for the uncertainty about both losses $L_{k,xO}$ and $L_{k,oX}$ when many of them are different.

We focus on cases in which K is so large that we cannot discuss or inspect each unit k and set its losses $L_{k,xO}$ and $L_{k,oX}$ individually. We refer to the case of constant pairs of losses, $L_{k,oX} = L_{oX} = 1$ and $L_{k,xO} = L_{xO} = R$ for all k as *trivial* loss structure, and to all other as nontrivial. In the simplest nontrivial structure, the K units are divided into (a small number of) H

categories, and trivial loss structure applies in each category. If the losses in category h are $L_{\text{oX}}^{(h)} = 1$ and $L_{\text{xO}}^{(h)} = R^{(h)}$, then the constant loss of one lossile for failure to discover is not natural, and an importance $m^{(h)}$ has to be introduced for each category. In this case, the categories can be treated as different countries and importances as exchange rates for conversion to the currency of one (reference) country, say $h = 1$, for which $m^{(1)} = 1$. Thus, the loss structure is described by $2H - 1$ positive constants $R^{(h)}$ and $m^{(h)}$, $h = 1, \ldots, H$, with $m^{(1)} = 1$. Unless H is small, introducing plausible ranges for these parameters can easily become unmanageable, even though the plausible expected losses are, in principle, easy to evaluate.

We thus face the task of minimising the importance-weighted sum of expected losses,

$$Q = \sum_{k=1}^{K} m_k Q_k, \tag{8.1}$$

where $Q_k = \min(Q_{k,\text{o}}, Q_{k,\text{x}})$, and $Q_{k,\text{o}} = \text{E}(L_{k,\text{oX}})$ and $Q_{k,\text{x}} = \text{E}(L_{k,\text{xO}})$. When the K decision problems fall into H categories, each with a trivial loss structure, then

$$Q = \sum_{h=1}^{H} m^{(h)} Q^{(h)},$$

and for each expected loss $Q^{(h)}$ we have the expression in equation (8.1), with superscripts (h) added.

When the decisions are autonomous, one not affecting any other, the overall expected loss Q is minimised by minimising each elementary expected loss Q_k. In practice, some decisions are of secondary importance, and we would prefer to drop them from consideration, so as to focus more sharply on the remainder. This idea can be implemented by attaching a cost C to each decision k, and dropping all decisions k for which less than C lossiles are at stake.

For a single decision, to be based on an unbiased normally distributed estimator $\hat{\theta}_1$ with sampling variance τ_1^2, we define the stake as the largest possible loss. With piecewise constant loss, the stake is $\max(R, 1)$. For a few decisions it is reasonable to add up the stakes, so that with a trivial loss structure and piecewise constant loss for M decisions, the overall stake is $M \max(R, 1)$. This is based on the assumption that we may get every one of the M decisions wrong, despite the aversion to the choice with more severe consequences.

When the number of decisions is large we can discount such a pessimistic scenario as implausible. We regard the M parameters θ_k, $k = 1, \ldots, M$, as a sample drawn from a population described by a distribution. Suppose this distribution is normal, $\mathcal{N}(\mu, \sigma^2)$. We can estimate its (super-) parameters μ and σ^2 by the method of moments. When each estimator $\hat{\theta}_k$ has the same sampling variance τ^2, there is no worthy competitor to the sample mean

$$\hat{\mu} = \frac{1}{M} \sum_{k=1}^{M} \hat{\theta}_k .$$

When the sampling variances (τ_k^2) are unequal the sample mean is replaced by the weighted mean

$$\hat{\mu} = \frac{1}{c_+} \sum_{k=1}^{M} c_k \hat{\theta}_k ,$$

where $c_k = 1/\tau_k^2$ is the reciprocal of the variance, known as the concentration, and $c_+ = c_1 + \cdots + c_M$ is their total. The variance σ^2 is estimated by adjusting the naïve estimator of σ^2,

$$\hat{\sigma}_\dagger^2 = \frac{1}{M-1} \sum_{k=1}^{M} \left(\hat{\theta}_k - \hat{\mu} \right)^2 , \tag{8.2}$$

assuming that the variances τ_k^2 are identical. This estimator is biased because it ignores the uncertainty about $\hat{\theta}_k$ and $\hat{\mu}$, but its bias can be adjusted for. The adjustment is derived in Appendix A.

We issue verdict $V_{k,x}$ when

$$\Phi(z_{k+}) - \Phi(z_{k-}) < \frac{1}{R+1} , \tag{8.3}$$

where $z_{k+} = (\theta - \hat{\theta}_k + \Delta\theta)/\tau_k$ and $z_{k-} = (\theta - \hat{\theta}_k - \Delta\theta)/\tau_k$. We issue verdict $V_{k,o}$ otherwise. We commit an error, and lose one lossile, when the inequality in (8.3) holds but $\theta - \Delta\theta < \theta_k < \theta + \Delta\theta$. We lose R lossiles when the inequality in (8.3) does not hold but $\theta_k < \theta - \Delta\theta$ or $\theta_k > \theta + \Delta\theta$. Without evaluating the estimates $\hat{\theta}_k$, $k = 1, \ldots, M$, we have two simple options: to issue $V_{k,x}$ for every k, and to issue $V_{k,o}$ for every k. The corresponding losses, integrated over the distribution of $\{\theta_k\}$ are

$$\mathrm{E}\left(L_{xO}; \mu, \sigma^2\right) = \int_{\theta-\Delta\theta}^{\theta+\Delta\theta} \varphi(x; \mu, \sigma) \, dx$$

$$= R\{\Phi(u_+) - \Phi(u_-)\}$$

$$\mathrm{E}\left(L_{oX}; \mu, \sigma^2\right) = 1 - \Phi(u_+) + \Phi(u_-) , \tag{8.4}$$

where $u_+ = (\theta + \Delta\theta - \mu)/\sigma$ and $u_- = (\theta - \Delta\theta - \mu)/\sigma$. Note that these two expectations are over the distribution of θ_k; they are not fiducial but for replications of draws from $\mathcal{N}(\mu, \sigma^2)$.

If we obtained each estimate $\hat{\theta}_k$, we could minimise the (fiducial) expected loss separately for each k. We estimate the average of these expected losses by simulations, drawing a (large) random sample $\{\eta_j\}$ from $\mathcal{N}(\mu, \sigma^2)$, contaminating each value with estimation error $e_j \sim \mathcal{N}(0, \sigma^2)$, establishing the verdict according to the rule in equation (8.3) applied to the estimate $\eta_j + e_j$, the consequent loss (0, 1 or R), and averaging these losses over the replicates $j = 1, \ldots, J$.

Example 15

We explore the following setting. The distribution of the (underlying) values θ_k is $\mathcal{N}(\mu = 0.5, \sigma^2 = 0.2)$. For each k, estimator $\hat{\theta}_k$ is unbiased for θ_k and has sampling variance $\tau^2 = 0.04$, the same for all k. Unit k is regarded as exceptional if $|\theta_k| > 0.75$, so that $\theta = 0$ and $\Delta\theta = 0.75$. The loss ratio is set to $R = 5$. In the simulations, we use 100 000 replications.

We obtained the following counts of replicates, cross-classified by the verdict (rows) and status (columns):

	State	
Verdict	O	X
O	69 052	12 312
X	1609	17 027

with the per-unit loss $(5 \times 1609 + 12\,312)/10^5 = 0.204$. These figures cannot be reproduced exactly because they involve randomness. For example, a replication of the entire simulation yielded per-unit loss 0.212. The tabulation indicates that an ordinary unit (O) is very likely to be classified correctly; there are only 1609 errors among 70 661 ordinary units (2.3%). In contrast, the performance on the exceptional units (X) might seem rather disappointing: there are 12 312 errors among 29 339 units (41.9%). However, this is the right strategy when errors committed on the exceptional units are far less serious than on ordinary units. If we alter the threshold $1/(R + 1) = \frac{1}{6}$ in equation (8.3), we alter the balance of the two kinds of error. For instance, for $R = 2.2$ we obtained the verdict-by-status table

	State	
Verdict	O	X
O	67 611	8 080
X	3533	20 776

Now the rate of false positives is $3533/(67\,611 + 3533) = 0.050$, the rate that would be aimed for by a hypothesis test. The rate of false negatives, 0.280, is much lower than with $R = 5$. However, the per-unit loss, evaluated with $R = 5$, is equal to 0.257. This is greater than the loss for the minimum expected loss decision, 0.204, by 26%. A kind of parity, when the numbers of errors of both kinds are approximately equal, arises for $R = 1.2$, when there are about 5800 errors of either kind. However, the numbers of units with states O and X are not even (70 700 vs. 29 300), so the error rates differ. The error rates are approximately equal for $R = 0.68$ (12.95%). These choices are distinctly inferior to the optimal decision rule, *if* we are committed to the loss ratio of $R = 5$. ☐

A special case arises when there are very few exceptional units (contaminants or gems), and we know that that is the case. Missing an exceptional unit (a false positive) in this case is a serious error; $L_{oX} \gg L_{xO}$. The selected units are then subjected to a refined search, so there is some incentive to obtain a small set of selected units. If we knew that the collection of K units contains M exceptions, we should nevertheless not discard any selection of size greater than M. It is better to admit to errors (false positives) than to retain the hope (or pretence) of correctness by selecting exactly M units, because we care about the expected loss, not the likelihood of no loss.

We can incorporate the incentive to select a small set of units by declaring a cost associated with a large (excessive) selection. This cost is a nondecreasing function $S(m)$. For $m \leq M$ there is no cost; $S(m) = 0$. For $m > M$, $S(m)$ is positive. The obvious choice is a fixed cost $s > 0$ for each unit in excess of M, that is, $S(m) = s(m - M)_+$. The plus sign in the subscript indicates the positive part, that is, $(h)_+ = h$ if $h > 0$ and $(h)_+ = 0$ otherwise.

We now minimise the total of the overall expected loss and the cost of excessive selection. This is implemented by sorting the units in the descending order of the balance $B_k = Q_{k,o} - Q_{k,x}$, $k = 1, \ldots, K$, and evaluating the partial (cumulative) totals $C_h = B^{(1)} + B^{(2)} + \cdots + B^{(h)} - S(h)$, where $B^{(j)}$ is the balance for the unit with rank j; $B^{(1)}$ is the largest balance, $B^{(2)}$ the second largest, and so on. We select the first m^* units in this list for which C_m is maximised. When there is a fixed cost s for each unit in excess of M, we select the units with the M largest balances, so long as their balances are all positive. If $B^{(M)} > 0$, then we select all units from the remainder, for which $B^{(m)} > s$.

8.2   Extreme selections

Verdict $V_{k,o}$ is not issued for any value of $\hat{\theta}_k$ when the ordinary range is too narrow (sufficiently small $\Delta\theta$), we are too averse to failure to discover (sufficiently small R) or the measurements are too imprecise, when $\tau_k^2 = \text{var}(\Delta\hat{\theta}_k)$, or their common value τ^2, are too large.

For fixed values of θ, $\Delta\theta$ and τ_k, the right-hand side of the inequality in equation (8.3) attains its maximum for $\hat{\theta}_k = \theta$. This is concluded by setting its partial differential

$$\frac{\partial \left\{ \Phi\left(z_{k_+}\right) - \Phi\left(z_{k_-}\right) \right\}}{\partial \hat{\theta}_k} = -\frac{1}{\tau_k} \left\{ \varphi\left(z_{k_+}\right) - \varphi\left(z_{k_-}\right) \right\}$$

to zero. Symmetry and unimodality of φ imply that the only solution is given

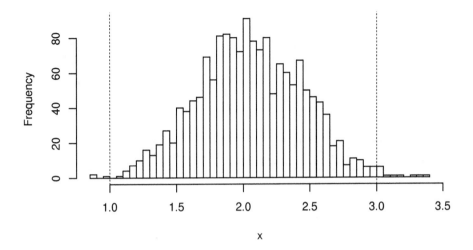

FIGURE 8.1
Histogram of the observations obtained by the first measurement.
The ordinary range is delimited by vertical dashes.

by the identity $z_{k_+} = -z_{k_-}$, and that is equivalent to $\hat\theta_k = \theta$. Therefore, the difference $\Phi(z_{k_+}) - \Phi(z_{k_-})$ is never greater than $1/(R+1)$ when

$$\Phi(z_0) - \Phi(-z_0) < \frac{1}{R+1},$$

where $z_0 = \Delta\theta/\tau_k$. The solution of this inequality for $\Delta\theta$ is

$$\Delta\theta < \tau_k\, \Phi^{-1}\left(\frac{R+2}{2R+2}\right). \tag{8.5}$$

The solution for R is

$$R < \frac{2 - 2\Phi(z_0)}{2\Phi(z_0) - 1}. \tag{8.6}$$

These evaluations have a natural interpretation. When all τ_k are identical, the strongest case for V_o arises when $\hat\theta = \theta$, when the measurement falls right in the middle of the ordinary range. When even such an observation would be declared as exceptional, then no observation would be declared as ordinary. The borderline R, given by the right-hand side of equation (8.6), is an increasing function of τ. Loosely interpreted, with more precise estimation, the selection is more discerning.

Example 16

Figure 8.1 displays the histogram of a set of $K = 1609$ observations, among which we want to identify exceptions. The observations are subject to estimation (or measurement) error with variance 0.040; they entail no bias. We

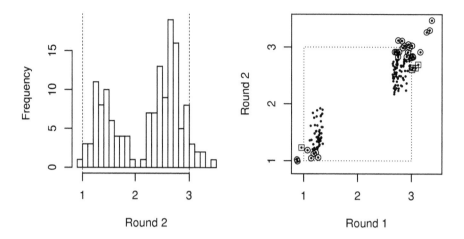

FIGURE 8.2
Candidates selected in the first round and their second-round measurements. Units selected in round 2 are marked by circles; units with round-1 measurements outside the ordinary range $(1.0, 3.0)$ that were not selected in round 2 are marked by squares.

distinguish between the underlying values θ_k and their estimates (or measurements) $\hat{\theta}_k$, $k = 1, \ldots, K$. Exceptions are defined as units with θ_k outside the ordinary range $(1.0, 3.0)$. The values of all the observations $\hat{\theta}_k$ are in the range $(0.873, 3.383)$, and only 13 of them are outside the ordinary range. Three values are smaller than 1.0 and ten exceed 3.0. The loss ratio is 0.04. A false exception is 25 times less serious an error than a failure to identify an exceptional unit.

There is a more precise method of measurement, with measurement error variance 0.011, but it cannot be afforded on all the observations. Therefore we proceed in three stages. First we select from the 1609 units candidates for exceptional status. Then we apply the more expensive and (more precise) measurement on the selected units, and finally we sub-select units based on these measurements.

The first stage concludes with 148 candidates; they are defined by the inequalities $\hat{\theta}_k < 1.353$ and $\hat{\theta}_k > 2.647$. Of course, they include the 13 observations with values outside the range $(1.0, 3.0)$. These 148 units proceed to the second round, where their values are measured with the more precise instrument.

Figure 8.2 displays a histogram of the values of the second measurements in the left-hand panel and a plot of the pairs of measurements for the candidate units in the right-hand panel. The second round of selection concludes with 28 units; seven of them have estimates smaller than 1.185 and the remaining 21 have estimates greater than 2.815. Of the 13 units with first measurements

TABLE 8.1
Units selected in round 2, sorted by their expected gains (balance).

Rank	Value	Balance	Rank	Value	Balance	Rank	Value	Balance
1	3.47	1.00	11	2.99	0.44	21	1.12	0.08
2	3.30	1.00	12	1.02	0.41	22	1.15	0.05
3	3.26	0.99	13	1.05	0.31	23	2.84	0.03
4	3.12	0.87	14	1.06	0.26	24	2.84	0.02
5	3.11	0.85	15	2.92	0.18	25	2.83	0.02
6	3.05	0.67	16	2.92	0.18	26	2.83	0.01
7	3.03	0.58	17	2.92	0.18	27	2.82	0.00
8	3.02	0.58	18	2.91	0.17	28	1.19	0.00
9	0.99	0.51	19	2.91	0.16			
10	2.99	0.45	20	2.90	0.15			

outside $(1.0, 3.0)$, five units are not included in this list: one of the three with small values and four with large values. They are marked in the plot by squares.

The histogram indicates that the round-2 measurements of the candidates with small and large values are less well separated than their round-1 measurements. This is a consequence of the selection (of extremes) in round 1, which has some preference for exceptional errors $\hat{\theta}_k - \theta_k$. The errors in round 2 are independent of them. Every unit with a round-2 measurement outside the ordinary range is selected, and several units that have both measurements within the ordinary range are also selected. This reflects our strong aversion to omitting exceptional units.

Table 8.1 lists the values of the selected units and their balances in round 2, in the descending order. A few units have very small balances; they have been included in the list by a narrow margin.

Suppose the selection of up to 12 units entails no cost, but the cost of selecting the next 12 units is 0.12 lossiles each, and the cost for each unit beyond the first 24 is 0.08 lossiles. The units ranked lower in the selection are the first candidates for discarding. Inclusion of units ranked $25-28$ costs 0.32 lossiles in total, whereas the expected gains over discarding them add up to only 0.03. Similarly, units $21-24$ are a poor value, since including them is associated with expected gains $0.02-0.08$ each, smaller than the cost of keeping them in the list, 0.12 each. Thus, we conclude with the selection of the units ranked $1-20$.

Selection of exceptional units is conditional on the values of the parameters and the assumption of normality. Sensitivity of the results can be assessed by altering the values of the parameters and using alternative distributions. The

obvious choice for the latter is a Student t distribution with a small number
of degrees of freedom.

Suppose the estimators $\hat{\theta}_k$ are such that each $\hat{\theta}_k/\hat{\tau}_k$ has noncentral t distribution with five degrees of freedom. Then the selection in round 1 concludes
with 142 units, six fewer than with the normality assumptions ($\mathcal{N}$). These
units are a subset of the 148 selected with $\mathcal{N}$. The six units for which the
selection is in discord have very small balances; for all of them the balances
are smaller than 0.01. Round 2 with measurement errors distributed according
to t_5 (scaled by $1/\sqrt{0.011}$) also yields a selection very similar to the selection
with $\mathcal{N}$. Twenty six units are selected; they coincide with the selection made
with the normality criterion, except for two units, both of them with very
small (positive) balances, 0.0025 and 0.0004, ranked 27 and 28 in the original
analysis (Table 8.1). By introducing the cost of selection, 0.12 for units ranked
$13-24$ and 0.08 for units ranked lower, we end up selecting the same set of
20 units as with the normality assumptions.

8.3   Grey zone

In a practical setting, there is an obvious value for the centre of the ordinary
range, θ, but there may be disagreement or ambiguity about the values of
$\Delta\theta$ and R. Plausible values have to be considered for these two parameters.
The pairs of equilibria, values of the estimates for which the expected losses
with the two verdicts coincide, are monotone functions of both $\Delta\theta$ and R, so
the plausible equilibria are delimited by the equilibria for the bounds of the
plausible ranges for $\Delta\theta$ and R, namely, $(\Delta\theta_-, R_-)$ and $(\Delta\theta_+, R_+)$.

An equilibrium E is found by the Newton-Raphson algorithm. By symmetry, its twin is $2\theta - E$. A pair of equilibria is a function of the standard error
τ_k. Figure 8.3 displays the continuum of plausible equilibria for the plausible
ranges $\Delta\theta \in (0.5, 0.8)$ and $R \in (0.04, 0.06)$. The grey zone that these equilibria delineate is bounded by the equilibria for ($R = 0.04, \Delta\theta = 0.5$) and
($R = 0.06, \Delta\theta = 0.8$). For the latter, equilibria do not exist for $\tau_k > 0.242$.
The curves for the negative and positive equilibria meet at the centre $\theta = 0$.

For any pair $(\hat{\theta}_k, \tau_k)$ in the grey zone we arrive at an impasse. For some
plausible values of R and $\Delta\theta$ one verdict, and for others the other verdict
is issued. With increasing standard error the ordinary range narrows until it
vanishes altogether at $\tau_k = 0.242$. The diagram implies that the uncertainty
about R introduces much less uncertainty about the equilibria than does the
uncertainty about $\Delta\theta$. It is easy to check that the equilibria for $R_- = 0.04$
and $R_+ = 0.06$ differ much less for any plausible value of $\Delta\theta$ than do the
equilibria for $\Delta\theta_- = 0.5$ and $\Delta\theta_+ = 0.8$ for any plausible value of R. This
suggests that it would be more useful to try and narrow down the plausible
range of $\Delta\theta$ than the plausible range of R.

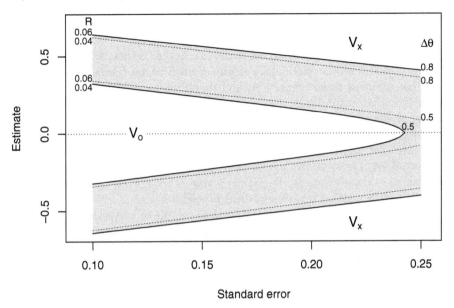

FIGURE 8.3
Grey zone. Equilibria for plausible ranges $\Delta\theta \in (0.5, 0.8)$ and $R \in (0.04, 0.06)$; $\theta = 0$.

8.4 Actions in a sequence

When we purchase an item or subscribe to a service, we are sometimes offered another item or service, often at a discount. The offer is a result of cursory or profound research of other customers' shopping 'careers'. We are told that many customers who in the past purchased A, just like we are purchasing it now, have returned later and purchased B. It is suggested that we, the present purchaser, are like them, and we can save some money by this offer, and spare ourselves the trouble of having to come back in the near future when we realise that item (or service) B would be useful to have.

A less frivolous example is from clinical medicine. A patient admitted to the emergency ward of a hospital undergoes procedure A, say, a surgery. The consultant involved recalls that past patients who undergo this procedure are often readmitted, also as emergency cases, and require procedure B, another surgery. It might amount to better care, and lower expenditure of resources, to apply procedures A and B together, as it is likely to save the patient's future episode of emergency when his or her life is in balance and an involved medical procedure has to be applied (again) at a short notice. Some of the preparatory steps that are common to surgeries of types A and B are applied only once when A and B are applied together, whereas they have to be repeated when A and B are applied after separate admissions.

The argument against applying A and B together is that, without reference to past incidents (other patients' histories of medical interventions), procedure B is not essential to accompany A, and therefore its application is in conflict with the ethical standards of applying no unnecessary procedures as well as with the general principle of being frugal with the available resources that could be applied more appropriately and effectively to other patients. The argument for applying A and B together is that, in longer term, resources are saved and better health care is provided by avoiding (another) emergency treatment and the stress on the patient's organs brought on by the emergency, together with another invasive medical procedure.

Which argument should prevail? In the example of marketing, when is the campaign worth pursuing and how large a discount would be appropriate to offer for the additional purchase? In the example of surgery, what kind of study should be conducted to arbitrate the issue?

Randomised clinical trials (Chapter 10) are the gold standard in the development of new treatments, including drugs and interventions, such as methods of surgery, and, more generally, guidelines to follow in the care for and treatment of patients. Their principal feature is random assignment of (recruited) patients to treatments. A key step in the recruitment of a patient, apart from the qualification of having the relevant condition, is informed consent, a conscious decision to enrol in the trial, arrived at without any coercion or threat of a sanction, after being informed of the two (or several) alternative treatments and made aware of the procedure of randomisation and other design features of the trial. Such a recruit is unlikely to develop a preference before the treatment is assigned to him or her and has no objection to leaving it to chance as to which treatment he or she will receive. Herein lies the main difficulty in conducting a clinical trial. Many patients either have a preference or prefer the choice to be made by the consultant—a competent professional. They may be enticed by the altruistic idea of contributing to the effort to establish which treatment is superior, but this is hardly the main motive for most of them (nor would be for any of us).

When the choice between two treatments is exercised, although without a regular or predictable pattern, an observational study is a credible alternative to a clinical trial. Informed consent in such a study is sought only for sharing the data about the episode of care: the treatment selected and applied, the associated factors, and the outcomes. This places a much lesser onus on the patient than giving consent to a treatment that would be assigned at random, outside any control by the patient and without being informed by the clinical attributes of the case.

It often takes a long time to recruit a sufficient number of patients for a randomised trial, given the many rejections in the attempts to recruit, and the need for recruiting agents to be present with the relevant information in a setting appropriate for requesting informed consent. Studies that use data collected in the past, preceding the commencement of the study (or even its design) are called retrospective. Not having to wait for the potential recruits is their distinct advantage. The advantage is further enhanced when the relevant variables, including descriptions of the outcomes of the treatment, have been recorded, together with an approval to use them for medical research not as yet specified when the data is recorded. With such registers, retrospective studies are much easier to design, organise and conduct, because the data is available; it can be extracted from an established database. If the register has been in existence for several years, then it contains data about a lot of patients.

The handicap of no control over the treatment assignment is addressed by selecting a subgroup from each treatment group so that for all purposes these two subgroups as a dataset could be regarded as arising from a trial with randomised treatment assignment, and could be analysed by a method suitable for such an experiment. The subgroups are formed by pairing up patients from one treatment group with patients from the other, so that the patients within a pair have as similar backgrounds as can be arranged. Such a procedure is referred to as matching. Details are moved to Appendix B.

Suppose a matching exercise concludes with n patients who underwent procedure AB (operations A and B applied at the same admission) and another n patients who underwent only procedure A. Suppose the outcome variable is the time-span between the procedure and the next admission for treatment, or death, for a related condition, with an appropriate arrangement when no such event took place. We look no further than two years after the treatment; that is the horizon of the study. The readmissions may be restricted to those in which A or B, or both, have been applied. An alternative to the time spanned is the binary variable indicating treatment-free survival for two years.

We assume that every patient is in acute need of procedure A, so the cost and other outlay, risks and impact on the patient that are associated with this procedure are common to the $2n$ patients, and can therefore be disregarded in any comparisons of AB vs. A. Application of the combined procedure AB is associated with cost C_B (in lossiles) in addition to the cost of A which we have discounted. When the element B of the procedure is redundant, there is a further cost C_R that can be interpreted as a penalty for breaching the principle of applying minimum necessary treatment.

If patient who received procedure A is readmitted within a given period of time, then the loss is C_E. This loss includes the cost of the procedure, C_B, the cost of the arrangements associated with emergency care (transport by ambulance, first aid and stabilisation), the harm caused to the patient and the danger to which the patient was exposed owing to having applied only A instead of AB, and the elements of the cost of applying B on its own that would have been spared by applying AB. They include recovery from the surgery if recovery from AB takes as long and requires as much of resources as recovery from A. More precisely, let R_{AB}, R_A and R_B be the resources required for recovery after the respective procedures AB, A and B. Reasonable assumptions are that $R_{AB} \geq R_A$, $R_{AB} \geq R_B$ but $R_{AB} < R_A + R_B$, and that $R_{AB} - R_A$ and $R_{AB} - R_B$ are small; every invasive treatment has its 'overhead' of placing the patient under stress and demanding a long recovery, not only in clinical care, but also for weeks or months that follow.

In brief, establishing the losses is a complex accounting exercise in which we tally the costs in terms of the resources expanded by the health-care services, their appropriateness, stresses, risks and incapacitation or disability to which the patient is exposed. Some of this we can postpone, but cannot avoid or ignore.

A simple and intuitive approach to the analysis entails within-pair contests. Each pair contains one patient from group AB and one from A, and every patient appears in at most one pair. The patients in each pair are matched on the background variables. For each matched pair, we arbitrate which patient had a better outcome. The possible outcomes are A followed by no admission, A followed by admission for B, AB followed by no admission, admission for A or B, and in all cases we distinguish between survival till the horizon of the study and death before reaching the horizon. Death is the worst outcome but in such cases we could tally the losses associated with treatment(s). Further, among survivors, A on its own is best without a readmission, followed by AB without readmission, then A with readmission and AB with readmission. Yet further, a readmission while still recovering from the previous procedure (surgery) may reasonably be regarded as worse than a readmission after full recovery.

For each pair, we have a comparison: treatment A is a winner, tie or loser. The ties can be ignored because they contain no information about which treatment is superior. Thus, the number of winners has a binomial distribution. The probability of being a winner is estimated by the sample proportion of winners. Its distribution can usually be well approximated by a normal distribution. The test based on this statistic is known as the sign test.

8.5   Further reading

This chapter is based on Longford (2014a) and (2013a). The problem of controlling the overall rate of false positive verdicts is a problem of considerable interest in hypothesis testing. Benjamini and Hochberg (1995) and Benjamini and Yekutieli (2001) present solutions that have been widely adopted. A Bayesian perspective on the problem is discussed by Berry and Hochberg (1999). The problem has received a new impetus with the expansion of research in genetics (e.g., Storey 2002), and disease mapping and surveillance. Imbens and Rubin (2015) is a comprehensive textbook on the potential outcomes framework. Rosenbaum (2017) presents a similar perspective.

8.6 Exercises

8.1. Review the issue of multiplicity in testing many hypotheses and the arguments against hypothesis testing in the context of making many decisions.

8.2. Explain why the correlation structure of the estimators $\hat{\theta}_1, \ldots, \hat{\theta}_K$ can be ignored when making decisions about units $1, \ldots, K$, but not when testing the corresponding set of hypotheses.

8.3. What is the difference between exceptional and outlying units? Find examples in which either property is important (and the other one is not).

8.4. What is the role of the importance factor in the expected loss of a collection of all decisions?

8.5. How is minimisation of the expected loss Q complicated when the decisions are not autonomous?

8.6. Derive the inequality in equation (8.3).

8.7. Discuss how the balance of the two expectations in equation (8.4) is altered when $\Delta\theta$ is reduced and when σ is reduced.

8.8. Reproduce Example 15.

8.9. Compile a list of approaches to making many decisions when it is not possible to define the loss separately for each elementary decision.

8.10. Discuss the conditions stated in the first paragraph of Section 8.2. Examine the rationale for labelling all units as exceptional in the setting introduced in the first paragraph of Section 8.2.

8.11. Derive the counterparts of the inequality in equation (8.3) for linear loss.

8.12. Derive the inequalities in equations (8.5) and (8.6).

8.13. How should the second stage of selection in Example 16 be adapted when obtaining an estimator $\hat{\theta}_k$ is associated with a (fixed) cost? Could the second-round estimator be improved by combining it with the first-round estimator?

8.14. It is known about a large collection of units that it contains a given number H of exceptional units. Why should we not, in general, reduce (or expand) the selection to H units? How is your answer altered when there is a cost associated with finding more than H exceptions?

8.15. Relate the problem of making many decisions to the problem in medical screening in Chapter 7.

8.16. In the secretary selection problem, there is a vacancy for a secretary. The director interviews candidates in a sequence and decides at the end of each interview whether to hire or reject the candidate. A rejected candidate cannot be recalled later. The director uses no information about the candidate other than the (oral) interview. Devise a selection procedure for the director. As factors, consider how well the director could cope without a secretary, the cost of advertising and arranging the interviews, the loss when failing to hire the best candidate, the loss associated with firing a secretary and with losing a secretary who leaves for another job.

8.17. Describe your attitude to impulse buying. While shopping according to a list of necessities you compiled before leaving home you notice some bargains that may be more expensive next time or items about which you are not certain whether you will need them. Relate it to the alternative treatments AB and A in Section 8.4.

8.18. Compare the status (fixed or random) of the outcomes and treatment assignment in the potential outcomes framework defined in Appendix B, and in modelling of the outcome by relating it to the treatment.

Appendix

A. Moment-matching estimator of the variance σ^2

We evaluate the expectation of the statistic $\hat{\sigma}_\dagger^2$ defined in equation (8.2), and then adjust $\hat{\sigma}_\dagger^2$ so that it would become an unbiased estimator of

$$\sigma^2 = \frac{1}{M-1} \sum_{k=1}^{M} (\theta_k - \mu)^2 .$$

Denote by e_k the error in estimating θ_k by $\hat{\theta}_k$; $e_k = \hat{\theta}_k - \theta_k$. Unbiasedness of $\hat{\theta}_k$ implies that $E(e_k) = 0$. Then

$$(M-1)\,E\left(\hat{\sigma}_\dagger^2\right) = E\left\{ \sum_{k=1}^{M} \left(\hat{\theta}_k - \hat{\mu}\right)^2 \right\}$$

$$= \sum_{k=1}^{M} E\left(\theta_k - \mu + e_k - \frac{1}{M}\sum_{l=1}^{M} e_l \right)^2 .$$

By expanding each square we obtain the squared deviations $(\theta_k - \mu)^2$, a squared term involving the estimation errors for which

$$E \left(e_k - \frac{1}{M} \sum_{l=1}^{M} e_l \right)^2 = \text{var} \left(e_k - \frac{1}{M} \sum_{l=1}^{M} e_l \right), \qquad (8.7)$$

and twice the crossproduct which has zero expectation. The variance in equation (8.7) is equal to

$$\tau^2 \left\{ \left(\frac{M-1}{M} \right)^2 + \frac{M-1}{M^2} \right\} = \frac{M-1}{M} \tau^2 .$$

Therefore $E \left(\hat{\sigma}_\dagger^2 \right) = \sigma^2 + \tau^2$, and hence the unbiased estimator $\hat{\sigma}^2 = \hat{\sigma}_\dagger^2 - \tau^2$.

B. The potential outcomes framework

The fundamental difference between clinical trials (experiments) and observational studies is in their design; the former control the treatment assignment, and the latter do not. By control we eliminate the influence of all the background variables that are not involved in the treatment assignment process, and accounting for the background variables that are involved is straightforward. For example, men may be randomised to treatments D and E with respective probabilities $\frac{3}{4}$ and $\frac{1}{4}$ and women with equal probabilities. Sex then has to be taken into account in the analysis, but age, socio-demographic background and medical history can be ignored. Given sufficiently large sample size, the two treatment groups are well balanced on these variables and factors; neither group is 'privileged' to have younger, better educated, healthier or more prosperous subjects. The imbalance on sex is as prescribed by the design. There may be some departures from the exact balance but they are unlikely to be large when the sample size of the study is large. These deviations can be attributed to chance; that is, they would vanish in averages over hypothetical replications of the study.

In the potential outcomes framework, we consider treatment-specific outcomes Y_{iD} and Y_{iE} for each subject i. Subject-specific treatment effects are defined as the differences of these outcomes, $\Delta Y_i = Y_{iE} - Y_{iD}$. None of these differences are realised when a subject can receive only one of the two alternative treatments. Denote by A_i the indicator of assigning treatment E. That is, $A_i = 1$ if subject i is assigned to receive treatment E. The outcomes Y_{iD} and Y_{iE} are assumed to be fixed and the treatment assignment is random. The distribution of the assignment is prescribed by the design in a clinical trial, whereas in an observational study it is not known. A hypothetical replication of the treatment assignment would yield the same outcome for patient i if the patient were assigned to the same treatment.

Randomisation enables us to estimate the average of these (within-subject) differences $\Delta \bar{Y}$ without bias by the difference of the mean outcomes within the treatment groups; $\bar{Y}_E - \bar{Y}_D$, with obvious adaptations when the assignment probabilities depend on some background variables. Without randomisation, one or a set of background variables provide an alternative explanation for the differences of the mean outcomes within the treatment groups. This problem can be addressed by finding subsets of the treatment groups that are well balanced on all the observed background variables. This can be interpreted as a search for a design in an already realised dataset, that is, *post-observational* design. Such a search must not involve any outcomes or other variables affected by the treatment. Formally, not only outcomes, but any other variable defined on the subjects in a study has well defined potential outcomes. For variables affected by the treatment, the two potential versions differ, for at least some subjects. For background variables, the two potential versions coincide for every subject.

Matched subgroups

If two subsets of subjects, one selected from each treatment group, are well balanced on the background variables, these variables have no role in the analysis that compares the outcomes of the two subsets. In contrast to randomised studies, concerns about confounding in an observational study are not allayed over background variables that are not observed. In this respect, an observational study is distinctly second-rate to a clinical trial. An observational study is more credible if its within-treatment subgroups are matched on a wider set of background variables, but it cannot reach the credibility of a clinical trial. However, observational studies have some strong points—their conduct requires less resources and can in some settings have sample sizes, or even study an entire population, that would be unthinkable in a clinical trial. Another advantage is that the balance achieved, admittedly only on the observed background variables, may be tighter than it would be in a clinical trial, even a hypothetical one that would have the same sample size.

Finding a matched pair of treatment subgroups is a computational exercise of increasing complexity as more background variables are involved (and the study becomes more credible). In brief, failure to design a study by controlling the treatment assignment is addressed by computational technology. The key theoretical result that makes matching feasible is that it suffices to balance the two subgroups on their propensities. Propensity is defined as the conditional probability of receiving one of the treatments (say, E) given the background profile. The propensity of a subject is the probability of assignment to the treatment given this (focal) subject's background, that is, the probability among the subjects who have the same background as the focal subject. Propensity has to be estimated, and the estimation entails a model search. However, the criterion in this search is not model validity but good balance. This search, called propensity modelling, involves only the treatment indicator and the background variables.

For a specific propensity model, such as the logistic regression (of the treatment indicator) on a set of background variables and some of their interactions, we evaluate the fitted propensities $\hat{p}_i$, $i = 1, \ldots, N$, and classify them to K categories (propensity groups). We form pairs of subjects so that each pair comprises a subject from either treatment group and the two subjects are from the same propensity group. For example, if propensity group k comprises n_{kD} and n_{kE} subjects from the respective treatment groups D and E, then $m_k = \min(n_{kD}, n_{kE})$ pairs can be formed in this propensity group. The $m = m_1 + \cdots + m_K$ matched pairs, formed in all the propensity groups, are reconstituted as two matched treatment subgroups of m subjects each. These two subgroups can be treated in the next step as if they arose in a randomised experiment. In particular, their outcomes can be compared without any (further) involvement of the background variables. Their potential confounding has been eliminated by matching. More precisely, such confounding would have been eliminated if the matching were perfect—if the two subgroups were perfectly balanced on their background profiles.

Balance

The balance on a continuous background variable X is assessed by comparing its means and standard deviations evaluated within the treatment subgroups. Let $\hat{\mu}_D$ and $\hat{\mu}_E$ be the means, $\hat{\sigma}_D$ and $\hat{\sigma}_E$ the standard deviations, and $\hat{\sigma}$ the pooled standard deviation of this variable. Then the balance is defined as $\Delta_X = |\hat{\mu}_E - \hat{\mu}_D|/\hat{\sigma}$ for the means and as $d|\log(\hat{\sigma}_E/\hat{\sigma}_D)|$ for the standard deviations; d is a suitable scaling constant. For a categorical variable, the balance is defined as the set of balances for its set of indicator variables ($H-1$ indicators for a variable with H categories).

The overall balance is defined as the average of the balances of the background variables. A convention is that every balance Δ_X should be smaller than 0.1. We suggest that the overall balance should be much smaller than 0.1, such as 0.025. An open-ended algorithm for finding a model is by trial and error. It comprises fitting a model (e.g., enhanced by one or a few newly added terms), defining propensity groups (e.g., by splitting the propensities according to their quantiles, such as deciles) and evaluating the balance. The conventional rules for model selection can be applied to speed up this process, but some level of improvisation is usually necessary. Be prepared to search through a large number of models, especially when there are many background variables.

Some adaptations

Proximity matching is an alternative to propensity matching. In proximity matching we define the distance between the background profiles of any two subjects and form matched pairs that are in a short distance from (that are similar to) one another. An upper bound can be set on the distance of a pair that qualifies for a match. The distance may be based on the fitted propensity;

it is more practical to use the logit of the propensity for this purpose. The Mahalanobis distance, and similar metrics, are alternatives.

Suppose there are fewer subjects in treatment group E than in D. We set an upper bound, such as 0.1, and search for each subject in E for a match from D that is closer than 0.1 on the scale of logit-propensity and is not matched (yet) with any other subject from E. In such pairing, preference is given to matches in shorter distance, but there is a danger that some 'popular' subjects from D are quickly used up, leaving no candidates for match for other subjects in E. This problem is addressed by sorting the subjects in E according to the number of candidates for a match (subjects from D that are within a set distance, called caliper, from the target subject), and prioritising matching for subjects who have fewer candidates.

These two methods of matching, within propensity groups and within calipers, have numerous adaptations and refinements. For example, we may insist on exact matching within the categories defined by one or a set of categorical background variables. This is implemented by splitting each propensity group according to these background groups, or defining propensity groups separately within each background group. For caliper matching, we add to the distance defined by log-propensity or by other means the distance of the background groups. The latter distance is defined so that two subjects are too far apart unless they fall into the same background group.

When there is an abundance of subjects in D, matched pairs can be replaced by matched cells that comprise one subject from E and a fixed number, say L, from D. These cells are then reconstituted as a matched group of Lm subjects from D and m subjects from E. Matching 2:1 has a strong advantage over 1:1 matching, so long as the number of triplets formed is not much smaller than the number of pairs would be. Matching 3:1, if feasible, presents further advantage, but this diminishes as larger cells are formed. Going beyond matching 5:1 is rarely of any advantage. Apart from forming fewer cells, the quality of the match should also be a concern.

Analysis of the outcomes

With a pair of matched subgroups, we can proceed to the analysis of the outcomes. In this analysis, the subjects in the pairs are regarded as if they arose in a study with randomised treatment assignment, and so the background can (and should) be ignored. The contrast of the within-subgroup means, $\hat{\Delta} = \bar{y}_{E'} - \bar{y}_{D'}$, is appropriate in most settings in which the outcome y is defined on a suitable scale, especially when the assumption of homoscedasticity (equal variances within the treatment subgroups) is palatable. The contrast $\hat{\Delta}$ may be transformed.

When we do not want to commit ourselves to any particular scale for the outcomes, a score may be assigned to each matched pair of subjects. In a simple version of this scheme, treatment group E is a winner in a matched pair, if its representative in the pair has an outcome superior to the outcome of its twin from D. Note that ties are also possible. In general, scores are assigned to each pair, such as 1 for a winner, 0 for a tie and -1 for a loser, and the scores are averaged or tabulated. In more complex schemes, the wins and losses are graded, such as ± 2 for a decisive difference and ± 1 for an insubstantial difference.

In inverse propensity weighting, matching is replaced by assigning weights w to every subject in the study in such a way that the two treatment groups would be represented by equal totals of weights within every propensity group or its refinement. For example, in the propensity group k with n_{kD} and n_{kE} subjects from the respective treatment groups D and E, each subject from treatment group T would be assigned weight $1/n_{kT}$. The weighted contrasts of the outcomes from the propensity groups are linearly combined with weights equal to the number of matched pairs that could be formed, that is, $\min(n_{kD}, n_{kE})$.

9

Performance of institutions

In developed countries, public and private institutions, such as schools, hospitals, providers of passenger transport, and some production units, operate in an environment where their performance is under constant scrutiny by customers (students, patients or passengers), their representatives and other stakeholders. This may take a variety of forms. At the most informal level, customers have an opportunity to comment on their experience, and these comments are compiled and evaluated by the management. The provider of the service (henceforth, the institution) may conduct a self-assessment. For example, a railway company may publish or be obliged to publish summaries, such as the proportion of the scheduled services that were delayed or cancelled in each calendar month. These summaries may be compared with standards set by the franchising authority or by the management of the institution, or be linked to future price adjustments and awards of franchise.

In this chapter, we consider the setting in which an outside body, such as an office for monitoring the standards, conducts a regular (e.g., annual) assessment of each institution in a particular domain (such as all the country's hospitals) and is charged with the task of classifying the institutions into a small number of ordinal categories or labels. An example of such a set of labels is 'unsatisfactory', 'satisfactory' and 'excellent'. When the assessment is conceived as a contest, the task may be to identify the best institution or a (small exclusive) set of institutions that are the best (or the worst).

We describe first a method closely related to hypothesis testing, then contrast it with a solution based on decision making, and conclude with an array of alternatives that can be closely tailored to the assessor's perspective. The strength of our approach is in its subjectivity with respect to the client (the assessor) and objectivity with respect to the institutions. By the former we mean that the method of analysis faithfully encodes the assessor's remit and intent, and by the latter that this intent is unambiguously formulated and disclosed, and is open to scrutiny by all parties as to whether it is relevant and whether the only prejudice it exercises is against performance that is poor according to the established standards. Of course, the uncertainty inherent in the assessment has to be addressed.

9.1 The setting and the task

Suppose there are n institutions, and a particular aspect of the performance of institution $k = 1, \ldots, n$ is characterised by a parameter θ_k. Suppose this parameter is estimated without bias by $\hat{\theta}_k$, with sampling variance τ_k^2, and the estimator $\hat{\theta}_k$ is normally distributed. In many contexts, θ_k is a probability, estimated by the proportion $\hat{\theta}_k = m_{k+}/m_k$, where m_k is the number of transactions made by institution k in the assessed period and m_{k+} the number of transactions that comply with a set norm. We assume first that the counts m_k and $m_{k+} \leq m_k$ are such that the distribution of $\hat{\theta}_k$ is well approximated by a normal distribution. We drop this assumption in Section 9.5.

In general, the assessor's remit is to identify institutions that are exceptional—that have the lowest and the highest rates θ_k. The problem is complex when the θ_k are estimated with appreciable uncertainty, so that $\hat{\theta}_k$ cannot be used as a straightforward substitute for θ_k in the intended comparison of θ_k with an adopted standard or the overall rate θ. Moreover, the institutions have caseloads m_k in a wide range, and so their rates θ_k are estimated with uneven precision. The rates themselves may be in a wide range.

9.1.1 Evidence of poor performance

Poor performance is often interpreted as being below an a priori set standard or the overall (domain) average θ, that is, $\theta_k < \theta$ for institution k. In the mindset of hypothesis testing, this naturally leads to testing the hypothesis that $\theta_k \geq \theta$ against the alternative $\theta_k < \theta$, separately for each institution k. Statements (verdicts) are generated for each institution in isolation, so no issues of multiplicity, testing many hypotheses, arise. The probability of a false positive is controlled separately for each institution because every institution is assessed on its own.

Rejecting the hypothesis for institution k constitutes evidence that k performed below par. Such evidence may trigger a sanction, such as a publicised letter of reprimand. Failure to reject the hypothesis is followed by no action, which might reasonably be interpreted as satisfactory performance; the institution has passed a set hurdle. This hurdle would seem to be fair—it is the same rule applied to every institution. This is easy to contradict when the institutions have caseloads m_k, and their estimators $\hat{\theta}_k$ sampling variances τ_k^2, in a wide range. The test has more power for institutions with larger caseloads m_k. Therefore, poor performance with a given deficit $\Delta = \theta - \theta_k > 0$ is easier to detect for a large institution than for a small one. A small institution is more likely to get away with a poor performance than a large institution.

Counting the number of significant comparisons is a poor practice for a multitude of reasons. When θ is (approximately) equal to the average of the values of θ_k, about one-half of the institutions fall below the standard of

θ but, when many θ_k are estimated with low precision, far fewer of them are highlighted as poor performers by hypothesis testing. If the assessment is meant to discover institutions that perform much worse than θ, then the deficit Δ that is on the borderline of 'satisfactory' and 'much worse' should be specified. Testing the hypothesis that $\theta_k \geq \theta - \Delta$ does not resolve the iniquity of hypothesis testing, nor does it help us count the number of poorly performing institutions.

An assessment is likely to lose credibility if it highlights many institutions. It is effective when it highlights a few of them that have performed very poorly. The term 'outlier' is sometimes used for such an institution, although it does not conform with its definition in mainstream statistics. Suppose an assessment comprises two hypothesis tests for each institution; one for θ_k, and another for ξ_k, both of them estimated without bias by normally distributed estimators. Suppose the sampling variance of $\hat{\theta}_k$ is smaller than of $\hat{\xi}_k$ for most institutions. Then it is likely that more hypotheses will be rejected for the θ_k than for ξ_k, suggesting to the institutions that they pay greater attention to their indicators θ_k; after all, nobody wants to be reprimanded. However, such attention may be misdirected if the processes or outcomes associated with θ_k are clinically more important.

9.1.2 Assessment as a classification

We regard assessment as a problem of classifying each institution to one of a small number K of categories, such as unsatisfactory, satisfactory and excellent. We assume that each institution has an underlying value of θ_k which would faithfully reflect the performance of institution k, if θ_k were observed. Instead, we have an estimate $\hat{\theta}_k$. All θ_k are on the same scale, such as the rate (percentage) of compliance with a set standard. We define the categories of assessment on this scale by $K - 1$ thresholds, $T_1, \ldots, T_{K-1}$. By specifying the $K \times K$ loss matrix $\mathbf{L}$ we are in the setting of Section 5.4, and are ready for the analysis. This would not do justice to the difficulties of establishing the values of the thresholds and the (off-diagonal) elements of the loss matrix.

9.2 Outliers

We regard one or a small set of institutions $\mathcal{O}$ as outliers if their values of θ_k stand out from the remainder. The extent of 'standing out' is specified by a gap $\Delta > 0$ between $\mathcal{O}$ and its complement. Thus, a set of institutions $\mathcal{O}$ are negative outliers if the set comprises fewer than half of all the institutions and there is a constant T such that $\theta_k < T - \frac{1}{2}\Delta$ for all $k \in \mathcal{O}$, and $\theta_k > T + \frac{1}{2}\Delta$ for all $k \notin \mathcal{O}$. We refer to T as the centre of the gap and to Δ as its width. A set of positive outliers is defined similarly; a set of institutions are positive outliers if they are negative outliers with respect to the values of $-\theta_k$, with centre at $-T$ and width Δ. Note that outlier status is a property of a collection of values of θ_k, together with a partitioning of this set of all values (or the corresponding institutions) to a set of outliers and their complement. Also, if a set $\mathcal{O}$ comprises outliers with a gap Δ, then this set comprises outliers also with any gap smaller than Δ. So the set of gaps for a given set of outliers has the form of an interval $(0, \Delta^*)$.

The fiducial probability that a given set of institutions $\mathcal{O}$ is a set of negative outliers with gap Δ centred at T is equal to

$$P(\mathcal{O}; T, \Delta) = \prod_{k \in \mathcal{O}} \mathrm{P}\left(\theta_k < T - \tfrac{1}{2}\Delta\right) \prod_{k \notin \mathcal{O}} \mathrm{P}\left(\theta_k > T + \tfrac{1}{2}\Delta\right), \qquad (9.1)$$

assuming that the fiducial distributions of θ_k are mutually independent, as they would be when the sampling distributions of $\hat{\theta}_j$, $j = 1, \ldots, n$, are mutually independent. It is advantageous to evaluate the logarithm of this probability, especially when there are many institutions. In principle, this probability should be evaluated on a fine grid of values of T and for all reasonable candidate sets $\mathcal{O}$. In practice, narrow ranges of candidate values of the centre T and of sets $\mathcal{O}$ are easily identified from a sorted list of values of $\hat{\theta}_k$ or their graphical display.

Instead of evaluating the probability in equation (9.1) for candidate sets of institutions, of which there may be many, we focus on locating the centre of a gap. With the assumption of independence of the fiducial distributions of θ_k, the probability of a gap of width Δ around a centre T is equal to

$$P(T, \Delta) = \prod_{k=1}^{n} \mathrm{P}\left\{\theta_k \notin \left(T - \tfrac{1}{2}\Delta, T + \tfrac{1}{2}\Delta\right)\right\}. \qquad (9.2)$$

When the fiducial distributions of θ_k are $\mathcal{N}(\hat{\theta}_k, \tau_k^2)$, each probability in this product is

$$1 - \Phi\left(z_{k,+}\right) + \Phi\left(z_{k,-}\right),$$

where $z_{k,+} = (T + \tfrac{1}{2}\Delta - \hat{\theta}_k)/\tau_k$ and $z_{k,-} = (T - \tfrac{1}{2}\Delta - \hat{\theta}_k)/\tau_k$. For a fixed centre T, this probability is a decreasing function of Δ. The width Δ can be interpreted as the standard for outlying; a narrow gap corresponds to a low standard and a wide gap to a high standard.

For a fixed $\Delta > 0$, we maximise $P(T, \Delta)$ as a function of T. We set a threshold P^*, and claim that there is a gap when $\max_T P(T, \Delta)$ exceeds P^*. We use the verb and noun 'claim' for statements that entail some uncertainty. When there is no uncertainty, we make statements that determine or identify (e.g., one or a set of institutions). Thus, statements (assertions) are valid or invalid, whereas a claim, apart from being valid or invalid, may be classified as appropriate or inappropriate. A claim is associated with a level of uncertainty, such as the probability P^*, and the claim is appropriate if the probability that it is valid exceeds this level. The terms 'claim' and 'verdict' are synonymous, although a verdict is used mainly for selection from a small set of options (a decision), whereas a claim need not be associated with such a set.

It remains then to claim which institutions, if any, are outliers. This can be done without any sophistication, regarding $\hat{\theta}_k$ as if they were θ_k, because there is likely to be a gap Δ wide also for the estimates. Note that $P(T, \Delta)$ is large for T way outside the range of values of $\hat{\theta}_k$, so the search for T that maximises $P(T, \Delta)$ has to be restricted to within the range of values of $\hat{\theta}_k$. Once a gap and its centre are located, we can revert to the evaluation of $P(\mathcal{O}; T, \Delta)$ given by equation (9.1), with the set $\mathcal{O}$ selected in the obvious way.

The meaning of the term 'outlier' is qualified by the magnitude of Δ. The gap Δ can be set by elicitation, and uncertainty about it represented by a plausible range of gaps, (Δ_-, Δ_+). Let $\hat{\theta}_{(1)} < \hat{\theta}_{(2)} < \ldots < \hat{\theta}_{(n)}$ be the ordered set of the estimates $\hat{\theta}_k$. If $\theta_{(j+1)} - \theta_{(j)} > \Delta$ and $j < \frac{1}{2}n$, then the institutions with the j smallest values of θ_k are a set of outliers. If Δ is set too small, then there may be several sets of institutions that are outliers. This would defeat the intention of the term 'outlier' to designate an exclusive (small) set of institutions that, as a group, deviate from the remainder.

In another approach, in which elicitation of Δ takes place only after data inspection and analysis, we find the smallest value of Δ, say Δ_{mn}, for which we would claim that there are no outliers. Then we ask the client whether a value $\Delta \leq \Delta_{mn}$ is plausible. If the answer is affirmative, then there are some outliers. Note that a plausible range for Δ is $(\Delta^*, +\infty)$; if a value is plausible, then so is any greater value.

In practice, the differences $\theta_{(k+1)} - \theta_{(k)}$ are not observed, and so we can make claims about outliers, or centres and gaps around them, only in terms of (fiducial or posterior) probabilities. To set a threshold P^* such that we issue the verdict of 'outlier' when $P(T, \Delta) > P^*$ and the verdict 'not outlier' otherwise, we introduce losses for the two kinds of inappropriate verdicts. Suppose an inappropriate claim that there is an empty gap (T, Δ), that is, a false positive, is associated with the loss of one lossile, and the failure to spot a gap (T, Δ), a false negative, is associated with the loss of R lossiles. Then the expected losses with the respective false negative and false positive verdicts are $1 - P(T, \Delta)$ and $RP(T, \Delta)$, so the threshold, where these two probabilities coincide, is $P^* = 1/(R + 1)$. We may have to consider a plausible range of thresholds P^* because of uncertainty about Δ and R.

A more elaborate scheme relates the loss to the size of the set of outliers. A false negative for a large set of outliers incurs greater loss, as does a false positive with the claim of a large set of outliers. That is, the number of institutions that have been misclassified, as outliers or non-outliers, is a factor contributing to the gravity of the error and the magnitude of the loss. A positive verdict may identify a set different from the genuine set of outliers, and a loss should be declared also for such an error. The loss may also be related to the width of the largest gap.

9.3   As good as the best

When the institutions' outcomes, or performances, θ_k, are on a continuous scale, a tie, $\theta_k = \theta_{k'}$ for some pair $k' \neq k$, is unlikely, and so there is a single institution that has the highest and another that has the lowest performance. Identifying this institution may be difficult, especially when there is considerable uncertainty about the values of θ_k. This uncertainty may be assessed informally by the overlap of the (95%) confidence intervals for the institutions with the highest values of $\hat{\theta}_k$. Declaring a single institution as the best (as the winner) is therefore problematic and may undermine the credibility of the assessment. However, selecting several institutions for an honour or a prize contradicts the assumption of no ties.

We address a problem more general than identifying the winner. Instead of the leader, or the back-marker, we form a list of institutions that are nearly as good as the best of them, or nearly as poor as the worst of them. We specify a positive value δ, called the tolerance. If all θ_k were known, we would include in the list of 'nearly as good as the best' all the institutions k for which $\theta_k > \max_j(\theta_j) - \delta$. Identifying the winner corresponds to setting δ to a very small (positive) value.

When θ_k are estimated, we evaluate the fiducial probability of institution k being within δ of the best institution. For a given maximum M, this probability is

$$
\begin{aligned}
P_k(M) &= \mathrm{P}\left(M - \delta < \theta_k < M \mid \hat{\theta}_k\right) \prod_{j \neq k} \mathrm{P}\left(\theta_j < M \mid \hat{\theta}_j\right) \\
&= \left\{ 1 - \frac{\mathrm{P}\left(\theta_k < M - \delta \mid \hat{\theta}_k\right)}{\mathrm{P}\left(\theta_k < M \mid \hat{\theta}_k\right)} \right\} \prod_{j=1}^{n} \mathrm{P}\left(\theta_j < M \mid \hat{\theta}_j\right) , \quad (9.3)
\end{aligned}
$$

assuming that the fiducial distributions of θ_j, $j = 1, \ldots, n$, are mutually independent. We evaluate this expression for a range of plausible maxima M and choose the value for which P_k attains its maximum. Values of M can be set separately for each institution k, although they are unlikely to differ much from one another. When the estimators $\hat{\theta}_k$ have respective distributions $\mathcal{N}(\theta_k, \tau_k^2)$, P_k is expressed in terms of the standard normal distribution function Φ as

$$
P_k = \left\{ 1 - \frac{\Phi(z_{k,-})}{\Phi(z_k)} \right\} \prod_{j=1}^{n} \Phi(z_j) , \quad (9.4)
$$

where $z_j = (M - \hat{\theta}_j)/\tau_j$ and $z_{j,-} = (M - \delta - \hat{\theta}_j)/\tau_j$, $j = 1, \ldots, n$. When n is large, numerical errors may accumulate in the evaluation of the product in equation (9.4). This can be avoided by evaluating $\log(P_k)$, which is a summation of log-probabilities. In an alternative way of evaluating $P_k(M)$, calculation is terminated if a partial product (or sum of logarithms) becomes too small because we are interested only in relatively large values of $P_k(M)$.

Instead of evaluating $P_k(M)$ on a grid of values of M, Newton-Raphson algorithm can be employed in its maximisation. It requires more intricate programming and the gains in reducing the amount of computing and higher precision of resulting probability $P_k(M_k^*)$ are only slight. Nevertheless, it is a useful practice in programming, using the expression

$$\frac{\partial \log\{\Phi(z_k)\}}{\partial M} = \frac{1}{\tau_k} \frac{\varphi(z_k)}{\Phi(z_k)},$$

where φ is the density of the standard normal distribution. Readers with greater computational ambitions and plans may invest in compiling a versatile R function in which the objective function and its differential are two arguments.

We can make claims about individual institutions that they are (nearly) as good as the best, about sets of institutions that all of them are (nearly) as good as the best, and about a set of institutions that they form the complete set of institutions that are as good as the best. These types of claims have to be distinguished carefully. They are different (probabilistic) events, have different probabilities, and are associated with different levels of uncertainty. For example, the latter claim for a set $\mathcal{O}$ has the probability

$$\prod_{k \in \mathcal{O}} P_k(M) \prod_{k \notin \mathcal{O}} \{1 - P_k(M)\} \,,$$

whereas the claim that every institution in $\mathcal{O}$ is (nearly) as good as the best, without claiming anything about the complement of $\mathcal{O}$ is equal to the first factor in this expression.

9.4 Empirical Bayes estimation

The collection of values θ_k can be regarded as a random sample from a distribution. This is motivated by conceiving the institutions as random draws from a large (infinite) superpopulation of institutions. We can posit two models, one for the parameters θ_k and another for the estimates $\hat\theta_k$. An example of the former is

$$\theta_k = \theta + \delta_k \,, \tag{9.5}$$

where δ_k, the deviations of the institutions from the average, are a random sample from a normal distribution $\mathcal{N}(0, \nu^2)$. The model for $\hat\theta_k$, $k = 1, \ldots, n$, is

$$\hat\theta_k = \theta_k + \varepsilon_k \,, \tag{9.6}$$

where ε_k, the errors in estimating θ_k by $\hat{\theta}_k$, are independent with respective distributions $\mathcal{N}(0, \tau_k^2)$. The set of $2n$ random variables, δ_k and ε_k, is also mutually independent.

The models can be generalised by replacing θ in equation (9.5) by a linear predictor, $\mathbf{x}_k \boldsymbol{\beta}$, where $\boldsymbol{\beta}$ is a vector of (regression) parameters and $\mathbf{x}_k$ a (row) vector of values of the covariates for institution k. Further, a linear model can be replaced by a generalised linear or a more complex model; see Exercise 9.12.

There is no contradiction in θ_k being a random variable in the model in equation (9.5) and a parameter (constant) in (9.6). The latter model is for the conditional distribution of $\hat{\theta}_k$ given θ_k. If the value of ν^2 were known, we could seemingly estimate each θ_k more efficiently than by $\hat{\theta}_k$ by the conditional expectation of θ_k given $\hat{\theta}_k$. The conditional distribution involved is derived from the joint distribution of $(\theta_k, \hat{\theta}_k)$, which is bivariate normal,

$$\mathcal{N}_2 \left\{ \begin{pmatrix} \theta \\ \theta \end{pmatrix}, \begin{pmatrix} \nu^2 & \nu^2 \\ \nu^2 & \nu^2 + \tau_k^2 \end{pmatrix} \right\} .$$

The conditional distribution of θ_k given $\hat{\theta}_k$ is also normal,

$$\left(\theta_k \mid \hat{\theta}_k \right) \sim \mathcal{N} \left\{ \theta + \frac{\nu^2}{\nu^2 + \tau_k^2} \left(\hat{\theta}_k - \theta \right), \frac{\nu^2 \tau_k^2}{\nu^2 + \tau_k^2} \right\} .$$

The parameters θ and ν^2 have to be estimated; details are given in Appendix A. Setting aside the issue of their uncertainty, we adopt the conditional expectation of θ_k as its estimator and the conditional variance, denoted by $\tilde{\tau}_k^2$, as a measure of uncertainty in estimating θ_k by this expectation. The conditional expectation is a convex combination of θ and $\hat{\theta}_k$,

$$\tilde{\theta}_k = \frac{\nu^2 \tau_k^2}{\nu^2 + \tau_k^2} \left(\frac{1}{\nu^2} \theta + \frac{1}{\tau_k^2} \hat{\theta}_k \right) . \tag{9.7}$$

The coefficients in this combination are $1/\nu^2$ and $1/\tau_k^2$, the respective concentrations (reciprocals of the variances) of θ and $\hat{\theta}_k$, which are regarded as alternative estimators of θ_k. The estimator $\tilde{\theta}_k$ is known as a shrinkage estimator, moving the unbiased $\hat{\theta}_k$ closer to the focus θ common to all the institutions k; θ is biased for all institutions (for which $\theta_k \neq \theta$), but has no variance, and a small variance when it is estimated. The shrinkage is governed by the relative magnitudes of ν^2 and τ_k^2, or the ratio $\nu^2/(\nu^2 + \tau_k^2)$. When $\nu^2 > 0$, larger variances τ_k^2 are associated with stronger shrinkage because θ is a relatively more credible alternative to $\hat{\theta}_k$. For a fixed value of τ_k^2, shrinkage is a decreasing function of ν^2. At an extreme, when $\nu^2 = 0$, we obtain $\tilde{\theta}_k = \theta$ for all k, with no uncertainty. In general, by shrinkage we exploit the similarity of the institutions. The term 'borrowing strength' is also used in the literature for this.

Next time you do not know which of two estimators to use, consider their convex combination. By choosing one of them, you aim to match the precision of the more efficient of the two estimators. In contrast, a linear combination of two estimators (of the same target) may be more efficient than either of the alternatives. Section 10.6.1 presents an application and Chapter 11 develops this idea further and in greater generality.

Regarding θ_k, $k = 1, \ldots, n$, as a set of parameters, the distribution $\mathcal{N}(\theta, \nu^2)$ may be viewed as their (Bayesian) prior. A Bayesian model specification would be complete only by specifying a prior also for θ and ν^2. Without it, θ and ν^2 are estimated; see below. If θ and ν^2 were known and $\mathcal{N}(\theta, \nu^2)$ were adopted as the prior for θ_k, then $\mathcal{N}\{\tilde{\theta}_k(\nu^2), \tilde{\tau}_k^2(\nu^2)\}$ would be the posterior distribution of θ_k. The shrinkage estimator of θ_k used in practice, $\tilde{\theta}_k(\hat{\nu}^2)$, is called empirical Bayes. It is not Bayesian, but its name acknowledges its Bayesian motivation. It is tempting to claim that this estimator is unbiased and that its variance is estimated well by $\tilde{\tau}_k^2(\nu^2)$. Both claims ignore the uncertainty about ν^2 and the nonlinear (functional) dependence of $\tilde{\theta}_k$ and $\tilde{\tau}_k^2$ on ν^2. Another contentious issue is the designation of the deviations δ_k as random, elaborated below.

For θ and $\hat{\theta}_k$ independent and unbiased estimators of θ_k, the shrinkage estimator in equation (9.7) is the convex combination with the smallest MSE, even when these estimators are not normally distributed. The convex combination is also unbiased for θ_k, and its sampling variance is

$$\frac{\nu^4 \tau_k^4}{(\nu^2 + \tau_k^2)^2} \left(\frac{1}{\nu^2} + \frac{1}{\tau_k^2} \right) = \frac{\nu^2 \tau_k^2}{\nu^2 + \tau_k^2}. \tag{9.8}$$

This derivation makes no use of any properties of the normal distribution. This suggests that we could dispense with the assumptions of normality altogether, and simply ask the question:

What is the best convex combination of $\hat{\theta}_k$ and θ,

$$(1 - b_k)\,\hat{\theta}_k + b_k\,\theta,$$

$0 \leq b_k \leq 1$, when $\hat{\theta}_k$ and θ are regarded as independent unbiased estimators of θ_k?

The result is not surprising; we combine the two estimators with weights proportional to their precisions in estimating θ_k, $1/\nu^2$ and $1/\tau_k^2$.

Note the licence we took in these derivations. We started by θ_k as a fixed quantity and switched its status to random. With θ and θ_k we operate at the level of institutions. If institutions are regarded as random entities, then indeed the deviations $\theta_k - \theta$ have zero expectation and variance ν^2. In contrast, $\hat{\theta}_k$ as an estimator refers to the sample of clients who attended a specific institution k. That is, in a replication of the annual business of the institution, on which its performance is assessed, there may be different clients, with different outcomes, but their average (or another summary) would be the same θ_k, which characterises the practices of the institution during the period of assessment, its management, application, attributes of the staff, and the like.

This perspective, together with the purpose of the assessment, suggest that each institution should be treated as a fixed entity that would perform at the same level in a replication—each θ_k is fixed because it is the target of the assessment. Values of θ_k that would be realised in replications of forming the institutions according to the model in (9.5) are not relevant.

Despite its lack of validity, we should not dismiss this model for θ_k altogether because it may be useful. We explore the properties of $\tilde{\theta}_k$ assuming that θ_k is fixed. The evaluations that follow are based on the expression

$$\tilde{\theta}_k = \frac{1}{\nu^2 + \tau_k^2}\left(\tau_k^2\theta + \nu^2\hat{\theta}_k\right),\tag{9.9}$$

derived from equation (9.7). First,

$$\mathrm{E}\left(\tilde{\theta}_k \mid \theta_k\right) = \theta_k + \frac{\tau_k^2}{\nu^2 + \tau_k^2}\left(\theta - \theta_k\right),$$

so $\tilde{\theta}_k$ is biased for θ_k unless $\theta_k = \theta$. Next, $\mathrm{var}(\tilde{\theta}_k \mid \theta_k) = \nu^4\tau_k^2/(\nu^2 + \tau_k^2)^2$, and hence

$$\begin{aligned}
\mathrm{MSE}\left(\tilde{\theta}_k\,;\theta_k\right) &= \frac{\tau_k^2}{(\nu^2 + \tau_k^2)^2}\left\{\nu^4 + (\theta - \theta_k)^2\,\tau_k^2\right\}\\
&= \frac{\nu^2\,\tau_k^2}{\nu^2 + \tau_k^2}\,\frac{\nu^2 + r_k^2\,\tau_k^2}{\nu^2 + \tau_k^2},
\end{aligned}\tag{9.10}$$

where $r_k^2 = (\theta_k - \theta)^2/\nu^2$ is the squared relative deviation of the performance of institution k from the average. According to the model (which we, admittedly dismissed earlier), $\mathrm{E}(r_k^2) = 1$. When $r_k^2 = 1$, the MSE reduces to the first factor in equation (9.10), which coincides with the averaged conditional MSE in equation (9.8). In this case, the invalid model yields the correct answer. When $r_k^2 = 0$, θ_k is equal to θ, and so $\tilde{\theta}_k$ is a convex combination of two unbiased estimators. The MSE in (9.10) is the $\nu^2/(\nu^2 + \tau_k^2)$-multiple of the model-based conditional variance. When $\tau_k^2 \gg \nu^2$, equation (9.8) is a gross overstatement.

In general, the value of r_k^2 has no upper limit, so the MSE could even exceed τ_k^2; in this case, shrinkage is counterproductive. This happens when

$$\frac{\nu^2\,\tau_k^2}{\nu^2 + \tau_k^2}\,\frac{\nu^2 + r_k^2\,\tau_k^2}{\nu^2 + \tau_k^2} > \tau_k^2,$$

that is, when

$$r_k^2 > 2 + \frac{\tau_k^2}{\nu^2}$$

or, equivalently, when $(\theta_k - \theta)^2 > 2\nu^2 + \tau_k^2$. Of course, this cannot be ruled out. To assess the likelihood of this, we revert to the assumption of θ_k being a random sample. If this sample is from a normal distribution, then $r_k = (\theta - \theta_k)/\nu$ has the standard normal distribution, and then $\mathrm{P}(|r_k| > \sqrt{2}) = 0.157$. This is an overstatement because it corresponds to $\tau_k^2 = 0$. But even when $\tau_k^2 = \nu^2$, this probability is 0.083. To appreciate fully the importance of these probabilities, recall that the assessment is concerned mainly with the institutions with the largest and smallest values of θ_k, and that is where shrinkage is counterproductive. This should not be interpreted as an outright dismissal of shrinkage, because it could be applied with a coefficient smaller than $b_k^* = \tau_k^2/(\nu^2 + \tau_k^2)$. It should be tailored to the purpose (the perspective) of the assessment.

9.5 Assessment based on rare events

Many outcome variables used in performance assessment are dichotomies, such as 0/1, Yes/No and Positive/Negative. They are summarised by within-institution counts which are assumed to have binomial distributions. Independence of the elementary (dichotomous) outcomes is supported by the idea of professional and detached conduct. Such conduct implies that the outcome associated with one customer does not affect the outcomes of the customers treated in the immediate future. In the longer term, an institution may alter its practices in response to self-assessment (which includes analysis of feedback, with praise, suggestions and complaints made by past customers) and by adopting new technology. A more insidious threat to the assumption of binomial distribution arises when an institution anticipates the external assessment (audit), and applies special measures to improve its practice towards the end of the audited period to avoid being sanctioned for poor performance. In this case, the sampling variance of the rate $\hat{\theta}_k$ is likely to be lower than the binomial variance $\theta_k(1 - \theta_k)/m_k$.

When the number of customers (or transactions) m_k is very large, and the classification of each transaction entails some nontrivial effort, the performance may be assessed, and θ_k estimated, from a random sample of the institution's transactions. Then it is less obvious how special measures for improving the institution's performance towards the end of the audited period can be assessed, motivated and implemented.

Estimators of the rates θ_k are distinctly non-normally distributed when the events counted, such as failures, complaints, deaths or other adverse events (e.g., infection during stay in a hospital), are rare. When the number of transactions m_k is large the assumption of Poisson distribution is appropriate. Note that even the Poisson distribution converges to the normal as the (Poisson) rate, its expectation, increases beyond all bounds.

The methods developed in the earlier sections carry over to the binomial and Poisson distributions with the obvious changes, replacing the normal distribution function Φ with the relevant discrete distribution function or the appropriate fiducial distribution. The granular nature of the cumulative probabilities is reflected in the verdicts of the assessment by the differences that would arise if the outcomes were altered by a single instance. The substantial uncertainty in the estimators is a disincentive to use rare outcomes in an assessment.

9.6   Further reading

This chapter is based on Longford (2018) and (2020), developed from Longford (2016c), where similar problems are addressed in the context of educational measurement. Similar methods are applied by Silber *et al.* (2014) in the assessment of hospitals. A landmark paper on the application of empirical Bayes methods to performance assessment is Goldstein and Spiegelhalter (1996). Spiegelhalter *et al.* (2002a) discuss a high-profile case of performance assessment in health care in the U.K. Spiegelhalter *et al.* (2012) develop a set of methods for health-care audit. Thomas, Longford and Rolph (1994) present an application of empirical Bayes logistic regression to comparing cause-specific mortality rates of U.S. hospitals. Shen and Louis (1998) show that different shrinkage is optimal for different purposes: for representing the variation of the institutions, for ranking them, finding their extremes and, generally, for targets that are nonlinear transformations of expectations δ_k. Rubin (2008) contains an in-depth discussion of matching and adjustment (see Appendix B), arguing firmly for matching.

9.7   Exercises

9.1. Do you commute to work by public transport? Find out how and by whom is the quality of the service assessed and how it is publicised.

9.2. For revision: Suppose $\hat{p}$ is a binomial proportion (the proportion of positive outcomes in a sequence of independent binary trials). When is its sampling distribution well approximated by a normal distribution? Confirm your statement by simulations on a few examples. Check how reasonable are related conventions suggested by elementary textbooks.

9.3. Consider two institutions, with identical rates of compliance with a standard. The rates are based on a dichotomous assessment ('satisfied' or 'failed') for each transaction. (For example, 'Did the customer receive a legible receipt for the transaction?') Suppose a rate lower than 96% is regarded as unsatisfactory, and the two institutions have the same (underlying) rate of 93.8%. In the assessed period, one institution had 600 customers and the other had 4000; the outcome is established for every customer. Evaluate the probabilities that the performance of one or the other institution would be found unsatisfactory. Discuss the result from the viewpoint of the larger institution.

9.4. Explain the difference between the probabilities in equations (9.1) and (9.2).

9.5. Suppose a set of 25 institutions have rates of compliance with a standard that are well described by a normal distribution with mean $\hat{\theta} = 65\%$ and standard deviation 5%. These 25 rates are estimated, based on finite samples. How many institutions are likely to be found to have their rates significantly lower than the overall rate θ? Which factors does the answer depend on?

9.6. Could there be a set of outliers that are well separated from the rest of the institutions, which, as a set, are neither negative nor positive outliers?

9.7. Discuss the proposal made in the introduction to Section 9.3 to check the overlaps of the confidence intervals as a way of assessing the extent of uncertainty about the best and worst performers.

9.8. Explain why we do not have to be concerned about capitalising on chance when evaluating the probability $P_k(M)$ in equation (9.4) for a range of values of M.

9.9. Follow up on the proposal to find the maximum of the probability $P_k(M)$ by the Newton-Raphson algorithm.

9.10. Elaborate on the differences of the claims that:

- a (small) set of institutions contains the best performer;
- every institution in the selected set is as good as the best;
- the set contains all the institutions that are as good as the best;
- the set contains all the institutions that are as good as the best, and contains no others.

Evaluate the fiducial probability of each of these claims.

9.11. Prove that the shrinkage estimator $\tilde{\theta}_k$ given by equation (9.7) is indeed the convex combination of $\hat{\theta}_k$ and θ that has the smallest MSE, and do so without any assumptions of normality of $(\hat{\theta}_k \mid \theta_k)$ or of the deviations $\theta_k - \theta$.

9.12. In the model in equation (9.5), θ could be replaced by a (linear) regression: $\theta_k = \mathbf{x}_k\,\boldsymbol{\beta} + \delta_k$, where $\mathbf{x}_k$ are the values of a set of covariates for unit k and $\boldsymbol{\beta}$ is a vector of regression parameters. Discuss the merits of such a model. How would the expression for the conditional distribution $(\theta_k \mid \hat{\theta}_k)$ have to be adjusted?

9.13. Suppose θ_k, $k = 1, \ldots, K$, are the rates of compliance with a particular standard in the institutions of a country. Suppose the institutions have caseloads (numbers of cases handled) $N_1, \ldots, N_K$. Discuss the difference between the national rate $\bar{\theta} = \frac{1}{N}\sum_k N_k\,\theta_k$ and the average rate $\theta = \frac{1}{K}\sum_k \theta_k$. Find condition(s) in which these two rates coincide.

9.14. In the survey sampling perspective, the outcomes as well as all the parameters are fixed, and the sampling process, which determines which members of the studied population form the sample, is the only source of uncertainty. Compare this perspective with the superpopulation perspective in which θ_k are a random sample. Which perspective do you find more natural, and why? Could the two perspectives be reconciled?

   Qualify the property of no bias of $\tilde{\theta}_k$. Conditionally on the value θ_k, estimator $\hat{\theta}_k$ is biased, unless $\theta_k = \theta$. Without conditioning on θ_k, $\tilde{\theta}_k$ is unbiased. Which perspective is (more) appropriate?

9.15. Construct an example of $n = 100$ institutions providing a particular service, with $200 < m_k < 3000$ cases each, and rates of failure (θ_k) in the range $0.5-1.5\%$. Estimate the rates of failure based on simulated binary outcomes. In replications, based on $\hat{\theta}_k$ or $\tilde{\theta}_k$, how frequently do you correctly identify the institution with the lowest and the highest rate θ_k? Using a suitable diagram, compare the ranks of θ_k, $\hat{\theta}_k$ and $\tilde{\theta}_k$. How frequently is the best institution (as judged by its value θ_k) found to be among the top ten institutions (as judged by its estimate)? Repeat this simulation with a few distinct distributions (e.g., normal and uniform), from which you generate, as random samples, the values of θ_k.

9.16. Consider a simulation study with the setting of Exercise 9.15. The sample sizes m_k differ among the institutions, but are fixed across the replications. In one scenario, the set of rates θ_k is fixed — not altered from one replication to the next. In the other, a fresh set of rates θ_k is drawn in each replication. Which scenario do you regard as more realistic? When is it meaningful to average the results across the replications?

9.17. Why (and when) is it reasonable to assume that the sampling variation of the estimator of the variance ν^2, in the context of equation (9.7), is small?

Appendix

A. Estimation of θ and ν^2

When $\hat{\theta}_k$ are the sample means (or proportions) within institutions k, then θ may be estimated by the overall average

$$\hat{\theta} = \frac{1}{m} \sum_{k=1}^{n} m_k \hat{\theta}_k \, ,$$

where m_k is the caseload of institution k and $m = m_1 + \cdots + m_n$ is their total. This estimator is biased because the target is $\theta = (\theta_1 + \cdots + \theta_n)/n$, in which each institution has the same weight. The bias is substantial when θ_k and m_k are strongly correlated. The equal-weight estimator, $\hat{\theta}_0 = (\hat{\theta}_1 + \cdots + \hat{\theta}_n)/n$, is unbiased but has a large variance, with a disproportionately large contribution to it from the institutions with the smallest caseloads and largest variances τ_k^2. There is no straightforward compromise of the two estimators, unless there is some prior information about the magnitudes of the ratios τ_k^2/ν^2 and the correlation of θ_k and m_k.

The uncertainty about the variance ν^2 has in general a stronger impact on estimation of θ_k because ν^2 is a key factor in the shrinkage coefficient. With n institutions, the upper bound on the efficiency is the variance of the hypothetical estimator based on the values θ_k, if they were known. With the assumptions of normality, this estimator would have the scaled χ^2 distribution with $n-1$ degrees of freedom. When each θ_k is estimated the number of degrees of freedom is bound to be lower, and the distribution is not directly related to a χ^2 distribution when the estimators $\hat{\theta}_k$ have uneven variances.

The naïve estimator,

$$S = \frac{1}{n-1} \sum_{k=1}^{n} \left(\hat{\theta}_k - \hat{\theta} \right)^2 \, ,$$

has a positive bias for ν^2 because the estimates $\hat{\theta}_k$ are dispersed more than the underlying quantities θ_k. We apply moment matching, adjusting S to remove its bias. Denote by ε_k the error in estimating θ_k by $\hat{\theta}_k$; $\varepsilon_k = \hat{\theta}_k - \theta_k$. Unbiasedness of each $\hat{\theta}_k$ implies that $E(\varepsilon_k) = 0$ for all k. The expectation of

S is

$$
E(S) = \frac{1}{n-1} \sum_{k=1}^{n} E\left(\theta_k - \theta + \varepsilon_k - \frac{1}{m} \sum_{j=1}^{n} m_j \varepsilon_j \right)^2
$$

$$
= \frac{1}{n-1} \sum_{k=1}^{n} (\theta_k - \theta)^2 + \frac{1}{n-1} \sum_{k=1}^{n} E\left(\frac{m - m_k}{m} \varepsilon_k - \frac{1}{m} \sum_{j \neq k}^{n} m_j \varepsilon_j \right)^2
$$

$$
= \frac{1}{n-1} \sum_{k=1}^{n} (\theta_k - \theta)^2 + \frac{1}{n-1} \left(\sum_{k=1}^{n} \tau_k^2 - 2 \sum_{k=1}^{n} \frac{m_j}{m} \tau_k^2 \right)
$$

$$
+ \frac{n}{n-1} \sum_{k=1}^{n} \frac{m_k^2}{m^2} \tau_k^2 .
$$

This implies the estimator

$$
\hat{\nu}^2 = S - \frac{1}{n-1} \left(\sum_{k=1}^{n} \tau_k^2 - 2 \sum_{k=1}^{n} \frac{m_k}{m} \tau_k^2 \right) - \frac{n}{n-1} \sum_{k=1}^{n} \frac{m_k^2}{m^2} \tau_k^2 .
$$

The concluding term is much smaller than $(\tau_1^2 + \cdots + \tau_n^2)/n$ because $m_k^2 \ll m^2$. Further, $(m_1/m + \cdots + m_n/m) = 1$, and so $(\tau_1^2 + \cdots + \tau_n^2)$ dominates in the expression in the parentheses. Therefore

$$
\hat{\nu}^2 \doteq S - \frac{1}{n-1} \sum_{k=1}^{n} \tau_k^2 ; \tag{9.11}
$$

the naïve estimator S is adjusted by a quantity approximated by the average of the sampling variances of the estimators $\hat{\theta}_k$. The approximation is close when n is large and τ_k^2 are not associated strongly with the caseloads m_j. When the institutions have identical caseloads, the estimator $\hat{\nu}^2$ has a form similar to the approximation in (9.11),

$$
\hat{\nu}^2 = S - \frac{1}{n} \sum_{k=1}^{n} \tau_k^2 .
$$

We have an unbiased estimator $\hat{\nu}^2$ of ν^2. It is not efficient, but both absence of bias and efficiency are not particularly valuable properties because they are not retained by nonlinear transformations, and we would like to use $\hat{\nu}^2$ as a substitute for ν^2 in the shrinkage estimator (9.9), or its adaptation. Arbitration about estimating θ_k is further obscured by how the estimates of θ_k are to be used. On balance, we therefore prefer working with $\hat{\theta}_k$.

B. Adjustment and matching on background

Institutions are bound to have customers with different distributions of background. For example, one institution may operate in a district with more deprivation (poor housing and unemployment), greater proportion of pensioners in the population, more (or all) people living in rural areas, and the like. The term *casemix* is used for the distribution of such attributes within the institutions. In some forms of assessment, and for some outcomes, it is desirable to take into account the difference in the casemix of the institutions. In this section, we discuss two generic methods for such accounting: adjustment for background variables and matching on them.

In a model-based adjustment, we formulate a model for the association of the outcomes on the background variables. An example of such a model is the regression

$$\mathbf{y}_k = \mathbf{X}_k \boldsymbol{\beta} + \delta_k \mathbf{1} + \boldsymbol{\varepsilon}_k, \qquad (9.12)$$

$k = 1, \ldots, m$, in which $\mathbf{y}_k$ is the $m_k \times 1$ vector of outcomes for customers in institution k, $\mathbf{X}_k$ the corresponding $m_k \times (p+1)$ regression matrix for p covariates and the intercept ($\mathbf{1}$, as its first column), δ_k the deviation specific to institution k and $\boldsymbol{\varepsilon}_k$ is a random sample from $\mathcal{N}(\mathbf{0}, \sigma^2 \mathbf{I})$. We do not call the elements of $\boldsymbol{\varepsilon}_k$ errors because they do not represent any departure from a value (of the outcome) presumed to be correct, from any norm, nor any mistake committed by any conceivable party. In most cases, incorrect is the expectation $\mathbf{X}_k \boldsymbol{\beta} + \delta_k \mathbf{1}$ because the process (mechanism) that underlies the generation of the outcomes is far more complex than any (linear) model could capture; it is grossly unfair to pass the buck from the 'regularity' and simplicity of the posited expectation $\mathbf{X}_k \boldsymbol{\beta} + \delta_k \mathbf{1}$ to $\boldsymbol{\varepsilon}_k$ or any other term, if there were one in the model.

The institutions would be compared on the values of δ_k, if these values were available. In practice δ_k are estimated. In one perspective, δ_k are fixed quantities because they refer to well identified entities (institutions), which are best represented in the model by parameters. Only their pairwise contrasts (differences) are identified, but they happen to be the quantities of interest. Identification is commonly resolved by selecting, arbitrarily, one institution, say $k = 1$, and setting its value of δ_1 to zero.

When the assessment involves many institutions the model contains many parameters, a well recognised handicap in estimation. An alternative is to regard the δ_k as a random sample and estimate its distribution, assumed to be normal with zero mean and an (institution-level) variance ν^2. We skip the discussion of how these two-level models are fitted because we want to put forward an alternative that involves no models for relating outcomes to background. Thus, a set of $n-1$ parameters is replaced by a single variance, which happens to have an interpretation relevant for the setting: it quantifies the variation (variety) of the institutions. Also, the normal distribution $\mathcal{N}(0, \nu^2)$ can be interpreted as a prior distribution for the deviations δ_k.

In summary, we regard the model in (9.12) as not valid, but with a potential to serve the cause of comparing the institutions. We regard each deviation δ_k as fixed, but acknowledge that the invalid assumption that they are a random sample may be constructive.

In an alternative approach, we seek a more direct answer to two questions posed after setting out the assumptions. Suppose we have a set of customers given by their backgrounds **x**. The customers do not have to be real (existing), they may be a result of our design and construction, informed by a consultation. We ask

What would be the outcomes of these customers if they were served by institution k?

This question, and the answer to it are referred to as direct standardisation and the set of customers as a reference set or template. If we obtain a credible answer to this question, we have a seemingly perfect basis for comparing the institutions, based on the same set of tasks. Such a comparison would be indisputably fair. Its drawback may be that it is not equally relevant because such a casemix may be typical in some institutions and (highly) atypical in some others. The latter may argue that they are not ready and well equipped for such customers because they rarely, if ever, have them. As a basis for assessing them, the task is not relevant.

The other question is

How would the customers of institution k fare if their custom were dispersed throughout the domain (all the institutions)?

This question and the answer to it are referred to as indirect standardisation. It is not fair because each institution is assigned a different task, its own caseload. However, the task is indisputably relevant because it refers to the institution's clientele. Indirect standardisation is suitable only for comparisons of the (focal) institution with the entire domain. A comparison of two institutions could be addressed by this approach, but it would have to be based on a set of customers who are equally usual or common in the two institutions.

The two outlined problems are solved by statistical matching. For the customers of an institution we find matches among the customers in the entire domain; that is, we form pairs of customers so that each pair comprises one customer from the focal institution and one from the domain and the two customers have as similar background as can be arranged. No customer can be included in more than one pair, although some customers may end up unmatched (unpaired). Suppose there are M pairs. They can be reconstituted as two groups of M customers each. Their outcomes can be compared by a method that disregards the background variables completely because the influence of these variables (potential confounding) has been eliminated by matching. Note the affinity of this approach with the potential outcomes framework (Appendix B, Chapter 8).

Matching is related to a hypothetical experiment in which customers are assigned completely at random to the focal institution or to the domain, which are regarded as treatments (interventions). In such an experiment, with sufficiently large sample size, the two treatment groups are well balanced on every conceivable background variable, irrespective of whether it is observed or not.

Matching can be interpreted as forming subsets of the observed dataset in such a way as to mimic the linked experiment. Matching is performed solely on the background, without drawing any information from the outcomes. In this respect, it can be related to the design of the experiment, where we have the freedom to prescribe the probabilities of assignment to the interventions according to the background. Matching is thus characterised as post-observation design.

In performance assessment we do not have the luxury of a random assignment, and only a limited set of background variables is usually recorded. Matching is more credible the more and a greater variety of background variables are recorded and used.

Matching is a nontrivial computational exercise when there are many background variables. A key result that reduces this task to matching on a single (continuous) variable is that it would suffice to match on the propensity (Rosenbaum and Rubin, 1983). The propensity is defined for a customer as the probability of being assigned to the focal institution given the set of background variables. The propensity is defined for every customer. It has to be estimated based on a suitable model. This propensity model has no role in inference; it is used solely as a vehicle to find a pair of within-intervention subgroups that are well matched on all the observed background variables.

The quality of the match is defined by the scaled differences of the means of the background variables within the subgroups. The imbalance for a variable x is defined as $|\bar{x}_1 - \bar{x}_2|/s$, where $\bar{x}_h$, $h = 1, 2$ are the within-subgroup means of x and s is the standard deviation of x pooled across the two subgroups. The overall imbalance is defined as the average of these imbalances.

Adjust or match?

In an observational study, such as performance assessment, we have to contend with two processes: assignment into treatment groups, and the effect of the treatment. The treatment is the target of inference, and the assignment process is a nuisance. That is, if we were privy to the details of the assignment, as we are in an experiment, estimation of the treatment effect would be (relatively) straightforward. Adjustment entails modelling in which these two processes are not separated in any way. Model uncertainty is an undesirable burden together with the assumptions such as normality, linearity and the homogeneity (constancy) of the treatment effect. The model diagnostics are in general unsatisfactory because they cannot confirm a model, merely find a contradiction with it. The more thorough the process of model checking and selection the more likely we are to get the final model wrong because of the compendium of many decisions based on hypothesis tests or similar criteria, in which the model being 'all right' is the default, even though it corresponds to failure to reject a (null) hypothesis.

In matching we have a single unambiguous diagnostic—the level of imbalance in the matched subgroups. There are no distributional assumptions for the outcomes, nor for the treatment effect. A conceptual strength of the analysis of matched pairs is the clear separation of the processes of treatment assignment (or selection) and application. Matching addresses assignment and the subsequent analysis of the outcomes is for the treatment effect.

Both adjustment and matching require background variables, and both of them are handicapped if the list of background variables is incomplete. With adjustment, we operate with a model that is not valid; with matching, there is the threat that an unobserved background variable is unevenly distributed in the two matched treatment subgroups. However, adjustment is problematic with many background variables because modelling is less likely to identify the valid parsimonious model, and validity is an imperative. In matching, irrelevant (redundant) background variables cause no problems because a good balance is desired for all background variables. The matching exercise becomes more complex with more background variables, but it is also more credible because is brings us closer to the (admittedly unattainable) ideal of matching on the entire background.

10

Clinical trials

Many studies are conducted with the purpose of comparing the outcomes of two distinct treatments (interventions, exposures or conditions). Each treatment can be regarded as a course of action or an option. At the conclusion of such a study, the analyst has to decide, on behalf of a client (the sponsor of the study) which of the alternative treatments yields superior outcomes in a specified population, such as patients with a particular disease. The treatment is applied to selected or recruited members of this population, called subjects. We consider first designs in which a subject can receive only one of the treatments, so the subjects form treatment groups. These groups are labelled by the treatment received, say, D and E. Other designs are discussed in Section 10.6.

If the subjects are assigned to the treatments in a haphazard fashion, according subjects' preferences (self-selection), or advice of a professional, the concern arises that the outcomes are superior on average in one treatment group not (only) as a result of the treatment applied but (also partly) as a result of the treatment that was selected or assigned. We say that in this case the processes of treatment application and assignment are *confounded*.

As an example, suppose the studied disease has stages of advance or a wide range of severity, and patients who are more severe cases tend to prefer treatment D, whereas milder cases tend to prefer E. Suppose the severity of the disease cannot be established at the point of administering the treatment. If the treatments merely alleviate the condition or its symptoms temporarily, and rarely or never cure it, then treatment E will have superior outcomes to treatment D, even if D may be more effective.

The term 'effective' in its comparative form refers to a comparison of like with like. If we had two identical patients (and therefore also identical severity of the disease at the outset), say, subjects α and β, then this term would have an unambiguous meaning: D is more effective if patient α, having received treatment D, has a superior outcome to patient β who received treatment E. In all but some esoteric cases, we cannot arrange for such near-identical pairs of subjects who would be administered the alternative treatments. We have to work hard, in design or analysis, to make a (usually expensive) study useful for drawing inferences about how the (average) outcomes would compare had the treatments been assigned alternately to the subjects within pairs of clones.

This example naturally leads to the consideration of *potential outcomes*, that is, the hypothetical outcomes that would be recorded if a particular

TABLE 10.1
Illustration of treatment heterogeneity. The realised values are underlined.

Subject (i)	Scale of y			Scale of $u = y^2$		
	y_D	y_E	$\Delta_i^{(y)}$	u_D	u_E	$\Delta_i^{(u)}$
1	2	<u>3</u>	1	4	<u>9</u>	5
2	5	<u>6</u>	1	25	<u>36</u>	11
3	<u>6</u>	7	1	<u>36</u>	49	13

treatment were administered to a subject. Suppose these outcomes are y_{iD} and y_{iE}, $i = 1, \ldots, n$. We assume that the outcome for a patient depends only on the treatment applied, never on the treatments applied (and outcomes observed) for other patients. This assumption is known as stable unit-treatment value assignment (SUTVA).

The treatment effect for subject i is defined as the difference $\Delta_i = y_{iE} - y_{iD}$. Instead of y_{iT}, T = D, E, an increasing transformation may be used. The first important consequence of this definition is that a constant treatment effect is a special case, with a vast array of alternatives. Therefore, statements related to superiority have to be qualified much more carefully than merely by stating that 'E is superior to D'. One qualification is 'on average'. A treatment may be more effective on some patients (whose Δ_i is greater) than on others; we say that the treatment effect is heterogeneous. At an extreme, E may be superior for some patients ($\Delta_i > 0$ for some i) and inferior for others ($\Delta_i < 0$).

When there is no obvious scale on which y should be recorded and analysed, treatment homogeneity (the complement of heterogeneity) is easy to dismiss. Suppose the potential outcomes y_{iD} are not constant, and neither are y_{iE}. If there is homogeneity, $\Delta_i = y_{iE} - y_{iD} = \text{const}$, on one scale, then there is heterogeneity on (almost) any other scale defined by a nonlinear increasing transformation. A small example is presented in Table 10.1.

There is treatment homogeneity on the scale of y, but after quadratic transformation there is treatment heterogeneity. In practice, only one of the two potential outcomes is realised, so the pattern that is obvious in Table 10.1 can be neither observed nor inferred. For example, if patients 1 and 2 receive treatment E and patient 3 receives D, then all we observe are the outcomes $y_{1E} = 3$, $y_{2E} = 6$ and $y_{3D} = 6$, underlined in the table. From these values we could not judge whether the treatment effect is constant or which transformation would promote treatment homogeneity. Nor could we observe or infer that the treatment effect Δ_i is positive for some and negative for other patients, if this were the case.

Let z_i be the treatment applied to subject i, so that we observe $y_i^{(\text{obs})} = y_{i,z_i}$. The established statistical instinct is to apply a transformation f to $y_i^{(\text{obs})}$ that promotes normality of $f(y_i^{(\text{obs})})$. This is inappropriate because the distributions that matter are for the outcomes within the treatment groups, and not for their mixture. This is obvious when the average treatment effect, $\bar{\Delta} = \mu_E - \mu_D$, is substantial. Also, clinical considerations should have a priority over statistical concerns in the choice of the transformation.

Our target, the average treatment effect, $\bar{\Delta}$, is the average of the individual treatment effects Δ_i that would be evaluated in a population or a subpopulation. Therefore, we should first establish that Δ_i, as a random variable, is defined meaningfully and that averaging of its values is a sensible operation. A definition of Δ is meaningful if any particular value of Δ, say 1.7, represents the same difference in substance for all possible potential outcomes y_D and y_E for which $y_E - y_D = 1.7$.

Averaging is a sensible operation when the average is a complete summary of the contributing elements. As a simple example, the average of 36.0 and 37.8 is the same as the average of 36.8 and 37.0, equal to 36.9. However, if these values are human body temperatures in degrees centigrade, then averaging hides the presence of an elevated temperature in one pair of values as opposed to none in the other pair. In summary, differencing and averaging are linear operations. They are appropriate only on scales that are linear in their substance.

10.1 Randomisation

The analysis of a study that compares two treatments has to disentangle the following effects:

- the effect of the treatment (its application, or administration);

- the effect of the treatment assignment;

- inexplicable effects (natural variation).

Background refers to the collection of variables whose values are not affected by the treatment. Potential values can be defined not only for outcomes but also for any other variable defined for the subjects in the study. A variable is said to be background if its two potential values, associated with treatments D and E, coincide for every subject. That is, the treatment effect on a background variable vanishes for every subject.

The target of the study is the treatment effect on the outcome variable. This is the effect we would like to establish. The effect of the treatment assignment is a nuisance; we would welcome if it were absent. The third effect, inexplicability, is an inherent characteristic of any population we care to study, and a human population in particular. The confusion between the first two effects is referred to as confounding. For example, if the within-treatment average $\bar{y}_D$ exceeds $\bar{y}_E$, but the two treatment groups have different means of a background variable, the difference in the average outcomes may well be attributed to the difference in the backgrounds instead of the quality (healing capacity or efficacy) of the treatment.

The influence of inexplicability is reduced by conducting larger studies, recruiting more subjects. The effect of the treatment assignment is eliminated by design, by controlling the assignment. Randomisation is an example of such control. It encompasses a variety of designs in which the treatment is assigned according to a random mechanism. That is, control is exercised not over the assignment as such, but over the *distribution* of assignment. In its simplest implementation, each subject has the same probability, equal to $\frac{1}{2}$, of being assigned to either treatment, and the assignments of the subjects are mutually independent. Simple adaptations of this design discard some extreme assignments, such as all subjects being assigned to the same treatment, or an a priori specified subset of the subjects (e.g., all women) being assigned entirely to one treatment. Such assignments have small probabilities, but the study operates with a single realisation of the treatment assignment, and that realisation has to be fit for the purpose of estimating the treatment effect.

In the frequentist perspective, we consider a large number of (hypothetical) replications of the treatment assignment. On average, the two treatment groups are well balanced on any conceivable background variable. That is, any summary of the background variables (e.g., the mean of one variable) has the same distribution over an infinite sequence of replications within one treatment group as within the other. This applies to both the observed (recorded) and unobserved variables. The symmetry, or good match, of the within-treatment background profiles suggests that these variables have no role to play in the analysis of outcomes. This can be confirmed in a more general case.

This property conveys a great advantage on randomised studies. No background variables have to be recorded, except those involved in the assignment process, (potentially) saving a lot of paperwork, reducing the response burden and simplifying the analysis. This advantage is lost, however, when the protocol is not adhered to. The most common deviations from the protocol are dropping out from the study (not completing the course of the assigned treatment), and its extreme form, refusal to accept the treatment despite the initial consent. After such deviations, background variables may be important as informants about what outcome would have been observed had the protocol not been breached.

Any imbalance in the treatment groups, interpreted as their asymmetry, is a threat to the validity of the analysis. Symmetry on average is in many contexts the best that can be arranged, and the threats of substantial asymmetry diminish with greater sample sizes. Some elements of symmetry can be enforced in the design. For example, randomisation may be constrained to having the same number of men (and women) in the two treatment groups. The assignment realised by the adopted randomisation scheme may be inspected for (substantial) imbalances, and rerun if necessary. An important feature of all such schemes is that they are not informed by the outcomes which are to be realised in the future. That is, randomisation is manipulated (e.g., its realisation is selectively rejected) using only background information.

In summary, randomisation eliminates the effect of the background, and makes the treatment assignment, as a process, ignorable. It enables us to analyse the outcomes with no regard for any variable other than the treatment that was assigned. The two treatment groups are by construction equivalent with respect to the background, and so the only reasons for differences in their average outcomes are the differential treatment effects and inexplicability. The latter is ameliorated by having sufficiently large treatment groups. Note however that the two treatment groups are equivalent with respect to background only on average over replications of the treatment assignment. This issue is commonly ignored but it becomes less acute with larger within-treatment sample sizes.

10.2 Analysis by hypothesis testing

The motivation for a typical clinical trial is to provide evidence that a proposed (novel) treatment is superior to the established treatment. The established treatment may be placebo when there is no approved treatment for the studied condition or disease. A clinical trial fits into the scheme of testing the hypothesis that the average treatment effect is zero or negative against the alternative that it is positive. Let n_D and n_E be the (sample) sizes of the treatment groups, $\bar{y}_D$ and $\bar{y}_E$ the sample means of the outcomes, and τ_D^2 and τ_E^2 the variances of the outcomes within the respective groups D and E. The average treatment effect Δ is estimated without bias by the difference

$$\hat{\Delta} = \bar{y}_E - \bar{y}_D ,$$

and its sampling variance $\sigma^2 = \text{var}(\hat{\Delta})$ by

$$\hat{\sigma}^2 = \frac{\hat{\tau}_D^2}{n_D} + \frac{\hat{\tau}_E^2}{n_E} .$$

Suppose the estimator $\hat{\Delta}$ is normally distributed. The hypothesis that $\Delta \leq 0$ is rejected for sufficiently large realised value of $\hat{\Delta}$. The threshold, called the critical value (of the test), is set so that under the presumption of the hypothesis, $\Delta \leq 0$, the probability of (inappropriate) rejection would be smaller than or equal to the size of the test, α, set by convention to 0.05. Values of 0.01 and 0.10 are used, rarely, as alternatives. The critical value is $c = \sigma \Phi^{-1}(1 - \alpha)$; if $\hat{\Delta} > c$ we reject the hypothesis, otherwise we do not reject it. In the expression for c, the distribution function of the standard normal, Φ, is replaced by its counterpart for the t distribution with the appropriate number of degrees of freedom when σ^2 is estimated, or another distribution function when $\hat{\Delta}$ is not normally distributed. The assumption that the variances τ_D^2 and τ_E^2 coincide is often adopted, and done so with good reason. Then also $\hat{\tau}_D^2 = \hat{\tau}_E^2$, and the same estimator (the pooled variance estimator) is used for both variances.

Not rejecting the hypothesis, when $\hat{\Delta} < c$, provides no support for the hypothesis, so such a failure is best interpreted as being in a state of ignorance. In this case, the study has been wasted because no conclusion about Δ is warranted. The practice is more pragmatic but problematic. Failure to reject the hypothesis is interpreted as a failure to prove the concept, so the treatment would be dealt with as if it were not superior to the established alternative (or placebo). This is at best a logical inconsistency because such action is not supported by the analysis.

Sample size calculation is an essential element of the design of a randomised study in which two treatments are compared. The purpose of this calculation is to set the sample sizes of the two treatment groups so that, presuming a specific alternative, $\Delta = \Delta_{\min} > 0$, the planned test would reject the hypothesis with a probability at least as high as a specified value. This value, called the power of the test and denoted by β, is usually set to 0.8 or 0.9. Sample size calculation ensures that resources are not wasted by having a study that is too small and thus likely to conclude with failure to reject the hypothesis even when the novel treatment is effective. At the same time, it ensures that the study is not too large, unnecessarily exposing some vulnerable subjects to experimentation. The verb 'ensure' has to be carefully qualified. In this statement it does not imply certainty but a probability greater than specified: $1 - \alpha$ for inappropriately concluded evidence when $\Delta \leq 0$ and β for failure to conclude with evidence of superiority of the proposed treatment when $\Delta > \Delta_{\min}$. However, there is no profound basis for the commonly selected values of α and β, other than a deeply ingrained convention.

Assuming that σ^2 is known, the condition for sufficient power,

$$P\left(\hat{\Delta} > c \mid \Delta_{\min}\right) \geq \beta,$$

is equivalent to

$$\sigma \leq \frac{\Delta_{\min}}{\Phi^{-1}(1 - \alpha) + \Phi^{-1}(\beta)}.$$

The sample sizes, involved in σ, are set by solving this inequality. Suppose $n_D = n_E \, (= n)$ and $\tau_D^2 = \tau_E^2 \, (= \tau^2)$, so that $\sigma^2 = 2\tau^2/n$. Then the solution is

$$n \geq 2\tau^2 \frac{\left\{\Phi^{-1}(1 - \alpha) + \Phi^{-1}(\beta)\right\}^2}{\Delta_{\min}^2}. \tag{10.1}$$

In summary, the sample sizes $n_D = n_E$ are a function of four factors, the variance τ^2, the smallest difference that is regarded as important, $\Delta_{\min}$, and the test-related probabilities α and β. The client's perspective is represented by $\Delta_{\min}$. There is some scope in setting the probabilities α and β, but this is rarely if ever exploited because there is no obvious way of balancing the hypothetical probabilities of the appropriate verdict, $1 - \alpha$ and β.

10.3 Electing a course of action—approve or reject?

The purpose of a clinical trial is to choose one of two available courses of action: to proceed towards production and distribution of the proposed (novel) treatment or to abandon the development. There may be some intermediate options but we ignore them to focus on the essence of our approach. Before discussing the client's perspective we have to establish who the client is. In fact, there may be several clients and they may have different perspectives. We refer to these clients as stakeholders. A key stakeholder is the patient community. It is not well defined because it includes not only the current patients but also future patients for whom the novel treatment is intended. The providers of health care, the clinical professionals (and their organisations) and institutions (e.g., hospitals and their management) are another stakeholder. The health-care insurance industry cannot be omitted because they are footing the bill. They demand a good return, in terms of good clinical outcomes and infrequent returns for further care. These stakeholders may be represented by a single body, the regulatory agency, such as the Food and Drug Administration in the U.S.A., and the European Medicines Agency in the European Union. The regulator formulates the perspective that is a compromise of, or is informed by, the perspectives of the stakeholders.

The developer and the prospective manufacturer or provider of the service associated with the treatment is another key stakeholder. Their perspective in general differs from the regulator's perspective. As a business, their priority is continual income and profit, and that would be assured by products that are in demand and that command prices sufficiently high in relation to the costs of research, development, the process of gaining approval, production, marketing and other operating costs. The least controversial way such demand may be generated is by products that are superior to alternative products made by competitors or have other attributes that are valued and sought by the purchasers.

The development of a pharmaceutical drug, a medical device or another health-care product is an expensive affair, afforded by a select few companies and corporations worldwide. When approved, some of the new products command high prices, claimed to be necessary to recoup the expenses on development, and also claimed to be justified by their quality—their contribution to health care. If the developer and producer are at liberty to set the price, they will aim to maximise their profits. This does not mean that the price would necessarily be exorbitant because if the product is unaffordable by many, sales, and hence profits, will diminish.

Suffice to say that the consequences of a failure to have a product approved if otherwise it would have earned income and good reputation are the loss of this income and failure to gain reputation. The consequences of having a proposed product approved which would then turn out to be harmful are potentially catastrophic. First, income would flow only until the harmful or otherwise unsatisfactory nature of the product is discovered and publicised, although even rumours and doubts may sometimes cause a reduction in the demand. When these doubts are substantiated various sanctions and penalties may follow, as a result of legal action or contractual obligations. Second, the reputation of the company may suffer, affecting the demand for its other products as well as the capacity to command high prices for them.

In the developed economies and markets for health-care products, the regulatory agency has the role of a gate-keeper. It receives proposals for new products and arbitrates between accepting and rejecting them. Developers submit to the regulator proposals for new products, together with protocols and plans for the analysis of studies intended to confirm the (claimed) good properties of the proposed product. The regulator, an officially sanctioned agency charged with a specific remit, has, or should have, a perspective and value judgement that it would willingly disclose to all other stakeholders, and the developers in particular. At present, such a perspective has no role in a typical plan of analysis which usually incorporates a randomised trial analysed by a hypothesis test. This we regard as a fundamental deficiency of regulatory process.

The developer and the regulator have in general different perspectives. That does not create any difficulties in our approach centred on decision making. The regulator may or may not approve a developer's proposal. A developer may or may not want to proceed with the production and marketing of the proposed product. The product's access to the market will be open if the regulator and the developer *both* elect the positive option. Therefore, two related problems have to be solved, one to find the developer's preferred course of action and the other the regulator's.

10.4   Decision about superiority

In this section, we address the problem of deciding whether the novel treatment is superior to the established treatment, and its distribution in the intended market is worth pursuing. The decision is based on a comparison of the two treatments. We assume that the outcome of an application of either treatment is measured or assessed by the same variable y (using the same measurement or assessment protocol) in a study in which the subjects are assigned to the treatments completely at random, possibly subject to set sample sizes within the two treatment groups and some subgroups defined by background variables. We assume that these sample sizes are not trivial.

Superiority is interpreted with respect to the scale on which the outcomes are recorded, possibly after a transformation, as $\Delta > \Delta^*$, where $\Delta^* > 0$ may be agreed by the developer and the regulator. If Δ were positive but small, that is, $\Delta \in (0, \Delta^*)$, the proposed treatment would still be regarded as a failure. Suppose first that the losses for the two kinds of error, approving the proposal (or proceeding with it) when $\Delta < \Delta^*$, and rejecting the proposal (or stopping its progress towards distribution) when $\Delta > \Delta^*$, are R lossiles and one lossile, respectively. One might reasonably assume that R is much greater than unity—distribution of an unsuitable or harmful product has far graver consequences than failure to introduce a useful product. The developer and the regulator may agree that $R \gg 1$, but may nevertheless disagree about the value of R. It is unlikely that a single value of R, or a narrow range of values, would be suitable for all clinical trials in regulatory submissions, or even their subset, such as trials for superiority or trials for a class of medical conditions.

Let $\hat{\Delta}$ be an unbiased estimator of the average treatment effect Δ and suppose it has normal distribution with known sampling variance τ^2. Then the expected losses for the two verdicts (a—accept and r—reject) are

$$Q_a = R \int_{-\infty}^{\Delta^*} \varphi\left(x; \hat{\Delta}, \tau\right) dx$$

$$= R\Phi(z^*)$$

and

$$Q_r = 1 - \Phi(z^*),$$

where $z^* = (\Delta^* - \hat{\Delta})/\tau$. We choose the option with the smaller expected loss, so we accept the proposal (issue verdict V_a) if

$$z^* < \Phi^{-1}\left(\frac{1}{R+1}\right),$$

that is, when $\hat{\Delta}$ exceeds the equilibrium

$$E = \Delta^* + \tau \Phi^{-1}\left(\frac{R}{R+1}\right),$$

and reject it otherwise. The equilibrium E is an increasing function of R, with $E = \Delta^*$ for $R = 1$. Greater loss ratio R corresponds to stronger aversion to inappropriate approval, and so the 'hurdle' E is set higher.

10.4.1 More complex loss functions

The loss structure introduced earlier in this section takes no account of the magnitude of the error; only the type of the error matters. In a reasonable perspective, the loss with verdict V_a is greater when $\Delta < 0$ than when $0 < \Delta < \Delta^*$. When $\Delta < 0$, the proposed treatment is worse than the established treatment; when $\Delta > 0$ but $\Delta < \Delta^*$, it is better but not by a margin deemed sufficient.

Suppose one lossile is incurred if the proposal is rejected (verdict V_r) inappropriately, so that $Q_r = 1 - \Phi(z^*)$, but R_0 lossiles are lost with verdict V_a

if $\Delta \in (0, \Delta^*)$ and $R_1 > R_0$ lossiles are lost if $\Delta < 0$. Now the expected loss with verdict V_a is

$$
\begin{aligned}
Q_a &= R_0 \int_0^{\Delta^*} \varphi\left(x; \hat{\Delta}, \tau\right) dx + R_1 \int_{-\infty}^0 \varphi\left(x; \hat{\Delta}, \tau\right) dx \\
&= R_0 \left\{ \Phi(z^*) - \Phi(z_0) \right\} + R_1 \Phi(z_0) ,
\end{aligned}
$$

where $z_0 = -\hat{\Delta}/\tau$. The balance equation,

$$
B(\Delta^*, R_0, R_1) = (R_0 + 1) \Phi(z^*) + (R_1 - R_0) \Phi(z_0) = 1 ,
$$

does not have a closed-form solution. The properties of the equilibrium as a function of Δ^*, R_0 and R_1 can be explored using the implicit function theorem. The partial differentials are:

$$
\frac{\partial B}{\partial \hat{\Delta}} = -\frac{1}{\tau} \left\{ (R_0 + 1) \varphi(z^*) + (R_1 - R_0) \varphi(z_0) \right\} \quad < 0
$$

$$
\frac{\partial B}{\partial R_0} = \Phi(z^*) - \Phi(z_0) \quad > 0
$$

$$
\frac{\partial B}{\partial R_1} = \Phi(z_0) \quad > 0 , \tag{10.2}
$$

so $\partial \hat{\Delta} / \partial R_0 = -\partial B / \partial R_0 / \partial B / \partial \hat{\Delta} > 0$, and therefore the equilibrium function is increasing in R_0. It is increasing also in R_1.

The linear loss function $R_0 + S(\Delta^* - \Delta)$ is an alternative to the piecewise constant loss. Here R_0 is the lower bound (infimum) for the loss with verdict V_a when it is elected inappropriately. The loss diverges to $+\infty$ as Δ diverges to $-\infty$. The expected loss with verdict V_a is:

$$
\begin{aligned}
Q_a &= \int_{-\infty}^{\Delta^*} \left\{ R_0 + S(\Delta^* - x) \right\} \varphi\left(x; \hat{\Delta}, \tau\right) dx \\
&= R_0 \Phi(z^*) + S\tau \int_{-\infty}^{z^*} (z^* - y) \varphi(y) dy \\
&= R_0 \Phi(z^*) + S\tau \, \Phi_1(z^*) .
\end{aligned}
$$

The loss associated with inappropriate rejection can be revised similarly. Defined as an increasing function of Δ for $\Delta > \Delta^*$, it would mirror the assumption that a more effective treatment would earn greater income (the developer's perspective) and its application would result in better outcomes (the regulator's perspective). The evaluations of the corresponding expected loss and the balance function are left for an exercise.

10.4.2   Trials for non-inferiority

A treatment may be proposed with a seemingly lower ambition, or weaker claim, to merely match the effectiveness of an established treatment. Such a

treatment may be proposed when its administration is simpler, more convenient, requires less supervision, is proven to have fewer and less severe side
effects or is less likely to be rejected by patients. The methods for analysis of
trials for superiority can be adapted for this setting with Δ^* set to zero, so
that $z^* = z_0$. Some trade-off of efficacy with other advantages of the proposed
treatment may be permitted. This is reflected in the analysis by setting Δ^*
to a negative value. In this case, $-\Delta^*$ is the largest permitted reduction in
effectiveness. A small loss may be introduced when Δ is in the range $(\Delta^*, 0)$.

10.5 Trials for bioequivalence

A developer may propose a treatment that could for all clinical purposes be
regarded as equivalent to an established treatment. For example, the proposed
drug may contain the same active compound but would be manufactured by
a different (less expensive or more reliable) process. The original developer
may hold a patent protecting its sole right to manufacture and market the
treatment (drug, device, service, or the like). After expiry of this patent, other
companies may enter the market, but first have to provide evidence that their
treatment is therapeutically nearly equivalent (or, identical) to the original.
Generic drugs, often cheaper substitutes for the original branded drugs, are
examples of such products.

Bioequivalence corresponds to the treatment effect $\Delta = 0$, but it is impossible to provide evidence of this or any other singular value of Δ. This is
resolved by allowing some leeway, and defining bioequivalence as $|\Delta| < \Delta^*$ for
an agreed small positive value of Δ^*.

Within the confines of hypothesis testing, evidence of bioequivalence is
provided by rejecting both hypotheses of superiority:

- that treatment D is superior to treatment E;

- that treatment E is superior to treatment D.

The sizes of the two tests are adjusted so that the probability of a false positive
would not exceed a threshold, set to 0.05 by convention. Some leniency may
be introduced by setting this probability to 0.10.

In the simplest version of the approach centred on electing one of the two
available options, accepting the claim of bioequivalence and rejecting it, we
associate inappropriate rejection with unit loss and inappropriate acceptance
with loss $R > 1$. The expected losses with the two verdicts are:

$$
\begin{aligned}
Q_{\mathrm{a}} &= R\left\{ \int_{-\infty}^{-\Delta^*} \varphi\left(x; \hat{\Delta}, \tau\right) \mathrm{d}x + \int_{\Delta^*}^{+\infty} \varphi\left(x; \hat{\Delta}, \tau\right) \mathrm{d}x \right\} \\
&= R\left\{ \Phi\left(z_-^*\right) + 1 - \Phi\left(z_+^*\right) \right\},
\end{aligned}
$$

and

$$Q_r = \int_{-\Delta^*}^{\Delta^*} \varphi\left(x; \hat{\Delta}, \tau\right) dx$$

$$= \Phi\left(z_+^*\right) - \Phi\left(z_-^*\right),$$

where $z_+^* = (\Delta^* - \hat{\Delta})/\tau$ and $z_-^* = (-\Delta^* - \hat{\Delta})/\tau$. The balance equation, $Q_a = Q_r$, is equivalent to

$$\Phi\left(z_+^*\right) - \Phi\left(z_-^*\right) = \frac{R}{R+1}.$$

Symmetry and unimodality of the normal distribution imply that the balance equation has either no solution, two solutions, or that $\hat{\Delta} = 0$ is the unique solution. The latter (anomalous) case arises when

$$\Phi\left(\frac{\Delta^*}{\tau}\right) - \Phi\left(-\frac{\Delta^*}{\tau}\right) = \frac{R}{R+1},$$

that is, for

$$\tau_0 = \frac{\Delta^*}{\Phi^{-1}\left(\frac{2R+1}{2R+2}\right)}. \tag{10.3}$$

The denominator is an increasing function of R, positive for all $R > 0$. For $\tau > \tau_0$, there is no equilibrium, and the verdict V_r (rejecting the claim of bioequivalence) is issued irrespective of $\hat{\Delta}$. With very little (or no) data or information, we reject the proposal of bioequivalence because the bioequivalence range $(-\Delta^*, \Delta^*)$ is too narrow in relation to τ and the ramifications of a false rejection are not as grave as of a false acceptance. The more rigorous the condition of bioequivalence (smaller threshold $\Delta^* > 0$) the more precise estimator $\hat{\Delta}$ is necessary to give the claim a chance.

Figure 10.1 illustrates the various scenarios. In the left-hand panel, the differences (probabilities) $\Phi(z_+^*) - \Phi(z_-^*)$ are drawn by solid lines for $\Delta^* = 0.1$ and values of τ indicated on the curves (from 0.06 to 0.12). Three values of R are represented by horizontal dashes drawn at $R/(R+1)$. So, the equilibria (E) occur where the difference $\Phi(z_+^*) - \Phi(z_-^*)$ intersects $R/(1+R)$, and does so at points $\hat{\theta} = \pm E$. There is no equilibrium when the curve lies entirely below $R/(1+R)$, for instance, for $\tau = 0.10$ and $R = 3.0$.

The right-hand panel of Figure 10.1 illustrates the dependence of $\Phi(z_+^*) - \Phi(z_-^*)$, and of the equilibria, on Δ^* and R. Here $\tau = 0.1$. The diagram confirms that narrower bioequivalence range $(-\Delta^*, \Delta^*)$ results in a lower and flatter balance function $\Phi(z_+^*) - \Phi(z_-^*)$, and therefore in the verdict of bioequivalence in a narrower range of values of $\hat{\theta}$. And, of course, the range may vanish altogether, as for $\Delta^* = 0.11$ and $R = 3.0$. Then the verdict of bioequivalence is not issued for any value of $\hat{\theta}$.

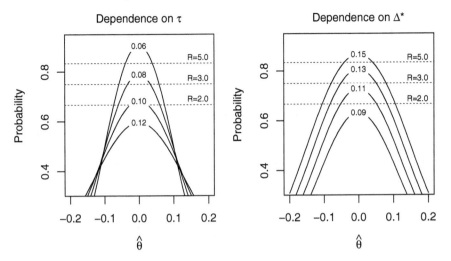

FIGURE 10.1
Balance equations and equilibria for bioequivalence as functions of Δ^*, τ and R.

When $\tau < \tau_0$, the equilibria can be found by line search or another iterative method. The Newton-Raphson algorithm is easy to implement using the differential of the balance function,

$$\frac{\partial B}{\partial \hat{\Delta}} = \frac{1}{\tau} \left\{ \varphi(z_-^*) - \varphi(z_+^*) \right\} .$$

The initial solution should not be set to zero because the differential vanishes there; $\hat{\Delta} = \Delta^*$ is suitable. The pair of equilibria are $\pm E$; bioequivalence is accepted for $|\hat{\Delta}| < E$.

10.6 Crossover design

When a subject receives only one of the two treatments that are compared in a clinical trial, the corresponding potential outcome is realised and the other potential outcome is not. Estimation of the treatment effect would be greatly simplified, at least in principle, if both potential outcomes of a subject were observed. This is impossible if one treatment irrevocably alters the health status of the subject. However, when a treatment has an effect that lasts for a limited period of time, subjects can be engaged in a clinical trial for two periods in which they receive the alternative treatments.

Thus, with treatments D and E, subjects are assigned to treatment regimes DE and ED in a crossover trial. Subject i in regime DE receives treatment D followed by measurement of the outcome y_{i1D}, a period of 'no treatment' called the washout period, in which the effect of D dissipates, and then treatment E, concluded with measurement of the outcome y_{i2E}. In regime ED, the roles of the treatments D and E are interchanged. Symmetry suggests that (approximately) the same number of patients should be included in either treatment-regime group. The rationale for randomisation to treatment regimes carries over from clinical trials with single-treatment administration.

Crossover design is suitable for treatments that have to be applied continually, and that have only palliative effects, alleviating the symptoms of a disease, so that each patient can be regarded, from a clinical point of view, as being in the same state at the beginning of each treatment period. Nevertheless, a wary analyst would be concerned about any deviation from this assumption. Is the washout period sufficiently long? Note that the length of a washout period may be limited by the need to provide some treatment to every subject.

The period effect is defined as the difference of an outcome following a treatment in period 2 and of the outcome following the same treatment in period 1. This difference cannot be observed, unless we introduce the repeat-regimes DD and EE. Without them, the randomised and balanced assignment to the regimes DE and ED helps us to eliminate this effect in the contrasts of the average outcomes $\bar{y}_E - \bar{y}_D = \frac{1}{2}(\bar{y}_{1E} + \bar{y}_{2E} - \bar{y}_{1D} - \bar{y}_{2D})$, where the averages are taken over the omitted indices, i for the subject in every case, and also for the period, 1 or 2, in $\bar{y}_D$ and $\bar{y}_E$. The elimination is exact when the period effect is constant. It may be variable, with each subject having a specific period effect. Problems arise when this effect is not additive, contrary to the assumption made below.

Despite the washout period, which separates the two treatment periods, there may be concern that a residue (carryover) of the effect of the treatment applied in the first period persists in the second treatment period. This problem, when addressed by the established model selection approaches, turns out to be deficient. The approach starts by formulating a linear or, more precisely, an additive model with constant period and carryover effects;

$$y_{ipt} = \mu + \lambda_p + \nu_t + \gamma_{pt} + \varepsilon_{itp}, \tag{10.4}$$

where λ_p, $p = 1, 2$, are the period effects, ν_t, $t = \mathrm{D}, \mathrm{E}$, are the treatment effects, γ_{pt} are the carryover effects and ε_{itp} is a random sample from a centred normal distribution with variance σ^2. To avoid some distracting complications, suppose the $2n$ subjects in a study are split evenly (and at random) to the two treatment regimes.

Without some constraints, this model is not well identified—it involves too many parameters. For example, in the presence of an intercept term μ, only the difference $\lambda_2 - \lambda_1$ is identified. The conventional choice is to set $\lambda_1 = 0$, $\nu_D = 0$, and three of the four terms γ_{pt} to zero, except for γ_{2E}. Then the observed outcomes can be summarised by the four period-by-treatment averages $\bar{y}_{pt}$, and the model for them has four parameters, μ, λ_2, ν_E and γ_{2E}.

The target is ν_E; the other three parameters are at best of secondary interest, mainly to assess, post hoc, the value of the crossover design. They are regarded as nuisance parameters. The maximum likelihood as well as the ordinary least squares estimator of ν_E is

$$\hat{\nu}^{(1)} = \bar{y}_{1E} - \bar{y}_{1D}.$$

It does not involve the data from the second period at all. That renders the exposure of subjects in the second period unnecessary. Such a wasteful exposure is unethical. It suggests that the crossover design should be used only when the carryover effect is absent, $\nu_{2E} = 0$, and this is confirmed at the design stage. In that case, ν_E is estimated, without bias, by

$$
\begin{aligned}
\hat{\nu}' &= \bar{y}_E - \bar{y}_D \\
&= \frac{\hat{\nu}^{(1)} + \hat{\nu}^{(2)}}{2} \\
&= \frac{\bar{y}_{1E} - \bar{y}_{1D} + \bar{y}_{2E} - \bar{y}_{2D}}{2}.
\end{aligned}
$$

Model selection might proceed by testing the hypothesis that $\nu_{2E} = 0$ and, depending on its outcome, using either $\hat{\nu}^{(1)}$ or $\hat{\nu}'$. The choice between $\hat{\nu}'$ and $\hat{\nu}^{(1)}$ can be interpreted as an issue of whether to use data from period 2 in addition to data from period 1. Note the insidious presence of model uncertainty is this plan.

We reject the categorical nature of this approach, which seeks to arbitrate as to whether the outcomes from period 2 should be used, on par with outcomes from period 1, or period-2 outcomes should be discarded altogether. We propose a compromise in which outcomes from period 2 are accorded less importance than outcomes from period 1. The extent of down-weighting the outcomes from period 2 depends on the prior information about how large the carryover effect could be.

10.6.1 Composition of within-period estimators

The within-period estimators of the treatment effect ν_E are $\hat{\nu}^{(p)} = \bar{y}_{pE} - \bar{y}_{pD}$, $p = 1, 2$; $\hat{\nu}^{(1)}$ is unbiased and, according to the model in equation (10.4), the bias of $\hat{\nu}^{(2)}$ is

$$E\left(\hat{\nu}^{(2)}\right) - \nu_E = \gamma_{2E};$$

the period effect is absent in $\hat{\nu}^{(2)}$ because it is present in $\bar{y}_{pD}$ and $\bar{y}_{pE}$ in equal measure. According to the model, $\hat{\nu}^{(1)}$ and $\hat{\nu}^{(2)}$ have identical variances $\tau^2 = 2\sigma^2/n$ and are independent because they involve non-overlapping sets of terms ε_{itp}.

Suppose we have an upper bound γ_{max} on the absolute value of the carry-over effect; $|\gamma_{2E}| < \gamma_{max}$. We seek the constant b for which the estimator

$$\tilde{\nu}_E(b) = (1-b)\hat{\nu}^{(1)} + b\hat{\nu}^{(2)}$$

has the smallest mean squared error (MSE) under the assumption of the largest plausible carryover effect. This is a simple problem of finding the minimum of the quadratic function

$$\text{MSE}\{\tilde{\nu}_E(b); \gamma_{2E} = \pm\gamma_{max}\} = \left\{(1-b)^2 + b^2\right\}\tau^2 + b^2\gamma_{max}^2.$$

The minimum is attained for

$$b^* = \frac{\tau^2}{2\tau^2 + \gamma_{max}^2}.$$

If we are certain that the carryover effect is absent, then $\gamma_{max} = 0$, and so $b^* = \frac{1}{2}$. In this case, the two periods yield equally valuable data for ν_E. With τ^2 fixed, the higher the upper bound γ_{max} the smaller the weight b^* accorded to the outcomes from period 2 and more faith is put in the data from period 1.

We refer to an estimator formed by a linear (or convex) combination of two or more estimators as *composite*, if their coefficients add up to unity. Thus, our proposal can be described as the composite estimator that would be efficient in the case of the largest plausible carryover. We explore next the behaviour of $\text{MSE}\{\tilde{\nu}_E(b^*); \gamma_{2E}\}$ when $|\gamma_{2E}| < \gamma_{max}$, evaluating the loss of precision owing to declaring γ_{max} too large (being too cautious), and the consequences of declaring γ_{max} too small (being reckless). The MSE,

$$\text{MSE}\{\tilde{\nu}_E(b^*); \nu_E\} = \frac{\tau^2}{2}\left\{1 + (1 - 2b^*)^2\right\} + b^{*2}\gamma_{2E}^2$$

$$= \frac{\tau^2}{2}\left\{1 + \frac{\gamma_{max}^4}{(2\tau^2 + \gamma_{max}^2)^2}\right\} + \frac{\tau^4\gamma_{2E}^2}{(2\tau^2 + \gamma_{max}^2)^2}, \qquad (10.5)$$

is an increasing (linear) function of γ_{2E}^2 and its slope on γ_{2E}^2 decreases with γ_{max}^2, from $\frac{1}{4}$ for $\gamma_{max} = 0$ to zero as $\gamma_{max}^2 \to +\infty$. For the largest plausible carryover,

$$\text{MSE}\{\tilde{\nu}_E(b^*); \nu_E, \gamma_{2E} = \gamma_{max}\} = \frac{\tau^2}{2}\left(1 + \frac{\gamma_{max}^2}{2\tau^2 + \gamma_{max}^2}\right).$$

This has the lower bound of $\frac{1}{2}\tau^2$, which is reached only when $\gamma_{max}^2 = 0$. Its upper bound of $\tau^2 = \text{var}(\hat{\nu}^{(1)})$ is the limit as γ_{max}^2 diverges to $+\infty$.

The partial differential of the MSE in equation (10.5) with respect to γ_{max}^2 is

$$\frac{2\tau^4\left(\gamma_{max}^2 - \gamma_{2E}^2\right)}{(2\tau^2 + \gamma_{max}^2)^3}. \qquad (10.6)$$

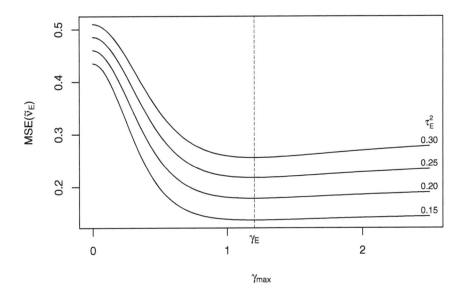

FIGURE 10.2
The MSE of the composite estimator $\tilde{\nu}_E$ as a function of $\gamma_{\max}$ for $\gamma_E = 1.2$ and τ^2 indicated at the right-hand margin.

As a function of γ_{2E}^2, the MSE decreases while $\gamma_{\max}^2 < \gamma_{2E}^2$ and increases for $\gamma_{\max}^2 > \gamma_{2E}^2$, so it attains its minimum for $\gamma_{\max}^2$. For a given deviation $\Delta = |\gamma_{\max}^2 > \gamma_{2E}^2|$, the differential in (10.6) has greater absolute value when $\gamma_{\max}^2 < \gamma_{2E}^2$ than when $\gamma_{\max}^2 > \gamma_{2E}^2$. So, the decrease of the MSE is steeper when $\gamma_{\max}^2 = \gamma_{2E}^2 - \Delta$ than the increase is when $\gamma_{\max}^2 = \gamma_{2E}^2 + \Delta$. Therefore, overstating $\gamma_{\max}^2$ results in a smaller inflation of MSE than understating it by the same amount.

This conclusion is confirmed on an example in Figure 10.2 in which the MSE in equation (10.5) is plotted as a function of $\gamma_{\max}$ for $\gamma_E = 1.2$ and $\tau^2 = 0.15, 0.20, 0.25$ and 0.30. The MSE is a flat function of $\gamma_{\max}$ for $\gamma_{\max} > \gamma_E$, and also in the left-hand vicinity of γ_E. However, when $\gamma_{\max}$ is set much lower than the (unknown) γ_E, the MSE is very large—the penalty for substantial understatement of $\gamma_{\max}$ is much harsher than for a similar overstatement.

Figure 10.3 compares the MSE functions for a set of composite estimators of ν_E defined by distinct values of $\gamma_{\max}$. The functions are drawn by solid lines in the intervals where $\gamma_{2E} < \gamma_{\max}$, and by hairlines further on, corresponding to understatement of $\gamma_{\max}$. The diagram confirms that lower $\gamma_{\max}$ results in smaller MSE while $\gamma_{2E} < \gamma_{\max}$, but the MSE increases steeply for values of γ_{2E} beyond $\gamma_{\max}$.

There is a small leeway for γ_{2E} in the right-hand neighbourhood of $\gamma_{\max}$, but one should not rely on it. Excessive understatement of $\gamma_{\max}$ exacts a severe

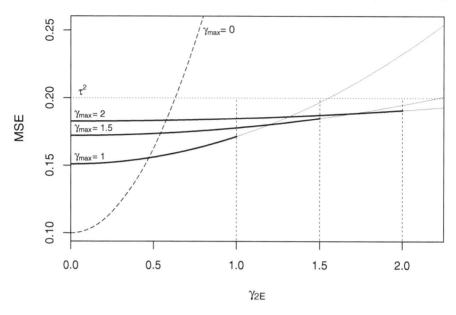

FIGURE 10.3
Mean squared errors of composite estimators with upper bounds γ_{max} set to 1.0, 1.5 and 2.0; $\tau^2 = 0.2$. Each MSE is drawn with solid line in the plausibility range, where $\gamma_{2E} < \gamma_{max}$.

inflation of the MSE. For example, the MSE when $\gamma_{2E} = 2.0$ and $\gamma_{max} = 1.0$ is 0.233, exceeding even $\tau^2 = 0.20$. The MSE of estimator $\hat{\nu}'$, defined by $\gamma_{max} = 0$, is drawn by long dashes. The estimator should be considered only when we are certain that γ_{2E} is very small, say $\gamma_{2E} < 0.2$. But even in that case, the composite estimator based on $\gamma_{max} = 0.2$ is more efficient than $\hat{\nu}'$.

The general idea of composite estimation is developed further in Chapter 11 as an alternative to model selection.

10.7 Further reading

This chapter is based on Longford (2016b) and (2017a). There is a vast literature on the design and analysis of clinical trials but most of it remains firmly committed to hypothesis testing. That applies even to experiments with novel adaptive designs. Schuirman (1987) is an example of a non-standard application of hypothesis testing in bioequivalence trials that overcomes the problem of providing evidence for a null hypothesis. Lindley (1998) presents a case for applying decision theory to bioequivalence studies, but his arguments apply much more generally. Kenward and Jones (1989) is an established text on design, estimation and hypothesis testing in crossover trials. Composite estimation of the treatment effect in crossover trials (Section 10.6.1) was introduced by Longford (2001).

10.8 Exercises

10.1. Find or construct a (realistic) example in which the assumption of SUTVA does not apply.
Hint: Consider a clinical trial in which patients know which treatment they receive, some are cared by the same medical staff, and discuss their state of health with one another during the trial.

10.2. Find in the literature (or in a textbook) arguments for and against involving background variables in the analysis of a clinical trial.

10.3. In smaller clinical trials, randomisation may yield treatment groups of very uneven sizes. For example, if 20 recruited patients are assigned to one of two treatments by tossing a fair coin for each (head – D; tail – E), the probability that 6 or fewer patients will be assigned to one and 14 or more to the other group is about 0.115. (Confirm this.)
Discuss the merits of the following adaptations. First, patients are assigned to pairs, in an arbitrary fashion, and within each pair a randomly selected patient is assigned to D and the other to E. Second, a threshold on the imbalance of the two treatment groups is set, such as having to have 7–13 patients in either treatment group. If randomisation, using (the computer version of) a fair coin, yields a sample with greater imbalance, with $0 - 6$ patients in one of the treatment groups, then it is rejected, and the randomisation is repeated. Third, a random sample of size 20 is drawn from a continuous distribution, one value for each subject. Subjects with the ten smallest values are assigned to treatment group D and the remainder to group E.

10.4. Search the literature for clinical trials in which the patients are assigned to treatments by randomisation with unequal probabilities, such as 1:2, for instance, aiming to have about 100 patients in treatment group D and 200 in group E. Compile a list of reasons why and in what circumstances such schemes are (or should be) applied.

10.5. Clinical trials are said to generate treatment groups with equal within-treatment distributions of all conceivable background variables, whether observed or not. But in any realisation, there is bound to be imbalance, however slight, on at least some background variables, and maybe even some nontrivial imbalance, especially when many background variables are considered. Qualify the statement of equal within-treatment distributions by reference to a replication scheme. Relate this scheme to the replication scheme considered in linear models for clinical trials.

10.6. A null-hypothesis test concludes with evidence against the hypothesis or failure to find such evidence. Discuss the various alternative statements, such as the finding of significant difference, finding of important difference, having found no difference, finding evidence that the difference is small, and so on. Discuss the interpretation(s) such statement(s) invite and their ambiguity.

One study concludes with no evidence against a hypothesis. Another study, testing the same hypothesis, concludes with evidence against the hypothesis. Are the two conclusions in a logical contradiction?

10.7. Explore the sample size required, given by the right-hand side of equation (10.1), as a function of α and β. Adapt the formula for designs with uneven assignment (e.g., 1:2.5) to the treatment groups.

10.8. A developer and a regulator consider a novel treatment in two settings: treatment that would compete with several established alternatives (e.g., for a common condition) and treatment intended for a condition for which there is no satisfactory alternative at present. How should the difference in the settings be reflected in the analysis?

10.9. Evaluate the expected loss Q_r and the balance function in Section 10.4.1.

10.10. In the setting with piecewise constant loss in Section 10.4.1, we have $R_1 > R_0$. The condition that $R_1 > 1$ is also reasonable. Discuss why R_0 should or should not be smaller than 1.0.

10.11. Derive the expected loss for the linear loss function for inappropriate rejection of superiority of the novel treatment. Derive the balance function when the loss for inappropriate acceptance is also linear. Set reasonable constraints on the parameters in the two loss functions.

10.12. Work out the details for the decision of superiority with a quadratic loss for inappropriate acceptance.

10.13. Collect in the literature on clinical trials in a specialty of your choice reports of trials for non-inferiority and classify them according to their motivation. For a selection of the reports, describe the methods they apply, and whether (and how) they could be adapted to deal with the problem of making decisions.

10.14. Find in the literature the distinction between average and population bioequivalence, and their link to treatment heterogeneity.

10.15. Check the steps that lead to the solution in equation (10.3). Discuss the difficulties in this derivation that would arise if the normal distribution were replaced by an asymmetric (but unimodal) distribution.

10.16. Work out the details for a piecewise linear loss for the problem of bioequivalence.

10.17. Discuss the merits of having a washout period before the first treatment, in addition to a washout period between the two treatments, in a crossover trial.

10.18. What are the two contemplated estimators of the treatment effect in a two-period crossover trial? Derive their biases, sampling variances and covariance.

10.19. Adapt the composite estimator $\tilde{\nu}_E$ developed in Section 10.6.1 for treatment groups of unequal size.

10.20. Work out the details of sample size calculation for a crossover trial in which an upper bound on the carryover effect γ_{2E} is to be used and the aim is to estimate ν_E with MSE lower than a given threshold.

10.21. Relate the sample size calculation for a crossover trial to the methods developed in Chapter 6.

10.22. Prove the statements made in the paragraph following equation (10.6) more directly, without using this expression.

11

Model uncertainty

Models play a pivotal role in statistics. A typical model is a simplified description of how a set of variables are inter-related in a specified population. A regression model describes one variable, the outcome or response, in terms of a set of others, called covariates. It states that a variable y has a conditional distribution given that a set of covariates $\mathcal{X}$ has a vector of values $\mathbf{x}$:

$$(y \mid \mathcal{X} = \mathbf{x}) \sim \mathcal{D}(\mathbf{x}; \boldsymbol{\theta}). \tag{11.1}$$

Here $\mathcal{D}$ is a class of distributions parametrised by $\mathbf{x}$ and $\boldsymbol{\theta}$; $\mathbf{x}$ is a row vector of length $K > 0$ (its dimensions are $1 \times K$) and $\boldsymbol{\theta}$ is a column vector of parameters, usually of length greater than K. By virtue of being a simplified description, the statement in equation (11.1) is usually not correct (or valid). Most phenomena we study defy any simple description but we would be satisfied even with a partial understanding of how they operate, what influences they exert and what would happen if we altered the values of some of the covariates (elements of $\mathbf{x}$) according to our designs. For that, a model that captures the most pronounced features and ignores the many less important ones would suffice. We are willing to sacrifice detail for simplicity and tractability.

This takes us into some murky waters in which clarity is difficult to establish. When is a model appropriate, adequate, useful or constructive? When are we on a firm ground with using a model for a particular purpose related to the studied variables? When can we claim that the properties of the model apply also to these variables, or to the phenomenon we want to study through them? Models are instruments to study reality. They become invaluable when we can reliably select from them a model that is a good match for the design of the study, its purpose, prior information and the data collected.

Formally, a model is a class of joint distributions for a set of variables defined in a population. The population is represented in the study by a sample on which the values of these variables are recorded. These records form a dataset, usually a rectangular array (matrix) of n subjects and K variables. We set aside cases in which this array contains some missing values, missing by design (in accordance with our plans or intention), or otherwise, by failure to adhere to the plan.

In most cases we contemplate several models, sometimes even a vast array of them, and would like to adopt one of them as a superior description of the data. All inferences would then be based on this selected model. The contemplated (or candidate) models may be partially ordered. That is, for some pairs of them, say models M_1 and M_2, M_1 is a submodel of M_2. This means that every distribution in M_1 is also included in M_2; M_2 is a wider set than M_1, $M_1 \subset M_2$. If the distributions in M_1 and M_2 are defined by values of a parameter vector $\boldsymbol{\theta}$, then we can identify a model, such as M_1, with a subset of the parameter space. For example, the parameter space may be $\mathcal{R}^p$, the p-dimensional Euclidean space, and M_1 may be defined by the constraint that the first three elements of $\boldsymbol{\theta} = (\theta_1, \theta_2, \ldots, \theta_p)^\top$ vanish; that is, $\theta_1 = \theta_2 = \theta_3 = 0$.

We say that a model is valid, if it contains the distribution from which the data at hand was generated. If M_1 is valid for a particular dataset, then any wider model M_2 is also valid, because it also contains the data-generating distribution. If both M_1 and M_2 are valid, and $M_1 \subset M_2$, then we prefer M_1 because the task of searching for the distribution that generated the data, defined by the parameter vector $\boldsymbol{\theta}$, is easier in a smaller set M_1, so long as the set contains the target of our search. Thus, there are two counteracting aims: validity, which provides an incentive to widen the model, and parsimony, defined as model reduction, within the constraints of validity. Validity would seem to be an imperative, whereas parsimony is desirable, although maybe not always essential. We will contest this view later in this chapter.

Validity and parsimony are often associated, respectively, with absence of bias and small sampling variance of an estimator. If we want to estimate a parameter with small MSE, then insisting on model validity reduces our attention to unbiased estimators, among which we seek, by pursuing model parsimony, the one with the smallest sampling variance. We show in this chapter that this strategy, although firmly established, is not optimal. We also argue that it is misguided because an estimator based on a selected model is biased and has an intractable distribution even in some simple settings. But first we establish that model selection is a decision problem, or a sequence of such problems.

11.1   Ordinary regression

In ordinary regression, the model $\mathcal{D}$ is the class of normal distributions, $\mathcal{N}$, and $\boldsymbol{\theta}$ a vector of length $K + 2$, comprising a (column) vector of regression parameters $\boldsymbol{\beta}$ of length $K + 1$, a $(K + 1) \times 1$ vector, and the residual variance σ^2, a positive scalar. The model is defined by the following assumptions:

- The conditional expectation $\mathrm{E}(y \mid \mathbf{X} = \mathbf{x})$ is a linear function of $\mathbf{x}$:

$$\mathrm{E}(y \mid \mathbf{X} = \mathbf{x}) = \mathbf{x}\boldsymbol{\beta}.$$

- The conditional variance $\mathrm{var}(y \mid \mathbf{X} = \mathbf{x})$ is a positive constant, denoted by σ^2.

- Conditionally on the vectors of covariates, the elements of the vector of outcomes, $\mathbf{y} = (y_1, y_2, \ldots, y_n)^\top$, are normally distributed and are mutually independent.

We rule out the degenerate case in which $\sigma^2 = 0$ because it leads to a trivial solution. It is expedient to introduce a compact matrix notation, in addition to $\mathbf{y}$. Let $\mathbf{X}$ be the $n \times (K+1)$ matrix composed of the vectors of covariates $\mathbf{x}_1, \mathbf{x}_2, \ldots, \mathbf{x}_n$ as its rows. A column of $\mathbf{X}$, denoted by X_r, $r = 0, \ldots, K$, contains the values of the variable r. By convention, index $r = 0$ is reserved for the intercept β_0; $X_0 = \mathbf{1}_n$, where $\mathbf{1}_n$ stands for the column vector of unities of the length n.

With this notation, the ordinary regression model is defined by the formula

$$\mathbf{y} = \mathbf{X}\boldsymbol{\beta} + \boldsymbol{\varepsilon}, \tag{11.2}$$

where $\boldsymbol{\varepsilon} \sim \mathcal{N}_n(\mathbf{0}_n, \sigma^2 \mathbf{I}_n)$; $\mathbf{0}_n$ stands for the $n \times 1$ vector of zeros and $\mathbf{I}_n$ for the $n \times n$ identity matrix. When the index of $\mathbf{0}$, $\mathbf{1}$ or $\mathbf{I}$ is obvious from the context, we omit it to reduce clutter in the notation. Throughout, we assume that $n > K + 1$ and $\mathbf{X}$ has the full rank $K + 1$. Otherwise, if $\mathbf{X}$ is singular, of rank $K' + 1 < K + 1$, we delete $K - K'$ of its columns (variables) $1, 2, \ldots, K$ for which the reduced matrix is nonsingular. Column 0, the intercept, with $X_0 = \mathbf{1}_n$, is always retained.

An alternative to the equation in (11.2) is in terms of the conditional expectation and variance matrix

$$\begin{aligned} \mathrm{E}\,(\mathbf{y} \,|\, \mathbf{X}) &= \mathbf{X}\boldsymbol{\beta} \\ \mathrm{var}\,(\mathbf{y} \,|\, \mathbf{X}) &= \sigma^2 \mathbf{I}, \end{aligned} \tag{11.3}$$

supplemented by the assumption of conditional normality of $\mathbf{y}$ given $\mathbf{X}$. More concisely,

$$(\mathbf{y} \,|\, \mathbf{X}) \sim \mathcal{N}\left(\mathbf{X}\boldsymbol{\beta}, \sigma^2 \mathbf{I}\right),$$

interpreted (or read) as: the conditional distribution of $\mathbf{y}$ given $\mathbf{X}$ is (multivariate) normal, with conditional expectation and variance matrix given in equation (11.3). Note the looseness in the notation we use for conditioning. By $(\mathbf{y} \,|\, \mathbf{X})$ we mean that the matrix of covariates has specific values—$\mathbf{X}$ is fixed. The looseness arises because we do not have separate notation for $\mathbf{X}$ as a (random) matrix and $\mathbf{X}$ as its realised (or set) values, a matrix of constants. The conditioning is always on the realised matrix $\mathbf{X}$.

Of the extensive theory of ordinary regression we refer to the results for estimating $\boldsymbol{\beta}$ and σ^2. The ordinary least squares (OLS) estimator of $\boldsymbol{\beta}$ is

$$\hat{\boldsymbol{\beta}} = \left(\mathbf{X}^\top \mathbf{X}\right)^{-1} \mathbf{X}^\top \mathbf{y}. \tag{11.4}$$

If the model is valid this estimator is unbiased, $\mathrm{E}(\hat{\boldsymbol{\beta}}) = \boldsymbol{\beta}$, and has the variance matrix $\mathrm{var}(\hat{\boldsymbol{\beta}}) = \sigma^2 (\mathbf{X}^\top \mathbf{X})^{-1}$. Further, an unbiased estimator of σ^2 is

$$\hat{\sigma}^2 = \frac{1}{n - K - 1} \mathbf{y}^\top (\mathbf{I} - \mathbf{P}) \mathbf{y}, \tag{11.5}$$

where $\mathbf{P} = \mathbf{X}(\mathbf{X}^{\top}\mathbf{X})^{-1}\mathbf{X}^{\top}$ is known as the projection matrix. The sampling distribution of $\hat{\sigma}^2$ is such that $(n - K - 1)\hat{\sigma}^2/\sigma^2$ has χ^2 distribution with $n - K - 1$ degrees of freedom. With these results, we could get a long way, but for one nontrivial hurdle, namely selection of the variables for the regression matrix $\mathbf{X}$. Validity corresponds to including all the relevant variables, possibly with some redundancy, and parsimony to excluding all redundant variables. We set aside the issues of whether the original matrix $\mathbf{X}$ contains all the relevant variables (whether they are all recorded), and whether the assumptions of linearity of the conditional expectation, conditional normality, independence and equal variance (homoscedasticity) are met, notwithstanding the onerous nature of this collection of issues.

In this section, we deal with a class of problems in which a parameter θ may have a 'special' value, such as zero. One course of action is appropriate for $\theta = 0$ and another course is advisable if θ is distant from zero. For example, when θ is a regression coefficient associated with a covariate X, it would be wise to discard it from the model when $\theta = 0$, when the covariate is redundant. In contrast, for a sufficiently large value of $|\theta|$ we would certainly retain X in the model because model validity is an imperative.

This rule is not formulated clinically because it does not state which action to take (exclude or retain the covariate X in the model) when θ is small or, more to the point, where is the threshold that separates small and large values of $|\theta|$. It is preferable to discard a covariate not only when its coefficient θ vanishes, $\theta = 0$, but also when $|\theta|$ is sufficiently small. In the latter case, with the covariate discarded, the model is not valid, but the bias of the estimator based on it is small in relation to the variance inflation that would result from retaining X in the model. The problem is further exacerbated by the value of θ not being known, and having to base model selection on its estimate $\hat{\theta}$ and related information.

In an established approach, the hypothesis that $\theta = 0$ is tested against the two-sided alternative that $\theta \neq 0$. With such a test, we control the probability of inappropriately including X in the model (false positive) by setting on it an upper limit of 0.05 (the test size), but leave ourselves exposed to a possibly large probability of inappropriate exclusion (false negative). Some control over the balance of the two probabilities may be a better solution than such a 'one-sided' control. Information about the relative magnitudes of the squared bias and variance inflation, as a function of θ, would help us define this balance.

In our perspective, the null hypothesis ($\theta = 0$) should be rejected outright, unless there is some prior information, external to the data, that supports the singular value of zero over all other values, even those extremely small, such as 10^{-19}. In the absence of such a support, zero is but one of uncountably many values, uncountably many of them in arbitrarily close distance from zero. Betting on any single one of these values is bound to be a losing proposition. But such a proposition is advanced by the null hypothesis and is applied in model selection. We presented a similar argument in the introduction to Chapter 5.

The two-stage procedure, in which we first select a model and then apply the estimator that would be appropriate if we knew that the selected model is parsimonious and valid, has two drawbacks. First, we pretend that the model selection is unerring—we ignore model uncertainty. The appropriate model is selected only with a limited probability; with the complementary probability inappropriate models are selected, and their influence on the selected-model based estimator is difficult to assess. Second, we rule out the possibility that an invalid model might yield an efficient estimator. Related to this is the possibility that some models may be useful for certain targets, which would imply that proceeding in two stages is a poor strategy. Instead of asking 'Which model?' and then placing all our faith in the model we select, we should pose the more direct question: 'Which estimator?' And, with this question, we might also ask whether selection is how to get the best out of a collection of estimators.

In the next sections we elaborate on this perspective. For the elementary problem of comparing two models, one being a submodel of the other, we assume that there are two thresholds, $T_1 < 0$ and $T_2 > 0$. One course of action (model reduction) is appropriate if $\theta \in (T_1, T_2)$, and the other (model retention) if $\theta \notin (T_1, T_2)$. That is, the two states are $\Theta_S = (T_1, T_2)$ and $\Theta_L = (-\infty, T_1) \cup (T_2, +\infty)$, referred to as 'small' (S) and 'large' (L), respectively. Denote $\bar{T} = \frac{1}{2}(T_1 + T_2)$. In most settings that we consider, $T_1 = -T_2$, and so $\bar{T} = 0$, but in the derivations that follow this is never assumed. In any case, such symmetry could be arranged by the linear transformation $\theta' = \theta - \bar{T}$.

11.1.1 Ordinary regression and model uncertainty

Suppose we are committed to a set of covariates with data in a $n \times (K+1)$ matrix $\mathbf{X}$, and we contemplate adding to it a variable with values in a $n \times 1$ vector $\mathbf{z}$. To simplify the discourse, suppose the residual variance σ^2 is known; usually it is estimated. Suppose further that the model based on the $n \times (K+2)$ matrix $\mathbf{Z} = (\mathbf{X} \ \mathbf{z})$ is valid, and so the estimator of the vector of regression coefficients β,

$$\hat{\beta} = \left(\mathbf{Z}^\top \mathbf{Z}\right)^{-1} \mathbf{Z}^\top \mathbf{y},$$

is unbiased, with variance matrix $\mathrm{var}(\hat{\beta}) = \sigma^2 (\mathbf{Z}^\top \mathbf{Z})^{-1}$. We assume that $\mathbf{Z}$ is nonsingular; that is, no column of $\mathbf{Z}$ can be reproduced by linearly combining the remaining columns of $\mathbf{Z}$. It implies that $\mathbf{X}$ is also nonsingular.

We focus on estimating a linear combination $\theta = \mathbf{c}^\top \beta$ for a given $(K+2) \times 1$ vector of constants $\mathbf{c}$. This includes a wide variety of problems, including estimation of a parameter β_k, which corresponds to an indicator vector $\mathbf{c}$. In this case, $\mathbf{c}$ comprises zeros except for a single unity in location k; $c_k = 1$ and $c_{k'} = 0$ for all $k' \neq k$. We address this problem by regarding retention and exclusion of $\mathbf{z}$ from the model as alternative courses of action. The consequences of committing an error in this problem are variance inflation if $\mathbf{z}$ is inappropriately retained and bias if $\mathbf{z}$ is inappropriately discarded. However, bias smaller than a certain threshold is preferable to variance inflation, so model validity (exclusion of $\mathbf{z}$ if $\beta_{K+2} = 0$) is a misleading imperative. The false dichotomy of β_{K+2} vanishing or being non-zero derails the effort to efficiently estimate θ by barring us from incurring a small bias in exchange for substantial variance reduction.

Denote by $\mathbf{c}_{\mathrm{x}}$ the subvector of the first $K+1$ elements of $\mathbf{c}$, which corresponds to $\mathbf{X}$, and by c_{z} the remaining element, which corresponds to $\mathbf{z}$;
$\mathbf{c} = \begin{pmatrix} \mathbf{c}_{\mathrm{x}} \\ c_{\mathrm{z}} \end{pmatrix} = (\mathbf{c}_{\mathrm{x}}^\top, c_{\mathrm{z}})^\top$. Split the vector $\boldsymbol{\beta}$ compatibly to $(\boldsymbol{\beta}_{\mathrm{x}}^\top, \beta_{\mathrm{z}})^\top$. Further, denote $\hat{\boldsymbol{\beta}}_{\mathrm{x}} = (\mathbf{X}^\top \mathbf{X})^{-1} \mathbf{X}^\top \mathbf{y}$. Note that $\hat{\boldsymbol{\beta}}_{\mathrm{x}}$ in general differs from the subvector of the first $K+1$ elements of $\hat{\boldsymbol{\beta}}$. The vector $\hat{\boldsymbol{\beta}}_{\mathrm{x}}$ would be an unbiased estimator of $\boldsymbol{\beta}_{\mathrm{x}}$ if variable $\mathbf{z}$ were redundant.

The two candidate models, one based of $\mathbf{X}$ and the other on $\mathbf{Z}$, are associated with two estimators of $\mathbf{c}^\top \boldsymbol{\beta}$: based on $\mathbf{X}$, the estimator is $\mathbf{c}_{\mathrm{x}}^\top \hat{\boldsymbol{\beta}}_{\mathrm{x}}$, which is biased unless $\beta_{\mathrm{z}} = 0$, and based on $\mathbf{Z}$ it is $\mathbf{c}^\top \hat{\boldsymbol{\beta}}$, which is unbiased, but has usually a larger (and never smaller) sampling variance than $\mathrm{var}(\mathbf{c}_{\mathrm{x}}^\top \hat{\boldsymbol{\beta}}_{\mathrm{x}})$. We would prefer the estimator that has smaller MSE, so the choice between $\mathbf{c}^\top \hat{\boldsymbol{\beta}}$ and $\mathbf{c}_{\mathrm{x}}^\top \hat{\boldsymbol{\beta}}_{\mathrm{x}}$ is a contest of the squared bias with variance inflation.

Denote

$$\mathbf{E}_{\mathrm{z}|\mathrm{X}} = \left(\mathbf{X}^\top \mathbf{X} \right)^{-1} \mathbf{X}^\top \mathbf{z}$$

$$D = \frac{1}{\mathbf{z}^\top \mathbf{z} - \mathbf{z}^\top \mathbf{X} \left(\mathbf{X}^\top \mathbf{X} \right)^{-1} \mathbf{X}^\top \mathbf{z}} . \tag{11.6}$$

These expressions are familiar from the matrix algebra for OLS, and they recur in the derivations that follow. Nonsingularity of $\mathbf{Z}$ implies that $D > 0$.

Since $\mathbf{c}^\top \boldsymbol{\beta} = \mathbf{c}_{\mathrm{x}}^\top \boldsymbol{\beta}_{\mathrm{x}} + c_{\mathrm{z}} \beta_{\mathrm{z}}$, the bias of $\mathbf{c}_{\mathrm{x}}^\top \hat{\boldsymbol{\beta}}_{\mathrm{x}}$ is

$$\begin{aligned} B_{\mathrm{x}} &= \mathbf{c}_{\mathrm{x}}^\top \left(\mathbf{X}^\top \mathbf{X} \right)^{-1} \mathbf{X}^\top (\mathbf{X} \boldsymbol{\beta}_{\mathrm{x}} + \mathbf{z} \beta_{\mathrm{z}}) - \mathbf{c}^\top \boldsymbol{\beta} \\ &= C \beta_{\mathrm{z}} , \end{aligned}$$

where $C = \mathbf{c}_{\mathrm{x}}^\top \mathbf{E}_{\mathrm{z}|\mathrm{X}} - c_{\mathrm{z}}$. With the model based on $\mathbf{Z}$, the variance of $\mathbf{c}^\top \hat{\boldsymbol{\beta}}$ is inflated by

$$\Delta v = \mathrm{var} \left(\mathbf{c}^\top \hat{\boldsymbol{\beta}} \right) - \mathrm{var} \left(\mathbf{c}_{\mathrm{x}}^\top \hat{\boldsymbol{\beta}}_{\mathrm{x}} \right) = C^2 D \sigma^2 ; \tag{11.7}$$

this expression is derived in Appendix A. If we knew β_{z}, the model choice would be:

- discard $\mathbf{z}$ if $|\beta_{\mathrm{z}}| < \sigma \sqrt{D}$;

- retain $\mathbf{z}$ if $|\beta_{\mathrm{z}}| > \sigma \sqrt{D}$.

The thresholds that divide the two states implied by these choices are $\pm \beta_{\mathrm{z}}^*$, where $\beta_{\mathrm{z}}^* = \sigma \sqrt{D}$. We use them as the thresholds T_1 and T_2 in other problems of choosing between 'small' and 'large' in Chapter 5.

With uncertainty about β_z, we resort to its fiducial distribution. By inverting the sampling-distributional identity

$$\hat{\beta}_z \sim \mathcal{N}\left(\beta_z, D\sigma^2\right),$$

we have the fiducial distribution

$$\beta_z \sim \mathcal{N}\left(\hat{\beta}_z, D\sigma^2\right).$$

Discarding $\mathbf{z}$, and estimating $\mathbf{c}^\top\boldsymbol{\beta}$ by $\mathbf{c}_x^\top\hat{\boldsymbol{\beta}}_x$, is the inferior choice when $\beta_z \notin (-\beta_z^*, \beta_z^*)$. The fiducial probability of this event is

$$Q_x = \int_{-\infty}^{-\beta_z^*} \varphi\left(x; \hat{\beta}_z, \beta_z^*\right) dx + \int_{\beta_z^*}^{+\infty} \varphi\left(x; \hat{\beta}_z, \beta_z^*\right) dx$$

$$= 1 - \Phi(u_2) + \Phi(u_1),$$

where $u_1 = (-\beta_z^* - \hat{\beta}_z)/\beta_z^* = -1 - u$, $u_2 = 1 - u$, and $u = \hat{\beta}_z/(\sigma\sqrt{D})$. The probability Q_x is equal to the expected loss. The condition $Q_x < \frac{1}{2}$, for when the submodel is preferred, is equivalent to

$$\Phi(u+1) - \Phi(u-1) > \tfrac{1}{2},$$

and this inequality is satisfied for $|u| < 0.933$.

This condition for model reduction differs substantially from the conventional model choice based on hypothesis testing, which corresponds to $|u| < 1.960$. Our criterion is derived without any reference to asymptotics. In contrast, the information criteria of Akaike (AIC; Akaike, 1974) and Schwartz (BIC; Schwartz, 1978) apply to the conventional model selection adjustments intended to have good asymptotic properties. Also, they are concerned with model validity, whereas our focus is on efficient estimation, without insisting on it being unbiased. On the flip side, our proposal is for ordinary regression only, whereas the information criteria are much more general.

The information criteria, as well as our proposal, have one distinct drawback, namely that the same dataset is used for model selection and the subsequent (post-selection) estimation. The distribution of the estimator derived by such a two-step process is a *mixture* of the distributions of the two candidate estimators,

$$I_z \mathbf{c}^\top\hat{\boldsymbol{\beta}} + I_x \mathbf{c}_x^\top\hat{\boldsymbol{\beta}}_x, \tag{11.8}$$

where I_z indicates retaining $\mathbf{z}$ and $I_x = 1 - I_z$ indicates dropping $\mathbf{z}$. The properties of such a *selected-model based* estimator (also known as post-selection estimator) are difficult to derive. It is left for an exercise to show, by an example, that such an estimator is biased and its MSE is understated if we pretend that the model was selected a priori. In brief, there is a penalty for selection in terms of inflation of MSE. Unlike the candidate estimators, the mixture in (11.8) is not normally distributed, unless one of the models is selected almost

surely (and the other with zero probability), when $P(I_z = 0)$ is equal to 0 or
1. In fact, examples can be constructed in which the mixture has a bimodal
distribution. The difficulty in establishing the distribution of the mixture is
compounded by the correlation of the selection indicators I_z and I_x with the
candidate estimators $\mathbf{c}^\top \hat{\boldsymbol{\beta}}$ and $\mathbf{c}_x^\top \hat{\boldsymbol{\beta}}_x$. Appendix B gives more background to
mixtures.

In our derivation, we adopted a symmetric piecewise constant loss function.
The symmetry is without contention, because the squared bias in excess of
the variance inflation is as detrimental to estimation as the variance inflation
in excess of the squared bias. However, there are some alternatives to the
constant loss function, the symmetric piecewise linear loss in particular. In
Appendix C we show that adopting it leads to a paradox.

11.1.2 Some related approaches

A limitation of any model selection approach is that it appears to dismiss any
estimator that is not based on a model. From the description of a selected-
model based estimator as a mixture in equation (11.8), it is obvious that the
properties of such an estimator depend on both candidate models. And in a
more complex modelling exercise, with many candidate models, the properties of the estimator based on the selected model depend on *all* the candidate
models, even those that were not involved in any comparisons, but would
be involved in some hypothetical replications of the model-selection process
applied to replicate datasets. This process is described by a multinomial distribution defined on the set of the candidate models.

In inferential statements that follow model selection (post-selection inference), the uncertainty brought about by model selection is commonly ignored.
This amounts to the pretence that the valid and parsimonious model has been
selected correctly with probability 1.0—that the selection process is perfect.
Not only is this probability smaller, but the potential of the models discarded
by selection is dismissed, often without a good justification. An invalid model
is superior for estimation of certain quantities if the squared bias incurred is
smaller than the variance inflation that would result from supplementing the
model to make it valid. Even when the valid model is identified with certainty,
its submodels should be considered, not because they may also be valid, but
because estimation based on them may incur bias that is small in relation to
the reduction of the sampling variance.

Another weakness of model selection is that it cannot incorporate some
information that may be available. The next section gives a generic example
related to the method developed in Section 10.6.1.

11.1.3 Bounded bias

Suppose we have an upper bound $\beta_{z,max}$ on $|\beta_z|$; that is, we are certain that
$|\beta_z| \leq \beta_{z,max}$. If we considered no other estimators of $\mathbf{c}^\top \boldsymbol{\beta}$ than $\mathbf{c}^\top \hat{\boldsymbol{\beta}}$ and

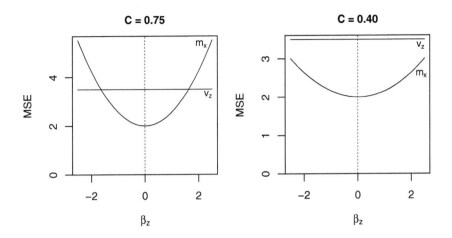

FIGURE 11.1
Mean squared errors of the estimators of $\mathbf{c}^\top\boldsymbol{\beta}$ based on the submodel (m_{x}) and the valid model (v_{z}), plotted over the plausible range for β_{z}.

$\mathbf{c}_{\mathrm{x}}^\top\hat{\boldsymbol{\beta}}_{\mathrm{x}}$, we might choose the estimator that is more efficient in the borderline (extreme plausible) case, when $\beta_{\mathrm{z}}^2 = \beta_{\mathrm{z,max}}^2$. The MSE of $\mathbf{c}^\top\hat{\boldsymbol{\beta}}$, denoted by v_{z}, does not depend on β_{z}, whereas the MSE of the submodel-based estimator, $m_{\mathrm{x}} = \mathrm{MSE}(\mathbf{c}_{\mathrm{x}}^\top\hat{\boldsymbol{\beta}}_{\mathrm{x}} ; \mathbf{c}^\top\boldsymbol{\beta}) = v_{\mathrm{x}} + B_{\mathrm{x}}^2$, is a quadratic function of β_{z}, with its minimum at $\beta_{\mathrm{z}} = 0$. Figure 11.1 presents two scenarios. In both of them $v_{\mathrm{x}} = 2$, $v_{\mathrm{z}} = 3.5$, $\beta_{\mathrm{z,max}} = 2.5$ and $\sigma^2 = 1$, but they differ in the value of $C = \mathbf{c}_{\mathrm{x}}^\top(\mathbf{X}^\top\mathbf{X})^{-1}\mathbf{X}^\top\mathbf{z} - c_{\mathrm{z}}$, given at the top of each panel, which controls how fast the squared bias B_{x}^2 increases with $|\beta_{\mathrm{z}}|$.

In the left-hand panel, the two MSE functions intersect at $\beta_{\mathrm{z}} = \pm1.63$, within the plausible range $(-\beta_{\mathrm{z,max}}, \beta_{\mathrm{z,max}})$ for β_{z}, whereas in the right-hand panel $v_{\mathrm{z}} > m_{\mathrm{x}}$ throughout the plausible range. When $v_{\mathrm{z}} \geq m_{\mathrm{x}}(\beta_{\mathrm{z}})$ for all $\beta_{\mathrm{z}} \in (-\beta_{\mathrm{z,max}}, \beta_{\mathrm{z,max}})$, including $\beta_{\mathrm{z}} = \pm\beta_{\mathrm{z,max}}$, the appropriate choice (the more efficient estimator of $\mathbf{c}^\top\boldsymbol{\beta}$) is $\mathbf{c}_{\mathrm{x}}^\top\hat{\boldsymbol{\beta}}_{\mathrm{x}}$. When $v_{\mathrm{z}} < m_{\mathrm{x}}(\beta_{\mathrm{z,max}})$, we achieve a lower maximum of the MSE with $\mathbf{c}^\top\hat{\boldsymbol{\beta}}$, even though the MSE of $\mathbf{c}_{\mathrm{x}}^\top\hat{\boldsymbol{\beta}}_{\mathrm{x}}$ is smaller for some plausible values of β_{z}.

Instead of selecting the estimator based on the model that fits better, we compose the two estimators with coefficient b that is optimal in the extreme case, when $\beta_{\mathrm{z}}^2 = \beta_{\mathrm{z,max}}^2$. This is a problem of minimising the quadratic function:

$$\mathrm{MSE}\left\{ (1-b)\mathbf{c}^\top\hat{\boldsymbol{\beta}} + b\,\mathbf{c}_{\mathrm{x}}^\top\hat{\boldsymbol{\beta}}_{\mathrm{x}} ; \mathbf{c}^\top\boldsymbol{\beta} \right\}$$
$$= (1-b)^2 v_{\mathrm{z}} + b^2\left(v_{\mathrm{x}} + B_{\mathrm{x}}^2\right) + 2b(1-b)\,\mathrm{cov}\left(\mathbf{c}^\top\hat{\boldsymbol{\beta}}, \mathbf{c}_{\mathrm{x}}^\top\hat{\boldsymbol{\beta}}_{\mathrm{x}}\right).$$

Owing to the identity $\text{cov}(\mathbf{c}^\top \hat{\boldsymbol{\beta}}, \mathbf{c}_x^\top \hat{\boldsymbol{\beta}}_x) = v_x$, this reduces to

$$b^2 \left(\Delta v + B_x^2 \right) - 2b\Delta v + v_z = v_x + (1-b)^2 \Delta v + b^2 B_x^2 ,$$

where $\Delta v = v_z - v_x = C^2 D \sigma^2$ and $B_x = C\beta_z$; see Appendix A. This function attains its minimum at

$$b^*_{mx} = \frac{\Delta v}{\Delta v + B_{x,max}^2} = \frac{1}{1 + \dfrac{\beta_{z,max}^2}{D\sigma^2}} ,$$

and the minimum attained with this coefficient is

$$v_z - \frac{(\Delta v)^2}{\Delta v + B_{x,max}^2} = v_x + \frac{B_{x,max}^2 \, \Delta v}{B_{x,max}^2 + \Delta v} .$$

Recall that $\beta_z^* = \sigma\sqrt{D}$ is the (positive) slope for which the estimators $\mathbf{c}^\top \hat{\boldsymbol{\beta}}$ and $\mathbf{c}_x^\top \hat{\boldsymbol{\beta}}_z$ have the same MSE for $\mathbf{c}^\top \boldsymbol{\beta}$; in this case $B_x^2 = \Delta v$. Denote $r = \beta_z^2 / \beta_z^{*\,2}$ and $r_{mx} = \beta_{z,max}^2 / \beta_z^{*\,2}$, so that $b_{mx}^* = 1/(1 + r_{mx})$. The MSE of the composition with this coefficient for arbitrary β_z is

$$v_x + \Delta v \, \frac{r + r_{mx}^2}{(1 + r_{mx})^2} . \tag{11.9}$$

This is an (increasing) linear function of r. Its slope is $\Delta v/(1 + r_{mx})^2$. In the plausible range of β_z, where $0 < r < 1$, this MSE is between v_x and v_z.

If a particular value of $\beta_{z,max}$ is an upper bound for β_z, then so is any larger value. However, there are some rewards for setting $\beta_{z,max}$ as small as possible, so long as it remains an upper bound. Declaring a smaller upper bound $\beta_{z,max}$ results in a uniformly smaller MSE for $|\beta_z| < \beta_{z,max}$. An undesirable consequence of a smaller upper bound is that the MSE increases faster with β_z^2. Therefore, dishonesty (or being misguided) in setting $\beta_{z,max}$ too low is punished more harshly the smaller the value of $\beta_{z,max}$. However, some small error is forgiven.

To explore the value of setting a tighter upper bound $\beta_{z,max}$, suppose r_2 is an appropriate upper bound for $r = \beta_z^2 / \beta_z^{*\,2}$ and $r_1 < r_2$ is not. That is, some values of r in the range (r_1, r_2) are plausible. We explore when the estimator based on r_1 is more efficient than the estimator based on r_2. The expression for the MSE in equation (11.9), in which v_x and Δv are constants, implies that this condition is equivalent to

$$\frac{r + r_1^2}{(1 + r_1)^2} < \frac{r + r_2^2}{(1 + r_2)^2} .$$

By elementary rearrangement we obtain the equivalent inequality

$$r < \frac{r_1 + r_2 + 2r_1 r_2}{r_1 + r_2 + 2} = r_1 + \frac{(r_2 - r_1)(r_1 + 1)}{r_1 + r_2 + 2} . \tag{11.10}$$

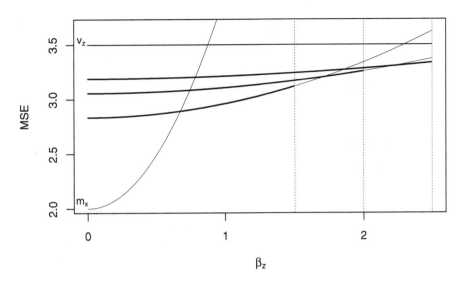

FIGURE 11.2

Mean squared errors of the estimators of $c^\top \beta$ based on the composition optimal for the largest plausible value of β_z; $\beta_{z,\max}$ set to 1.5, 2.0 and 2.5.

Apart from $r \leq r_1$ this is satisfied also for some values $r > r_1$, but not for $r = r_2$, and neither for $r = \frac{1}{2}(r_1 + r_2)$, so the threshold, up to where the estimator that is optimal at r_1 is more efficient, is below $\frac{1}{2}(r_1 + r_2)$.

Figure 11.2 illustrates this approach on the setting with $v_x = 2.0$, $v_z = 3.5$ and $C^2 \beta_z^{*2} = 1.96$. We consider three values of $\beta_{z,\max}$, 1.5, 2.0 and 2.5, to explore the importance of this factor. The MSEs of the two basis estimators, $c^\top \hat{\beta}$ and $c_x^\top \hat{\beta}_x$, are indicated as m_x and v_z. All the MSEs we consider are symmetric functions of β_z, so the plot can be restricted to $\beta_z > 0$; this restriction improves the resolution of the plot. Clearly, selecting $c_x^\top \hat{\beta}_x$ is unwise because m_x, its MSE, is very large for some plausible values of β_z; for example, it is equal to 5.92 at $\beta_z = 2.0$.

The MSE functions for the three settings of $\beta_{z,\max}$ are drawn by solid lines up to the assumed upper bound (1.5, 2.0 and 2.5), and are continued by hairlines. For the largest value, $\beta_{z,\max} = 2.5$, the estimator is uniformly more efficient than $c^\top \hat{\beta}$, but its performance for small values of β_z is not much better.

By setting $\beta_{z,\max}$ to 2.0, we obtain an estimator more efficient in almost the entire range of plausible values of β_z. The MSE would be only slightly greater for β_z close to 2.5. By setting $\beta_{z,\max}$ to 1.5, we make greater gains *if* β_z is indeed smaller than this assumed upper bound, but the MSE is greater than with the more conservative setting of $\beta_{z,\max} = 2.5$ for $\beta_z > 1.86$. In fact, the estimator is less efficient than $\mathbf{c}^\top \hat{\boldsymbol{\beta}}$ for $\beta_z > 2.30$.

In summary, integrity in the specification of the upper bound $\beta_{z,\max}$ is important, but understating its value by a small margin is not punished severely. However, a substantial understatement, for example, by setting $\beta_{z,\max} = 1.5$ when $\beta_{z,\max} = 2.5$ would be appropriate, exposes us to the risk of large MSE. The MSE may then exceed even v_z, the variance of the unbiased estimator.

11.2   Composition

Selection between two estimators or, more generally, from a larger set of estimators, can be regarded as a search for an optimal estimator in a set defined by a basis and a class of operations on the elements of this basis. We use the term 'basis' as it is used in algebra. For example, a K-dimensional linear space is defined by a set of $K \times 1$ vectors that are not linearly dependent, and the operations of vector addition and scalar multiplication. The space is defined as all the elements that can be generated from the basis by these operations. We, and our more distant colleagues in other branches of mathematics, like to work with linear spaces because they are tractable.

The class of estimators that arise in model selection is defined by a family of selection processes, and they correspond to mixtures; see Section 7.3 and Appendix B of this chapter. The basis is defined by the models that we consider or, more precisely, by the corresponding estimators. This viewpoint clarifies the difficulty—identifying a process that would yield the optimal estimator, one with the smallest MSE. Establishing the properties of a mixture is difficult because mixture is an inherently nonlinear operation. Finding an element in this space that has some desirable properties is impossible without an intimate understanding of mixtures, and that we are lacking. The problem is compounded by the dependence (correlation) of the selection indicator (which model is selected) with the estimators.

As an alternative, we consider composition. For two estimators, $\hat{\theta}_0$ and $\hat{\theta}_1$, of the same target θ, the family of compositions is defined as all the linear combinations $\tilde{\theta} = (1 - b)\hat{\theta}_1 + b\hat{\theta}_0$ for any real b. We write $\tilde{\theta}(b)$ to indicate the specific value of b on which $\tilde{\theta}$ is based. Convex combinations, which correspond to $b \in [0, 1]$, are a subset of these compositions. Denote the respective sampling variances of $\hat{\theta}_0$ and $\hat{\theta}_1$ by V_0 and V_1 and their covariance by ρ_{01}. We assume that V_0, V_1 and ρ_{01} are known and that $\hat{\theta}_1$ is unbiased, being based on the widest model that is contemplated; $\hat{\theta}_0$ is based on a submodel. The bias of $\hat{\theta}_0$ is denoted by B_0. The estimators may have arbitrary distributions; in particular, they need not be normally distributed. Nor do they have to be derived from or be related to any models.

The principal advantage of composition over selection, and the mixtures it entails, is that the properties of composite estimators are easy to explore. Composition $\tilde{\theta}$ has bias $E\{\tilde{\theta}(b)\} - \theta = bB_0$ and variance,

$$\text{var}\left(\tilde{\theta}\right) = (1-b)^2 V_1 + 2b(1-b)\rho_{01} + b^2 V_0.$$

The MSE of $\tilde{\theta}$, equal to $\text{var}(\tilde{\theta}) + b^2 B_0^2$, is a quadratic function of the coefficient b. The MSE is nonnegative for all b, so it has a unique minimum, except when the MSE is constant, when $V_0 = V_1 = \rho_{01}$. This case, when the two estimators are perfectly correlated and differ only by a constant, is of no importance, so we discount it. The minimum is attained for

$$b^* = \frac{V_1 - \rho_{01}}{V_0 + V_1 - 2\rho_{01} + B_0^2}. \tag{11.11}$$

In practice, b^* has to be estimated because B_0 is unknown. If its value were established, the ideal estimator $\hat{\theta}(b^*)$ would have MSE:

$$V_1 - \frac{(V_1 - \rho_{01})^2}{V_0 + V_1 - 2\rho_{01} + B_0^2} = V_0 + B_0^2 - \frac{(V_0 - \rho_{01} + B_0^2)^2}{V_0 + V_1 - 2\rho_{01} + B_0^2}. \tag{11.12}$$

This is smaller than both $V_0 + B_0^2$ and V_1. But we are bound to lose some efficiency (incur some inflation of the MSE) by using an estimator $\hat{b}^*$ in place of the optimal constant b^*.

We apply this method to estimating the linear combination $\mathbf{c}^\top\boldsymbol{\beta}$ where $\boldsymbol{\beta}$ is the $(K+2) \times 1$ vector of regression parameters. We consider the same two models as in the problem of selection treated in Section 11.1.1. From there we have the following identities:

$$V_0 = \sigma^2 \mathbf{c}_x^\top \left(\mathbf{X}^\top\mathbf{X}\right)^{-1}\mathbf{c}_x$$

$$V_1 = \sigma^2 \mathbf{c}^\top \left(\mathbf{Z}^\top\mathbf{Z}\right)^{-1}\mathbf{c}$$

$$= V_0 + C^2 D\sigma^2$$

$$\rho_{01} = \mathbf{c}^\top \left(\mathbf{Z}^\top\mathbf{Z}\right)^{-1}\mathbf{Z}^\top\mathbf{X}\left(\mathbf{X}^\top\mathbf{X}\right)^{-1}\mathbf{c}_x = V_0, \tag{11.13}$$

and $B_0 = C\beta_z$. The optimal coefficient is

$$b^* = \frac{\sigma^2 DC^2}{\sigma^2 DC^2 + \beta_z^2 C^2}$$

$$= \frac{1}{1 + \dfrac{\beta_z^2}{D\sigma^2}}.$$

The corresponding MSE, derived by substituting the identities in (11.13) in both sides of equation (11.12), is

$$\sigma^2\left\{\mathbf{c}^\top\left(\mathbf{Z}^\top\mathbf{Z}\right)^{-1}\mathbf{c} - \frac{D^2C^2}{D\sigma^2 + \beta_z^2}\right\} = \sigma^2\left\{\mathbf{c}_x^\top\left(\mathbf{X}^\top\mathbf{X}\right)^{-1}\mathbf{c}_x + \frac{D\sigma^2\beta_z^2}{D\sigma^2 + \beta_z^2}C^2\right\}.$$

So, we would estimate $\mathbf{c}^\top \boldsymbol{\beta}$ optimally by the composition

$$\tilde{\theta}(b^*) = (1 - b^*)\,\mathbf{c}^\top \hat{\boldsymbol{\beta}} + b^* \mathbf{c}_{\mathbf{x}}^\top \hat{\boldsymbol{\beta}}_{\mathbf{x}}\,,$$

if the bias $\beta_{\mathbf{z}}$, on which b^* depends, were known. Of course, $\beta_{\mathbf{z}}$ has to be estimated. It may be estimated efficiently by $\hat{\beta}_{\mathbf{z}}$, but the nonlinear transformation
involved in b^* given by (11.11) undermines this efficiency.

Since $\mathrm{var}(\hat{\beta}_{\mathbf{z}}) = D\sigma^2$, we can absorb the variance $D\sigma^2$ within the estimator
$\hat{\beta}_{\mathbf{z}}$ and the target $\beta_{\mathbf{z}}$. The ideal composite estimator is

$$\begin{aligned}
\tilde{\theta} &= \hat{\theta}_0 + b^* \left(\hat{\theta}_1 - \hat{\theta}_0 \right) \\
&= \hat{\theta}_0 + C b^* \hat{\beta}_{\mathbf{z}}\,,
\end{aligned}$$

and no generality is lost by assuming that $\hat{\theta}_0$ and $\hat{\theta}_1 - \hat{\theta}_0$ are independent.
This makes the problem essentially univariate, to estimate $1/(1 + \beta_{\mathbf{z}}^2)$ using
an unbiased estimator $\hat{\beta}_{\mathbf{z}}$ with unit variance. The factor C^2 can be treated
similarly when we have an unbiased estimator of $\beta_{\mathbf{z}}$ with unit variance. We
assume now that this estimator is normally distributed. We consider the class
of shrinkage estimators $\tilde{b}(\gamma) = 1/(1 + \gamma\hat{\beta}_{\mathbf{z}}^2)$, where γ is a positive constant. We
identify the value of γ for which $\tilde{b}$ has the smallest MSE.

The problem is not tractable analytically, so we apply simulations. For
each value $\beta_{\mathbf{z}}$ on a grid of values $0.0, 0.2, \ldots, 3.0$ and γ set to $0.01, 0.02, \ldots, 1$,
we simulate a random sample of large size H of values of $\hat{\beta}_{\mathbf{z}}$, drawn from
$\mathcal{N}(\beta_{\mathbf{z}}, 1)$, and evaluate the empirical MSE of $\tilde{b}(\gamma)$, defined as

$$\widehat{\mathrm{MSE}}\left\{ \tilde{b}(\gamma); b^* \right\} = \frac{1}{H} \sum_{h=1}^{H} \left(\frac{1}{1 + \gamma\hat{\beta}_{\mathbf{z}}^2} - \frac{1}{1 + \beta_{\mathbf{z}}^2} \right)^2.$$

Denote this function by $G(\gamma; b^*)$. For each b^* we find the minimum of G
and explore the function defined by these minima. We seek a single value of
γ, denoted by γ^* for which $G(\gamma^*; b)$ is as close as can be arranged to the
pointwise minima.

Another class of shrinkage estimators is given by $(0.4 + \eta)/(1 + \hat{\beta}_{\mathbf{z}}^2)$. We
conduct for it simulations by the same approach. The constant 0.4 is added
to simplify the comparisons we make below. Figure 11.3 plots the root-MSEs
of the two classes of shrinkage estimators with respective coefficients γ and η
on the horizontal axis. We are at liberty to choose one coefficient that would
be used universally. The estimator should be as efficient as possible, but it
cannot be efficient uniformly, for all values of $\beta_{\mathbf{z}}$.

The diagram is not well suited for the task of selecting the best shrinkage.
The curves are 'reorganised' in Figure 11.4, where a panel each compares the
estimators within the two classes; with shrinkage in the denominator in the
left-hand panel, and in the numerator in the right-hand panel.

For the estimators with shrinkage in the denominator, the optimal coefficient is $\gamma^* = 0.52$ for which the root-MSEs are in the range $0.211 - 0.329$.

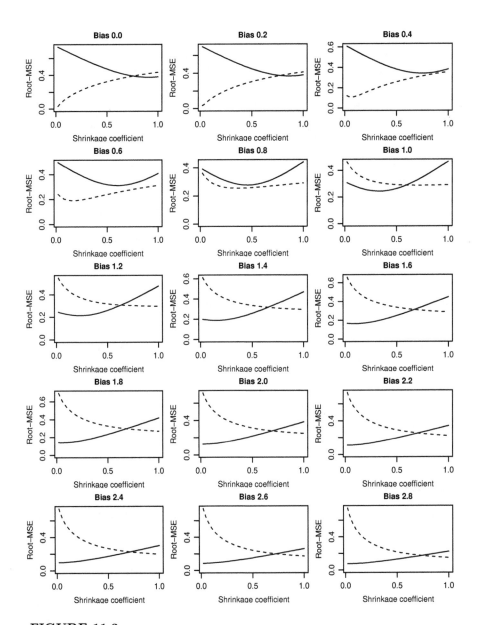

FIGURE 11.3

Root-MSEs of the shrinkage estimators of the coefficient $1/(1 + \beta_z^2)$ for values of β_z given in the panel titles. Solid line – shrinkage in the denominator; dashes – shrinkage in the numerator.

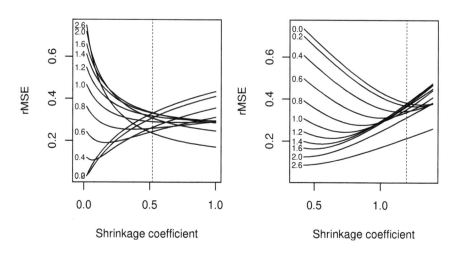

FIGURE 11.4
Root-MSEs of the estimators of $1/(1 + \beta_z^2)$; shrinkage in the denominator (left-hand panel) and of the numerator (right-hand panel).

With this choice, the root-MSE is smaller than or equal to 0.329. This upper bound is greater for any other choice of γ. For the estimators with shrinkage in the numerator, the optimal shrinkage is $\eta = 0.8$ (1.2 on the horizontal axis in the right-hand panel) and the corresponding root-MSEs are in the range 0.184 – 0.381. The upper bound is greater, so we prefer to apply shrinkage in the denominator, with $\gamma^* = 0.52$.

11.3   Composition of a complete set of candidate models

In this section, we develop and explore the properties of a composite estimator for the setting of an ordinary regression with K covariates (in addition to the intercept), in which every one of the 2^K models formed by the whole set and any subset of the covariates is a candidate. The target is a linear combination of the regression parameters, $\theta = \mathbf{c}^\top \boldsymbol{\beta}$. To simplify the discourse, we assume that each covariate is either dichotomous or ordinal (in common parlance, continuous), so that every covariate is represented by one column of the regression matrix $\mathbf{X}$. From the derivations that follow we conclude that, with an approximation, we can reduce our attention to the composition of the model with all the covariates and the simplest model, which contains only the intercept. Then we search for some improvements on this composition.

We assume that the model with all the K covariates is valid, although it may contain some redundant covariates—it need not be parsimonious. As a basis for composition, the 2^K models contain a lot of redundancy. An equivalent basis that has no redundancy is formed by the models that contain no covariate, covariate 1, covariates 1 and 2, and so on, till the model with all the covariates $1, 2, \ldots, K$. Denote this increasing sequence of models as $M_0, M_1, \ldots, M_K$, and the OLS estimators based on them as $\hat{\theta}_k = \mathbf{c}^\top \hat{\boldsymbol{\beta}}_{[k]}$, $k = 0, \ldots, K$. That is, the covariates are in a particular order, such as their position as columns in $\mathbf{X}$, and model M_k contains the first $k + 1$ columns of $\mathbf{X}$. No generality is lost by assuming that the variables in $\mathbf{X}$ are orthogonal.

The OLS estimator of θ based on a single covariate $k > 1$ (and the intercept) can be composed from the basis as $\hat{\theta}_0 + \hat{\theta}_k - \hat{\theta}_{k-1}$. The coefficients of this combination are 1, 1 and -1, so it is a composition. Suppose the basis can be used for composing the OLS estimator $\hat{\theta}$ based on a set of covariates that do not include covariate k. Then the OLS estimator with this covariate added to the model can be composed as $\hat{\theta} + \hat{\theta}_k - \hat{\theta}_{k-1}$. This proves, by mathematical induction, that $M_0, M_1, \ldots, M_K$ generates by composition the same space of estimators as the set of all 2^K estimators.

Denote by $\hat{\boldsymbol{\theta}} = (\hat{\theta}_0, \hat{\theta}_1, \ldots, \hat{\theta}_K)^\top$ the vector of estimators in the basis, and by $\mathbf{b}$ and $\mathbf{V}$ their vector of biases for θ and variance matrix, respectively. Model M_K is valid, so $\hat{\theta}_K$ is unbiased and its variance, as well as its MSE for $\mathbf{c}^\top \boldsymbol{\beta}$, is $V_K = \sigma^2 \mathbf{c}^\top (\mathbf{X}^\top \mathbf{X})^{-1} \mathbf{c}$. By removing a covariate k from the model, the bias of the resulting OLS estimator is changed by adding $c_k \beta_k$ to it. By such a model reduction, the variance is *always* reduced, whereas the squared bias may move in either direction. In particular, by removing a covariate that is not redundant, we may reduce both the squared bias *and* the variance. This suggests that some invalid models may yield more efficient estimators of θ than the parsimonious valid model.

We search for an efficient estimator of $\theta = \mathbf{c}^\top \boldsymbol{\beta}$ in the space of estimators spanned by the basis $\hat{\boldsymbol{\theta}} = (\hat{\theta}_0, \hat{\theta}_1, \ldots, \hat{\theta}_K)^\top$, that is,

$$\tilde{\theta} = \mathbf{d}^\top \hat{\boldsymbol{\theta}}, \tag{11.14}$$

where $\mathbf{d}$ is a vector of constants such that $\tilde{\theta}(\mathbf{d})$ is a composition; that is, $\mathbf{d}^\top \mathbf{1}_{K+1} = 1$. In the derivations that follow, we drop the subscript $K + 1$ for **1**. The bias of $\tilde{\theta}$ in estimating θ is $\mathrm{B}(\tilde{\theta}; \theta) = \mathbf{d}^\top \mathbf{b}$, and its sampling variance is $\mathrm{var}(\tilde{\theta}) = \mathbf{d}^\top \mathbf{V} \mathbf{d}$. Hence

$$\mathrm{MSE}\left\{\tilde{\theta}(\mathbf{d}); \theta\right\} = \mathbf{d}^\top \mathbf{V} \mathbf{d} + \left(\mathbf{d}^\top \mathbf{b}\right)^2. \tag{11.15}$$

This is a quadratic function of $\mathbf{d}$, so its minimum is unique and can be found by the method of Lagrange multipliers that enforces the constraint $\mathbf{d}^\top \mathbf{1} = 1$. The differential of the function $m(\mathbf{d}) = \mathrm{MSE}\{\tilde{\theta}(\mathbf{d}); \theta\} + \lambda(\mathbf{d}^\top \mathbf{1} - 1)$ is

$$\frac{\partial m}{\partial \mathbf{d}} = 2\left(\mathbf{V} + \mathbf{b}\mathbf{b}^\top\right)\mathbf{d} + \lambda \mathbf{1},$$

so the solution, denoted by $\mathbf{d}^*$, has to be a scalar multiple of $(\mathbf{V} + \mathbf{b}\mathbf{b}^\top)^{-1}\mathbf{1}$. The constraint $\mathbf{d}^\top\mathbf{1} = 1$ implies that

$$\mathbf{d}^* = \frac{1}{\mathbf{1}^\top \left(\mathbf{V} + \mathbf{b}\mathbf{b}^\top\right)^{-1} \mathbf{1}} \left(\mathbf{V} + \mathbf{b}\mathbf{b}^\top\right)^{-1} \mathbf{1}. \tag{11.16}$$

Numerical inversion of $\mathbf{V} + \mathbf{b}\mathbf{b}^\top$, which may be a large matrix, can be avoided by applying the identity

$$\left(\mathbf{V} + \mathbf{b}\mathbf{b}^\top\right)^{-1} = \mathbf{V}^{-1} - \frac{1}{1 + \mathbf{b}^\top\mathbf{V}^{-1}\mathbf{b}} \mathbf{V}^{-1}\mathbf{b}\mathbf{b}^\top\mathbf{V}^{-1}. \tag{11.17}$$

There are several ways of proving this identity; see Appendix A. We show below that $\mathbf{V}$ has a particular pattern and its inverse $\mathbf{V}^{-1}$ has a closed form.

Denote $B_0 = \mathbf{1}^\top\mathbf{V}^{-1}\mathbf{1}$, $B_1 = \mathbf{b}^\top\mathbf{V}^{-1}\mathbf{1}$, $B_2 = \mathbf{b}^\top\mathbf{V}^{-1}\mathbf{b}$, and further, $B = B_0(1 + B_2) - B_1^2$. The expression for the inverse of $\mathbf{V} + \mathbf{b}\mathbf{b}^\top$ leads to

$$\begin{aligned}
\mathbf{d}^* &= \left(B_0 - \frac{B_1^2}{1 + B_2}\right)^{-1} \left(\mathbf{V}^{-1}\mathbf{1} - \frac{B_1}{1 + B_2}\mathbf{V}^{-1}\mathbf{b}\right) \\
&= \frac{(1 + B_2)\mathbf{V}^{-1}\mathbf{1} - B_1\mathbf{V}^{-1}\mathbf{b}}{B}.
\end{aligned}$$

By substituting $\hat{\theta}_K\mathbf{1} + \hat{\mathbf{b}}$ for $\hat{\boldsymbol{\theta}}$ we obtain

$$\mathbf{d}^{*\top}\hat{\boldsymbol{\theta}} = \hat{\theta}_K + \frac{(1 + B_2)\mathbf{1}^\top\mathbf{V}^{-1}\hat{\mathbf{b}} - B_1\mathbf{b}^\top\mathbf{V}^{-1}\hat{\mathbf{b}}}{B}.$$

If the constants B_0, B_1 and B_2 were known, we would have from equation (11.15)

$$\begin{aligned}
\mathrm{MSE}\left(\mathbf{d}^{*\top}\hat{\boldsymbol{\theta}};\theta\right) &= \frac{\left\{(1 + B_2)\mathbf{1}^\top - B_1\mathbf{b}^\top\right\}\mathbf{V}^{-1}\{(1 + B_2)\mathbf{1} - B_1\mathbf{b}\}}{B^2} + \frac{B_1^2}{B^2} \\
&= \frac{1 + B_2}{B}. \tag{11.18}
\end{aligned}$$

This expression understates $\mathrm{MSE}(\tilde{\theta};\theta)$ because it is based on knowing the constants B_0, B_1 and B_2. The remaining problem is that the vector of biases $\mathbf{b}$ is unknown, although its elements are estimated without bias by the differences $\hat{\theta}_k - \hat{\theta}_K$. We deal with this problem by replacing $\mathbf{b}$ with $\hat{\mathbf{b}}$ throughout, including in B_1 and B_2. This leads to the estimator

$$\tilde{\theta} = \hat{\theta}_K + \frac{\hat{B}_1}{\hat{B}_0\left(1 + \hat{B}_2\right) - \hat{B}_1^2}. \tag{11.19}$$

Next, we obtain expressions for B_0, B_1 and B_2 that exploit the association of the estimators in the basis. Since the models in the basis form an increasing sequence, $M_0 \subset M_1 \subset \ldots \subset M_K$,

$$\mathrm{cov}\left(\hat{\theta}_k, \hat{\theta}_h\right) = \mathrm{var}\left(\hat{\theta}_{\min(h,k)}\right). \tag{11.20}$$

The simplest way of proving this identity is by assuming that the columns of $\mathbf{X}$ are orthogonal, that $\mathbf{X}^{\top}\mathbf{X}$ is a diagonal matrix. No generality is lost by this assumption because orthogonality can be arranged by linear transformations (Gram-Schmidt orthogonalisation), and the models and estimators associated with the transformed matrix $\mathbf{X}$ are not altered.

Thus, the variance matrix $\mathbf{V}$ has $K + 1$ distinct elements $v_0 < v_1 < \ldots < v_K$, where $v_k = \mathrm{var}(\hat{\theta}_k)$. The smallest variance, v_0, appears in the first row and column; the second smallest, v_1, throughout the second row and column, except for elements $(1,2)$ and $(2,1)$; v_K appears only in the element $(K + 1, K + 1)$. The inverse of such a matrix is tridiagonal. Before giving an example, we introduce some notation. Let $\Delta u_0 = 1/v_0$, $\Delta u_{K+1} = 0$, and $\Delta u_k = 1/(v_k - v_{k-1})$ for $k = 1, \ldots, K$. Then $\mathbf{V}^{-1}$ has diagonal entries $V_{kk} = \Delta u_k + \Delta u_{k-1}$, $k = 1, \ldots, K + 1$, and entries on the first off-diagonal $V_{k,k+1} = V_{k+1,k} = -\Delta u_k$, $k = 1, \ldots, K$. For example, for $K = 4$, the inverse of the 5×5 matrix

$$\mathbf{V} = \begin{pmatrix} v_0 & v_0 & v_0 & v_0 & v_0 \\ v_0 & v_1 & v_1 & v_1 & v_1 \\ v_0 & v_1 & v_2 & v_2 & v_2 \\ v_0 & v_1 & v_2 & v_3 & v_3 \\ v_0 & v_1 & v_2 & v_3 & v_4 \end{pmatrix}$$

is

$$\mathbf{V}^{-1} = \begin{pmatrix} \frac{1}{v_0} + \Delta u_1 & -\Delta u_1 & 0 & 0 & 0 \\ -\Delta u_1 & \Delta u_1 + \Delta u_2 & -\Delta u_2 & 0 & 0 \\ 0 & -\Delta u_2 & \Delta u_2 + \Delta u_3 & -\Delta u_3 & 0 \\ 0 & 0 & -\Delta u_3 & \Delta u_3 + \Delta u_4 & -\Delta u_4 \\ 0 & 0 & 0 & -\Delta u_4 & \Delta u_4 \end{pmatrix}. \tag{11.21}$$

The row totals of $\mathbf{V}^{-1}$ vanish except for the first row, which has the total $1/v_0$. So, $\mathbf{V}^{-1}\mathbf{1} = (1/v_0, 0, \ldots, 0)^{\top}$, and therefore $B_0 = 1/v_0$ and $B_1 = b_0/v_0$. Further, $B_2 = R + b_0^2/v_0$, where $R = \Delta b_1^2 \Delta u_1^2 + \cdots + \Delta b_K^2 \Delta u_K$ and $\Delta b_k = b_k - b_{k-1}$. Substituting these expressions in equation (11.19) yields the formula

$$\tilde{\theta} = \hat{\theta}_K + \frac{\hat{b}_0}{1 + \hat{R}}.$$

Owing to the identity $\hat{\theta}_0 = \hat{\theta}_K + \hat{b}_0$, equivalent expressions are

$$\tilde{\theta} = \hat{\theta}_0 - \frac{\hat{R}\hat{b}_0}{1 + \hat{R}} = \frac{\hat{\theta}_0 + \hat{R}\hat{\theta}_K}{1 + \hat{R}}.$$

Setting aside the one instance of approximation in equation (11.19), we conclude that the optimal composition of the $K + 1$ basis estimators in $\hat{\theta}$ is a composition of merely two basis estimators, one based on the simplest model, and the other on the valid model. Admittedly, R involves the biases of all the basis estimators, but R can now be estimated directly, for the composition of $\hat{\theta}_0$ and $\hat{\theta}_K$. We should be concerned only about the composition, not about R, which is an intermediate quantity.

This result takes us back to the composition of two estimators, one based on the model M_K with all the covariates considered, and the other with no covariates, M_0. As a function of R, this estimator, $\tilde{\theta} = (\hat{\theta}_0 + R\hat{\theta}_K)/(1 + R)$, has MSE

$$\frac{1}{(1 + R)^2} \left(v_0 + 2Rv_0 + R^2 v_K + b_0^2 \right).$$

Its minimum is found as the root of the differential of its logarithm,

$$\frac{\partial \log \left\{ \mathrm{MSE} \left(\tilde{\theta}; \theta, R \right) \right\}}{\partial R} = 2 \left(\frac{v_0 + R v_K}{v_0 + b_0^2 + 2Rv_0 + R^2 v_K} - \frac{1}{R + 1} \right).$$

It yields the coefficient

$$R^* = \frac{b_0^2}{v_K - v_0},$$

which can be interpreted as a 'contest' between the squared bias and the variance reduction. If $\hat{\theta}_0$ is unbiased, then $\tilde{\theta} = \hat{\theta}_0$, and $\hat{\theta}_K$ is not useful. In contrast, if the squared bias b_0^2 is much greater than the variance reduction, then $\hat{\theta}_0$ is not useful, as its weight in the composition, $1/(1 + R^*)$, is small. In this case, we cannot improve substantially on the large-variance unbiased estimator $\hat{\theta}_K$. The MSE of the composition based on R^* is

$$\mathrm{MSE} \left(\tilde{\theta}; \theta, R^* \right) = \frac{b_0^2 v_K + v_0 (v_K - v_0)}{b_0^2 + v_K - v_0}. \tag{11.22}$$

In practice, R^* has to be estimated, and so this MSE is somewhat optimistic, and so is its estimator. The variances v_K and v_0 are scalar multiples of σ^2, such as $u\sigma^2$, where $u = \mathbf{c}^\top (\mathbf{X}^\top \mathbf{X})^{-1}\mathbf{c}$. By absorbing σ^2 in b_0, writing

$$R^* = \frac{a_0^2}{u_K - u_0},$$

where $a_0 = b_0/\sigma$, $u_K = v_K/\sigma^2$ and $u_0 = v_0/\sigma^2$, we convert the model selection problem into a unidimensional affair, estimating the constant a_0^2. This enables us to assess how much the naïve estimator of the MSE in equation (11.22) underestimates the MSE of the estimator $\tilde{\theta}(\hat{R}^*)$ and, indeed, how to estimate R^* to achieve efficiency of $\tilde{\theta}(\hat{R}^*)$. Since $\tilde{\theta}$ is a nonlinear function of R^*, this task is not equivalent to efficient estimation of R^*, and certainly not to efficient estimation of the bias b_0. There is no analytical solution, but a simulation study can be devised which explores estimation of $b_0/(1 + R)$.

In a practical problem, there is no argument about the estimator $\hat{\theta}_K$, or the underlying model M_K, although at the design stage the list of covariates to be recorded may (and always should) be compiled with great care, balancing their relevance with feasibility and cost of recording them. The 'empty' model M_0 may be replaced by a model with one or a few a priori important covariates. In hindsight, this suggests an ordering of the covariates, in the descending order of their a priori relevance. Dropping a few models from the original basis then corresponds to truncating the basis from the left, to $M_{K_0}, M_{K_0+1}, \ldots, M_K$, and retaining the most important (principal) variables $X_1, \ldots, X_{K_0}$ in every contemplated model.

Larger space of composite estimators implies a more efficient composition. This suggests that we should not attempt to reduce the basis $\hat{\boldsymbol{\theta}}$. However, the optimum in the space spanned by $K + 1$ estimators is close to the optimum in the substantially smaller space spanned by $(\hat{\theta}_0, \hat{\theta}_K)$. Thus, there is little value in speculating that some of the intermediate models might be excluded a priori. In contrast, dropping $\hat{\theta}_0$, or more generally the first $K_0 < K$ models may be useful. This is equivalent to including the first K_0 covariates in the model unconditionally, according them the same status as we have done to the intercept in the original formulation.

Such a reduction results in a substantially more efficient composition when the 'new' simplest-model based estimator $\hat{\theta}_{K_0}$ has a small bias and its variance is not much greater than for the (discarded) model with no covariates. Suppose $K_0 = 1$. Then we have to resolve whether, and in what circumstances, is the MSE in equation (11.22) reduced when b_0 and v_0 are replaced by the respective bias and variance b_1 and v_1 of a different estimator $\hat{\theta}_1$, based on a model M_1 wider than M_0. By rearranging the terms in the inequality

$$\frac{a_0^2 u_K + u_0 (u_K - u_0)}{a_0^2 + u_K - u_0} > \frac{a_1^2 u_K + u_1 (u_K - u_1)}{a_1^2 + u_K - u_1},$$

we obtain the condition

$$a_0^2 \nu - \frac{a_1^2}{\nu} > u_1 - u_0, \tag{11.23}$$

where $\nu = (u_K - u_1)/(u_K - u_0)$. Note that ν is a known constant, a function of $\mathbf{c}$, $\mathbf{X}$ and the models involved, M_0, M_1 and M_K. Since $u_K \geq u_1 \geq u_0$, $0 \leq \nu \leq 1$. The condition in equation (11.23) implies that if $a_0^2 < (u_1 - u_0)/\nu$, then search for an intermediate model M_1 is futile; even if $\hat{\theta}_1$ were unbiased ($a_1 = 0$), its composition with $\hat{\theta}_K$ would be less efficient than with $\hat{\theta}_0$. This result is not surprising: if the mean $\hat{\theta}_0$ has a small bias for θ, then it is nearly efficient because every other basis estimator has variance greater than $\text{var}(\hat{\theta}_0)$.

11.3.1   Summary

In this chapter, we presented a case against searching (with imperfect tools) for the valid model (to eliminate bias) that is parsimonious (to reduce the variance). If we are committed to estimation with minimum MSE, then we should find the balance of the squared bias with the narrowest submodel and the variance inflation with the widest model, which we have assumed to be valid. A commitment to estimators based on a single (selected) model is misguided on two counts. First, compositions of model-based estimators have a much greater potential, and second, the properties of a selected-model based estimator depend not only on the model selected but on all the models that have taken part in the model-selection contest. In model selection, the candidate estimators take part in a contest, and the best one can hope for is that the winner is the most efficient estimator. In composition, the candidate estimators 'cooperate' in estimation of the target, aiming to estimate it with efficiency greater than the most efficient of the candidate model-based estimators.

Composite estimation of a sequence of models reduces to a composition of only two estimators, based on the simplest and the most complex model. The properties of such a bivariate composition of OLS estimators are relatively easy to explore because $\hat{\theta}_K$ and $\hat{b}_0 = \hat{\theta}_0 - \hat{\theta}_K$ are independent. Further, the only factor subject to uncertainty is the chi-squared-like statistic $\hat{b}_0^2/\hat{\sigma}^2$. Its properties can be explored by simulations. The problem of selecting from a multitude of estimators is replaced by a univariate problem of composing two estimators.

11.4   Further reading

This chapter is based on Longford (2017c). Burnham and Anderson (2002) give a comprehensive account of model selection. Some of the weaknesses of model selection are addressed by model averaging (Kass and Raftery, 1995; Hoeting *et al.*, 1999), which can be described as composition with the same vector of coefficients **b** for all targets. Claeskens and Hjort (2003 and 2008) constructed the focused information criterion which selects a model according to the target. They approximate the MSE of a post-selection estimator. An overview of the general problem is given by Efron (2014). Lasso (Tibshirani, 1996; Li and Shao, 2015) is an alternative to methods of model selection based on information criteria (Akaike, 1974; Schwartz, 1978; and Spiegelhalter *et al.*, 2002b).

11.5 Exercises

11.1. Prove the following properties of the projection matrix $\mathbf{P}$: it is idempotent (i.e., $\mathbf{P}^2 = \mathbf{P}$) and its eigenvalues are 0 and 1 with respective multiplicities $n - K - 1$ and $K + 1$. What about the matrix $\mathbf{I} - \mathbf{P}$?

11.2. Derive the expression for the OLS estimator $\hat{\boldsymbol{\beta}}$ in equation (11.4). Show that it is also the maximum likelihood estimator. Derive the distribution of $\hat{\boldsymbol{\beta}}$.

11.3. Derive the expression for the estimator $\hat{\sigma}^2$ in equation (11.5). How does it differ from the maximum likelihood estimator of σ^2? Derive the distribution of $\hat{\sigma}^2$.

11.4. Suppose the regression matrix $\mathbf{X}$ has only two columns, the intercept $\mathbf{1}$ and a covariate X; $\mathbf{X} = (\mathbf{1} \ X)$. Let $\bar{x}$ be the mean of X; that is, $\bar{x} = \frac{1}{n} X^\top \mathbf{1}$. Show that, irrespective of any details of the response variable $\mathbf{y}$, the ordinary regression model with $\mathbf{X}$ is equivalent to the ordinary regression model with $\mathbf{X}' = (\mathbf{1} \ X - \bar{x}\mathbf{1})$.

11.5. Extend the result of the Exercise 11.4 as follows. Show by construction that for any ordinary regression model with regression matrix $\mathbf{X}$ there is an equivalent ordinary regression model with regression matrix $\mathbf{X}'$ that has orthogonal columns; that is, $\mathbf{X}'^\top \mathbf{X}'$ is a diagonal matrix. How is the regression parameter vector $\boldsymbol{\beta}'$ in the model with $\mathbf{X}'$ related to its counterpart $\boldsymbol{\beta}$ in the model with $\mathbf{X}$? How are the corresponding OLS estimators $\hat{\boldsymbol{\beta}}$ and $\hat{\boldsymbol{\beta}}'$ related?

11.6. In the simple regression (with $K = 1$), based on a dataset of your choice or an artificially constructed vector X, find the borderline value $\beta^* > 0$ of the slope on X such that if $|\beta| < \beta^*$, then the constant zero has smaller MSE for β than the OLS estimator based on $(\mathbf{1} \ X)$; otherwise the $\bar{y}$ has greater MSE (is less efficient). Hint: Take advantage of the equivalence proved in Exercise 11.4.

11.7. Simulate the setting of Exercise 11.6 on the computer with your choices of β and σ^2. Apply a model selection procedure of your choice and record how many times each model is selected. Explain why no generality would be lost by constraining the intercept β_0 to zero and σ^2 to unity. (Then only one parameter, β_1, has to be set.) Collect the simulated values of $\hat{\boldsymbol{\beta}}$, with $\hat{\beta}_1 = 0$ when X is eliminated from the model, evaluate a linear combination $\hat{\theta} = \mathbf{c}^\top \hat{\boldsymbol{\beta}}$ with $\mathbf{c}$ of your choice, and draw a histogram of the simulated values of $\hat{\theta}$. Relate the observed distribution of $\hat{\theta}$ to the distribution of $\hat{\theta}$ that you would infer based on a realised dataset, or a single replication. Repeat the entire exercise with values of β_1 selected with intent to obtain both models (with and without X) in a nontrivial fraction of

replications. Relate your conclusions to equation (11.8). Show that
the choice made (the selected model) is related to $\hat{\beta}$.

11.8. Read and discuss the paper by Box (1976).

11.9. Read and discuss the editorial by the author (Longford, 2005).

11.10. Section 11.1.1 contains the claim that $\hat{\beta}_{x} = (\mathbf{X}^{\top}\mathbf{X})^{-1}\mathbf{X}^{\top}\mathbf{y}$ differs,
in general, from the subvector of the 'full-model' estimator $\hat{\beta} = (\mathbf{Z}^{\top}\mathbf{Z})^{-1}\mathbf{Z}^{\top}\mathbf{y}$. Describe the cases in which this does not hold, when
$\hat{\beta}_{x}$ is a subvector of $\hat{\beta}$.

11.11. Explain in detail why the denominator of D in equation (11.6) is
positive for any non-zero vector $\mathbf{z}$.
Hint: Recall the definition and the properties of the projection matrix.

11.12. A matter for revision (Appendix A): the inverse and the determi-
nant of a (symmetric) partitioned matrix $\mathbf{D} = \begin{pmatrix} \mathbf{A} & \mathbf{B} \\ \mathbf{B}^{\top} & \mathbf{C} \end{pmatrix}$.
Check the expression for $\mathbf{D}^{-1}$ numerically on a few examples, start-
ing with one in which $\mathbf{A} = \mathbf{I}$ and $\mathbf{C}$ is a scalar.
How can the identity for $(\mathbf{V} + \mathbf{b}\mathbf{b}^{\top})^{-1}$ be related to (or derived
from) the inverse of the partitioned matrix?

11.13. Compare Figures 11.2 and 10.3 and discuss the similarity of the two
settings.

11.14. Prove the identity $\text{cov}(\mathbf{c}^{\top}\hat{\beta}, \mathbf{c}_{x}^{\top}\hat{\beta}_{x}) = \text{var}(\mathbf{c}_{x}^{\top}\hat{\beta}_{x})\ (= v_{x})$, claimed
in Section 11.1.3.

11.15. The geometric average of a set of positive numbers x_{i}, $i = 1, \ldots, n$,
is defined as $\exp\{\frac{1}{n}\sum_{i=1}^{n}\log(x_{i})\}$, that is, $\sqrt[n]{\prod_{i=1}^{n}x_{i}}$. Show that
the inequality in (11.10) holds for $r < \sqrt{r_{1}r_{2}}$ and does not hold
otherwise. Why is this result important?

11.16. Linear algebra for revision: definition of linear spaces and their gen-
eralizations; operations defined on the elements of a space; definition
of a basis and the space it generates.

11.17. Prove the identity in equation (11.20) following the hints in the
text.

11.18. Check that the expression for the inverse $\mathbf{V}^{-1}$ in equation (11.21)
is correct and that it can be generalised to any dimension $K + 1$.
How is $\mathbf{V}$ related to a Markov chain?

11.19. Consider the set of estimators based on the empty model (intercept only) and all the simple regression models, which comprise
the intercept and a single covariate. Prove that this set, as a basis,
generates the same set of estimators as the basis of increasing models. Construct the corresponding variance matrix $\mathbf{V}$, find its inverse

and derive the optimal composition. Why is the result the same as $\tilde{\theta}$ obtained by composition based on the set of increasing models?

11.20. Explore the following idea. If we reduce the basis by discarding the empty model M_0 and perhaps a few of the simplest models $M_1, \ldots, M_L$, then the optimal composition of the remainder (the reduced basis) is a composition of M_{L+1} and M_K. If the estimator based on M_{L+1} has both small variance and small bias, then the optimal composition of M_{L+1} and M_K is likely to be much more efficient than the optimal composition of M_0 and M_K. Discuss the implications of this idea.

11.21. Prove that the matrices $\mathbf{A}$, $\mathbf{C}$ and $\mathbf{C} - \mathbf{B}^\top \mathbf{A}^{-1} \mathbf{B}$ introduced in Appendix A are positive definite.

11.22. Complete the derivation of the inverse of the partitioned matrix $\mathbf{D}$ in Appendix A by suitable pre- and post-multiplication. Follow up the derivations numerically, using a partitioned matrix of your choice.

11.23. Prove the statements made about mixtures in the first paragraph of Appendix B. Extend them to mixtures of more than two (but finitely many) components.

11.24. Implement the EM algorithm for fitting a finite mixture of normal distributions with unrelated expectations and variances.

11.25. Explore by simulation estimators of the ratio $b_0/(1 + R)$, as suggested in Section 11.3.

Appendix

A. Inverse of a partitioned matrix

Suppose a symmetric positive definite matrix $\mathbf{D}$ is partitioned as

$$\mathbf{D} = \begin{pmatrix} \mathbf{A} & \mathbf{B} \\ \mathbf{B}^\top & \mathbf{C} \end{pmatrix}$$

Then $\mathbf{A}$, $\mathbf{C}$ and $\mathbf{E} = \mathbf{C} - \mathbf{B}^\top \mathbf{A}^{-1} \mathbf{B}$ are also positive definite (and symmetric), and the inverse of $\mathbf{D}$ is

$$\mathbf{D}^{-1} = \begin{pmatrix} \mathbf{A}^{-1} & \mathbf{0} \\ \mathbf{0} & \mathbf{0} \end{pmatrix} + \begin{pmatrix} \mathbf{A}^{-1}\mathbf{B} \\ -\mathbf{I} \end{pmatrix} \mathbf{E}^{-1} \begin{pmatrix} \mathbf{A}^{-1}\mathbf{B} \\ -\mathbf{I} \end{pmatrix}^\top$$

where $\mathbf{0}$ stands for a matrix of zeros of the appropriate dimensions and $\mathbf{I}$ is the identity matrix. The identity can be proved by checking that the matrix product of the expressions for $\mathbf{D}$ and $\mathbf{D}^{-1}$ is indeed the identity matrix.

More insight can be gained by applying standard operations to derive the inverse of $\mathbf{D}$. Set side by side the matrices $\mathbf{D}$ and $\mathbf{I}$, and apply operations of pre- or post-multiplying both matrices so that $\mathbf{D}$ would be altered to $\mathbf{I}$. The result of the same operations on the other matrix (originally $\mathbf{I}$) is the inverse of $\mathbf{D}$. Start this process by post-multiplying $\mathbf{D}$ by

$$\begin{pmatrix} \mathbf{A}^{-1} & \mathbf{0} \\ \mathbf{0} & \mathbf{I} \end{pmatrix},$$

so that its (1,1) block is transformed to $\mathbf{I}$. Then post-multiply by

$$\begin{pmatrix} \mathbf{I} & -\mathbf{B} \\ \mathbf{0} & \mathbf{I} \end{pmatrix},$$

which transforms the (1,2) block to $\mathbf{0}$. Post-multiplication by

$$\begin{pmatrix} \mathbf{I} & \mathbf{0} \\ -\mathbf{E}^{-1}\mathbf{B}^{\top}\mathbf{A}^{-1} & \mathbf{I} \end{pmatrix}$$

transforms the (2, 1) block to $\mathbf{0}$. The rest is left for an exercise. This includes following up on the transformations applied to $\mathbf{I}$, which yield the inverse of $\mathbf{D}$.

As a special case, suppose $\mathbf{A} = \mathbf{X}^{\top}\mathbf{X}$, $\mathbf{B} = \mathbf{X}^{\top}\mathbf{z}$ and $\mathbf{C} = \mathbf{z}^{\top}\mathbf{z}$; see equation (11.6). Then

$$\begin{pmatrix} \mathbf{X}^{\top}\mathbf{X} & \mathbf{X}^{\top}\mathbf{z} \\ \mathbf{z}^{\top}\mathbf{X} & \mathbf{z}^{\top}\mathbf{z} \end{pmatrix}^{-1} = \begin{pmatrix} \left(\mathbf{X}^{\top}\mathbf{X}\right)^{-1} & \mathbf{0} \\ \mathbf{0} & \mathbf{0} \end{pmatrix}$$

$$+ D \begin{pmatrix} \left(\mathbf{X}^{\top}\mathbf{X}\right)^{-1}\mathbf{X}^{\top}\mathbf{z} \\ -1 \end{pmatrix} \begin{pmatrix} \left(\mathbf{X}^{\top}\mathbf{X}\right)^{-1}\mathbf{X}^{\top}\mathbf{z} \\ -1 \end{pmatrix}^{\top},$$

where $D = 1/\{\mathbf{z}^{\top}\mathbf{z} - \mathbf{z}^{\top}\mathbf{X}(\mathbf{X}^{\top}\mathbf{X})^{-1}\mathbf{X}^{\top}\mathbf{z}\}$. By pre- and post-multiplying by $\mathbf{c}$ we obtain the expression in equation (11.7).

A related identity, used in equation (11.17), is

$$\left(\mathbf{V} + \mathbf{b}\mathbf{b}^{\top}\right)^{-1} = \mathbf{V}^{-1} - \frac{1}{1 + \mathbf{b}^{\top}\mathbf{V}^{-1}\mathbf{b}} \mathbf{V}^{-1}\mathbf{b}\mathbf{b}^{\top}\mathbf{V}^{-1},$$

where $\mathbf{V}$ is a $K \times K$ matrix and $\mathbf{b}$ a $K \times 1$ vector. Apart from checking by multiplication, this identity can be derived by guessing that the inverse of $\mathbf{V} + \mathbf{b}\mathbf{b}^{\top}$ is bound to be a matrix of the form $\mathbf{V}^{-1} - u\mathbf{V}^{-1}\mathbf{b}\mathbf{b}^{\top}\mathbf{V}^{-1}$, and finding the scalar u.

B. Mixtures

A simple example of a mixture is given by the following process. Suppose A_0 and A_1 are two random variables and B is a binary random variable. Then a mixture of A_0 and A_1 is defined as A_B. That is, the value of the mixture is defined as the value (realisation) of A_0 or A_1, depending on the value of B. In most applications, B is independent of both A_0 and A_1. In this case, the mixture has the expectation $p_0 E_0 + p_1 E_1$ and variance $p_0 V_0 + p_1 V_1 + p_0 p_1 (E_1 - E_0)^2$, where E_b is the expectation and V_b the variance of A_b, $b = 0$ or 1, and $p_b = \mathrm{P}(B = b)$.

If we relax the assumption that B and A_b, $b = 0, 1$, are independent, the properties of the mixture A_B are much more difficult to establish. In selection between two models, the selection variable is independent of the two model-based estimators only in some unrealistic settings.

Finite mixture models are defined from a sequence of random variables $A_1, A_2, \ldots, A_K$ and a multinomial variable B defined on the integers (indices) $1, 2, \ldots, K$ as A_B. A wide class of distributions is obtained by such a mixture for a set of (basis-) distributions of A_k, $k = 1, \ldots, K$, and the class of all multinomial distributions on the indices $1, \ldots, K$, even when B is independent of all the basis variables A_k. If the assumption of independence is relaxed, the class of mixture distributions is much wider yet.

EM algorithm

In many problems, observations are made of the mixture A_B, without B being observed. That is, we conjecture that the random sample we observe is from a finite mixture of distributions, and we would like to recover the number of components K, distributions of $A_1, \ldots, A_K$ and the mixture probabilities $p_k = \mathrm{P}(B = k)$. The basis distributions may have no interpretation related to the substance of the problem. In that mode, these distributions are used merely as a vehicle for defining a wider class of distributions that adequately describes the recorded data.

Finite mixture models (with the assumption of independence) can be fitted by the EM algorithm (Dempster, Laird and Rubin, 1977). This algorithm provides a general prescription for fitting models for which we do not have an off-the-shelf procedure. The EM algorithm is connected to and motivated by the missing-data principle. Instead of the recorded dataset $\mathbf{X}_-$, called the incomplete data, we consider a larger dataset $\mathbf{X}$ for which we have an off-the-shelf procedure for maximum likelihood estimation. Denote by $\mathbf{X}_{\mathrm{miss}}$ the missing data—the part of $\mathbf{X}$ that is absent from $\mathbf{X}_-$. The log-likelihood for $\mathbf{X}$, $l_{\mathbf{X}}$, depends on $\mathbf{X}_{\mathrm{miss}}$ through a set of sufficient statistics $\mathbf{S}_{\mathrm{miss}}$. Such a vector is called linear, if $l_{\mathbf{X}}$ is a linear function of $\mathbf{S}_{\mathrm{miss}}$.

The EM algorithm proceeds by iterations, each comprising two steps, E (expectation) and M (maximisation). In the E step, the conditional expectation of $\mathbf{S}_{\mathrm{miss}}$, $\hat{\mathbf{S}} = \mathrm{E}(\mathbf{S}_{\mathrm{miss}} \mid \mathbf{X}_-, \boldsymbol{\theta})$, given the data and the current (provisional) estimates of all the parameters, is evaluated. In the M step, $l_{\mathbf{X}}$ is maximised, with $\mathbf{S}_{\text{miss}}$ replaced by its conditional expectation $\hat{\mathbf{S}}$ evaluated in the preceding E step. The iterations are stopped when the estimates obtained in two consecutive iterations differ by less than a prescribed margin.

For fitting a finite mixture, the missing data are the realisations of the multinomial (mixing) variable. Their conditional (E-step) expectations are the conditional probabilities of belonging to each category,

$$\hat{p}_{ik} = \frac{\pi_k f_k\left(\mathbf{x}_i, \hat{\boldsymbol{\theta}}\right)}{\displaystyle\sum_{h=1}^{K} \pi_h f_h\left(\mathbf{x}_i, \hat{\boldsymbol{\theta}}\right)}, \tag{11.24}$$

where $i = 1, \ldots, n$ is the index of the observation, f_k is the density of component $k = 1, \ldots, K$, and $\hat{\boldsymbol{\theta}}$ is the current estimate of the parameter vector $\boldsymbol{\theta}$; π_k is the prior probability of component k. When the K component distributions are in the same family of distributions, such as the normal, the ordering of the components is immaterial, and then the only meaningful choice of the prior probabilities is $\pi_1 = \ldots = \pi_K = \frac{1}{K}$. This amounts to dropping the factors π_h from the expression in equation (11.24). With the updated values of $\hat{p}_{ik}$, the parameters associated with each component are estimated in the next iteration by maximising the log-likelihood, with $\hat{p}_{ik}$ substituted for the unobserved indicators of the component. For instance, when component k has normal distribution, its mean is estimated by the weighted mean of the observations, with the weights set to $\hat{p}_{ik}$, $i = 1, \ldots, n$.

In the problem of selecting between two models, the EM algorithm is not useful because we have only one observation, comprising the two estimates and the selection made. Further, the selection is correlated with the two estimators. However, the EM algorithm helps us to elucidate the problem. It suggests that the estimates of the candidate models should be combined, with weights equal to the conditional probabilities of the models being appropriate. These probabilities cannot be estimated but selecting one estimator, by setting $p_k = 1$, is clearly inappropriate because it corresponds to an extreme case ($0 \leq p_k \leq 1$).

The invention of the EM algorithm set off a wide-ranging research in its applications to settings in which established methods were deficient or none were available. Much of it entails ingenuity in defining what should be regarded as the missing information. The EM algorithm is relatively easy to implement if a method for fitting the complete-data model is available. A distinct weakness of the EM algorithm is its slow convergence but this has over the years also been addressed in many settings. Meng and van Dyk (1996) describe several improvements on this general scheme.

C. Linear loss

In this Appendix we explore linear loss for the elementary model selection problem.

If we exclude **z** from the model, the expected loss with a linear loss function is $Q_x = Q_{x-} + Q_{x+}$, the sum of expected losses for large positive and large negative β_z, and

$$Q_{x-} = \int_{-\infty}^{-\beta_z^*} \left(x^2 - \beta_z^{*2}\right) \varphi\left(x; \hat{\beta}_z, \beta_z^*\right) dx.$$

First, we apply the transformation $y = (x - \hat{\beta}_z)/\beta_z^*$.

$$Q_{x-} = \beta_z^{*2} \int_{-\infty}^{-1-z} \left\{(y+z)^2 - 1\right\} \varphi(y)\, dy,$$

where $z = \hat{\beta}_z/\beta_z^*$. Next, we apply integration by parts, integrating $\varphi(y)$ and differentiating $(y+z)^2 - 1$. For $y = -1 - z$, the first term vanishes, and so

$$Q_{x-} = -2\beta_z^{*2} \int_{-\infty}^{-1-z} (y+z)\, \Phi(y)\, dy.$$

Another application of integrating by parts yields

$$\begin{aligned}
Q_{x-} &= 2\beta_z^{*2} \left\{ \Phi_1(-1-z) + \int_{-\infty}^{-1-z} \Phi_1(y)\, dy \right\} \\
&= 2\beta_z^{*2} \left\{ \Phi_1(-1-z) + \Phi_2(-1-z) \right\}.
\end{aligned}$$

We express this identity in terms of $\varphi(1+z)$ and $\Phi(1+z)$, using the relations

$$\begin{aligned}
\Phi_1(-u) &= -u + u\Phi(u) + \varphi(u) \\
2\Phi_2(-u) &= u^2 + 1 - \left(u^2 + 1\right)\Phi(u) - u\varphi(u)
\end{aligned}$$

derived directly from the definitions of Φ_1 and Φ_2 in respective Sections 3.2 and 3.3. From them we obtain the identity

$$\begin{aligned}
Q_{x-} &= \beta_z^{*2} \left\{ -2(1+z) + 2(1+z)\Phi(1+z) + 2\varphi(1+z) \right. \\
&\quad \left. + 2 + 2z + z^2 - \left(2 + 2z + z^2\right)\Phi(1+z) - (1+z)\varphi(1+z) \right\} \\
&= \beta_z^{*2} \left\{ z^2 - z^2\Phi(1+z) + (1-z)\varphi(1+z) \right\}.
\end{aligned}$$

The second part of Q_x is evaluated similarly, although integration by parts cannot be applied directly.

$$\begin{aligned}
Q_{x+} &= \int_{\beta_z^*}^{+\infty} \left(x^2 - \beta_z^{*2}\right) \varphi\left(x; \hat{\beta}_z, \beta_z^*\right) dx \\
&= \beta_z^{*2} \left[\int_{-\infty}^{+\infty} \left\{(y+z)^2 - 1\right\} \varphi(y)\, dy - \int_{-\infty}^{1-z} \left\{(y+z)^2 - 1\right\} \varphi(y)\, dy \right].
\end{aligned}$$

The former integral is equal to z^2 and for the latter we proceed by two-fold integration by parts. This leads to the result

$$
\begin{aligned}
Q_{\mathrm{x}+} &= \beta_{\mathrm{z}}^{*\,2}\left\{z^2 + 2\Phi_1\left(1 - z\right) - 2\Phi_2\left(1 - z\right)\right\} \\
&= \beta_{\mathrm{z}}^{*\,2}\left\{z^2 + \left(1 - z^2\right)\Phi(1 - z) + (1 + z)\varphi(1 - z)\right\}.
\end{aligned}
$$

There is no reduced expression for $Q_{\mathrm{x}} = Q_{\mathrm{x}-} + Q_{\mathrm{x}+}$.

The expected (linear) loss incurred by retaining $\mathbf{z}$ in the model, if $|b| < \sqrt{\Delta v}$, can be evaluated similarly. This expected loss is

$$
\begin{aligned}
Q_{\mathrm{z}} &= \int_{-\beta_{\mathrm{z}}^{*}}^{\beta_{\mathrm{z}}^{*}}\left(\beta_{\mathrm{z}}^{*\,2} - x^2\right)\varphi\left(x; \hat{\beta}_{\mathrm{z}}, \beta_{\mathrm{z}}^{*}\right)\mathrm{d}x \\
&= \int_{-\infty}^{+\infty}\left(\beta_{\mathrm{z}}^{*\,2} - x^2\right)\varphi\left(x; \hat{\beta}_{\mathrm{z}}, \beta_{\mathrm{z}}^{*}\right)\mathrm{d}x + Q_{\mathrm{x}} \\
&= \hat{\beta}_{\mathrm{z}}^2 + Q_{\mathrm{x}}.
\end{aligned}
$$

This implies that $Q_{\mathrm{z}} > Q_{\mathrm{x}}$, and therefore unconditional choice of the model without $\mathbf{z}$.

12

Postscript

With a bit of immodesty and a tinge of obsession, I would like to refer to the principal theme of this book as a paradigm, ascribing to it as much importance and distinction as to the frequentist and Bayesian paradigms in more established discourse in statistics. Statistics deals with errors, as cleaners do with dust and dustmen do with our domestic refuse. Our shared purpose is not to eradicate errors, dust or refuse, but to manage it in a way that is effective and economic. Statistical errors are not made equal; we all agree that they differ in magnitude and frequency, and the characterisation of these features, by standard errors, false discovery rates, and similar quantities is an important preoccupation in both theory and practice of statistics.

This volume presents arguments against regarding the magnitude and frequency as the objectively preeminent features of errors in statistics. They ignore the (client's) purpose of the analysis which introduces the consequences (ramifications) of the errors as another feature to reckon with. The purpose is subjective; if the client changed their perspective, value judgements or aims and priorities, the consequences would be altered, and that may result in a different solution being optimal. Analyses are (or should be) for clients, just as horses are for courses. Clients do have a lot in common, but it is rather presumptuous to have a single (objective) analysis, yielding a solution that would be offered to all and sundry, advertised as equally good (efficient) for whoever the client may be. That is a statistical restaurant hoping to be busy with a single item on the menu.

Not every analysis that is conducted has a well identified client. But studies and their analyses that matter, have, and the analyst's service is greatly enhanced by integrating the client's perspective and purpose or remit in the analysis. That implies a dismissal of most established formats of inferential statements. Efforts at estimation with minimum mean squared error (MSE) are misdirected if positive and negative errors of the same magnitude $|\hat{\theta} - \theta|$ have consequences of different gravity. In many cases, they undeniably do; for instance, when it is preferable to err on the side of caution, protecting the client from exposure to higher cost, greater risk of harm or inconvenience.

The main target of this volume is the inappropriate application of hypothesis testing for the purpose of choosing one or the other of two alternative courses of action. I categorically dismiss any application of hypothesis test because it is oblivious to the consequences of the two kinds of error. If the consequences do not matter, then the analysis is inconsequential. Such an analysis is hardly worth conducting because it demeans the analyst and the profession of statistics as a whole.

In general, a hypothesis is subjected to a test because there would be some profit or benefit if we knew whether the hypothesis or the alternative is valid. I have assumed in this volume that such an issue arises because the client would without hesitation follow one course of action if it were known that the hypothesis is valid, and would follow another if the alternative were known to be valid. I regard the management of the error in this setting and context, by limiting (controlling) the probability of one type of error, as wholly inadequate. First, the hypothesis and the alternative are treated asymmetrically; in the absence of evidence against it, the hypothesis is promoted, even though there is no evidence to support it. The hypothesis is treated as a default, applicable until it is contradicted by evidence. Second, the calculus of error probabilities of a set of hypothesis tests is too complex and ignores the relative importance of the hypotheses. And third, the probabilities handled in hypothesis tests are hypothetical, evaluated under temporarily adopted assumptions, an exercise abstract even to the most hardened theoretician.

These problems are well appreciated and widely discussed, but response to the associated criticism has been by appeal to the inexactness of statistics as a science that operates with models and other assumptions that render our handling of data tractable and the results of their analyses interpretable or constructive. The view prevails that until there is a better edifice to move into, we are staying put in the house of hypothesis testing—it accommodates everybody and, at least for the time being, is the established norm.

Indeed, hypothesis testing was a great statistical innovation in the 1950s, and its subsequent promotion into the mainstream statistical practice is one of the great success stories in science. However, times have moved on since, and however hard we (try to) innovate within the confines of hypothesis testing, the imperviousness to the consequences of the errors is not overcome because they are not taken into account directly. By leaving it to others to sort it out, the output of the statistical analyst's effort is at best half-baked, leaving a nontrivial evaluation to a party that is unlikely to be equipped with the relevant skills as a statistical analyst is, or should be.

The decision-theoretical approach presented in this volume manages the errors by comparing expected losses. Expected loss is additive, so the calculus for combining the expected losses in several studies or analyses (applications) is simple, although the evaluation of an expected loss may involve integration. In most common settings this is analytically tractable. The result of the analysis, a comparison of the expected losses, requires no interpretation; it is a proposal for what the client should do—which course of action to follow.

The method requires the same inputs as its established alternatives (estimation and hypothesis testing), with some additions. First, the consequences have to be quantified, by means of a loss function. The scale on which the loss function is defined has to have the property of additivity, akin to a monetary currency, and has to be such that the client, when thinking and acting rationally, would want to be frugal with it—spend as little of it as possible. Second, the domains in which each contemplated course of action (option) is preferable have to be defined more carefully than their counterparts in hypothesis tests (the hypothesis and the alternative). For example, the counterpart of the two-sided null hypothesis is the pair of thresholds that separate small and large values of a parameter.

I regard specifying these two inputs, losses and thresholds, as a much more difficult hurdle than the analytical (computational) complexity or, more precisely, lack of familiarity. It requires an intimate understanding of the client's perspective, similar to the ideal in the relationship of a lawyer and client, medical consultant and patient, political representative and (voter) constituency. This hurdle is not technical but cultural.

These difficulties may be discouraging because they require substantial investment in skills that are poorly appreciated and not well developed at present, by both analysts and (potential) clients. However, the suggestion that consequences of the errors that we as analysts make in our practice are not material and do not deserve detailed scrutiny and integration in the analysis is not sustainable. Leaving the client to take these consequences into account amounts to abrogation of responsibility and a strategic error of providing incomplete service by failing to respond to a more complete remit and failing to exploit the full potential of statistics as a science.

In conclusion, I want to thank all the readers who got this far for their perseverance, interest in my work and toleration of all the usual (and other) deficiencies of the presentation. An apology is in order for the terseness of the examples, some of which are at best sketchy on their background, especially of the elicitation. My humble explanation is that I managed to engage my clients in thorough elicitation rather infrequently. The examples represent my ideas about a particular aspect of operating as a statistical analyst, not the realities of my and others' operations in the past. They imply a kind of client I would like to have, not the kind I usually have had. By this statement I do not mean to denigrate clients in any way, but to invite another author to address (prospective) clients directly and promote transparency and openness about their perspectives, value judgements and priorities or remits because they are the gateway to a better service from a statistical analyst.

References

Akaike, H. (1974). A new look at the statistical model identification. *IEEE Transaction on Automatic Control* **19**, 716–723.

Baker, S.G. (2003). The central role of receiver operating characteristic (ROC) curves in evaluating tests for the early detection of cancer. *Journal of the National Cancer Institute* **95**, 511–515.

Benjamini, Y., and Hochberg, Y. (1995). Controlling the false discovery rate: A practical and powerful approach to multiple testing. *Journal of the Royal Statistical Society* Ser. B **57**, 289–300.

Benjamini, Y., and Yekutieli, D. (2001). The control of the false discovery rate in multiple testing under dependence. *Annals of Statistics* **29**, 1165–1188.

Berger, J.O. (1985). *Statistical Decision Theory and Bayesian Analysis*, 2nd ed. Springer-Verlag, New York.

Berry, D.A., and Hochberg, Y. (1999). Bayesian perspectives on multiple comparisons. *Journal of Statistical Planning and Inference* **82**, 215–227.

Box, G.E.P. (1976). Science and statistics. *Journal of the American Statistical Association* **71**, 791–799.

Burnham, K.P., and Anderson, D.R. (2002). *Model Selection and Multimodel Inference: A Practical Information-Theoretic Approach*, 2nd ed. Springer-Verlag, New York.

Claeskens, G., and Hjort, N.L. (2003). The focused information criterion. *Journal of the American Statistical Association* **98**, 900–945.

Claeskens, G., and Hjort, N.L. (2008). *Model Selection and Model Averaging*. Cambridge University Press, Cambridge, MA.

DeGroot, M.H. (1970). *Optimal Statistical Decisions*. McGraw-Hill, New York.

Dempster, A.P., Laird, N.M., and Rubin, D.B. (1977). Maximum likelihood from incomplete data via the EM algorithm. *Journal of the Royal Statistical Society* Ser. B **39**, 1–38.

Efron, B. (2014). Estimation and accuracy after model selection. *Journal of the American Statistical Association* **109**, 991–1007.

Ferguson, T.S. (1967). *Mathematical Statistics: A Decision Theoretic Approach*. Academic Press, New York.

Fisher, R.A. (1955). Statistical methods and scientific induction. *Journal of the Royal Statistical Society* Ser. B **17**, 69–78.

Fluss, R., Faraggi, D., Reiser, B., and Hu, J. (2015). Youden index and its associated cutoff point. *Biometrical Journal* **47**, 458–472.

Garthwaite, P.H., Kadane, J.B., and O'Hagan, A. (2005). Statistical methods for eliciting probability distributions. *Journal of the American Statistical Association* **100**, 680–701.

Goldstein, H., and Spiegelhalter, D.J. (1996). League tables and their limitations: statistical issues in comparisons of institutional performance. *Journal of the Royal Statistical Society* Ser. A **159**, 385–443.

Hanea, A.M., McBride, M.F., Burgman, M.A., and Wintle, B.C. (2018). Classical meets modern in the IDEA protocol for structured expert judgement. *Journal of Risk Research* **4**, 417–433.

Hannig J. (2009). On generalised fiducial inference. *Statistica Sinica* **19**, 491–544.

Harville, D.A. (1997). *Matrix Algebra from a Statistician's Perspective.* Springer-Verlag, New York.

Hedges, L.V., and Olkin, I. (1985). *Statistical Methods for Meta-Analysis.* Academic Press, Boston, MA.

Helenius, K., Longford, N., Lehtonen, L., Modi, N., and Gale, C., (2019). Association of early postnatal transfer and birth outside a tertiary hospital with mortality and severe brain injury in extremely preterm infants: observational cohort study with propensity score matching. *British Medical Journal* **367**, l5678.

Hoeting, J.A., Madigan, D., Raftery, A.E., and Volinsky, C.T. (1999). Bayesian model averaging: a tutorial. *Statistical Science* **14**, 382–401.

Holland, P.W. (1986). Causal inference and statistics. *Journal of the American Statistical Association* **81**, 945–960.

Imbens, G.W., and Rubin, D.B. (2015). *Causal Inference for Statistics, Social, and Biomedical Sciences. An Introduction.* Cambridge University Press, New York.

Jones, B., and Kenward, M.G. (1989). *Design and Analysis of Cross-over Trials.* Chapman and Hall, London, UK.

Kadane, J.B., and Wolfson, L.J. (1998). Experiences in elicitation. *Journal of the Royal Statistical Society* Ser. D **47**, 3–19.

Kass, R.E., and Raftery, A.E. (1995). Bayes factors and model uncertainty. *Journal of the American Statistical Association* **90**, 773–795.

Kish, L. (1987). *Statistical Design for Research.* Wiley, New York.

Lange, K. (1998). *Numerical Analysis for Statisticians.* Springer-Verlag, New York.

Lehmann, E.L., and Romano, J.P. (2005). *Testing Statistical Hypotheses*, 3rd ed. Springer-Verlag, New York.

Li, Q., and Shao, J. (2015). Regularizing lasso: a consistent variable selection method. *Statistica Sinica* **25**, 975–992.

Liese, F., and Miescke, K.-J. (2008). *Statistical Decision Theory. Estimation, Testing and Selection.* Springer-Verlag, New York.

Lindley D.V. (1958). Fiducial distribution and Bayes' theorem. *Journal of the Royal Statistical Society* Ser. B **20**, 102–107.

Lindley, D.V. (1985). *Making Decisions.* Wiley, Chichester, UK.

Lindley, D.V. (1998). Decision analysis and bioequivalence trials. *Statistical Science* **13**, 136–141.

Longford, N.T. (2001). Synthetic estimators with moderating influence: carry-over in cross-over trials revisited. *Statistics in Medicine* **20**, 3189–3203.

Longford, N.T. (2003). An alternative to model selection in ordinary regression. *Statistics and Computing* **13**, 67–80.

Longford, N.T. (2005). Editorial: Model selection and efficiency. Is 'Which model . . . ?' the right question? *Journal of the Royal Statistical Society* Ser. A **168**, 469–472.

Longford, N.T. (2007). Playing consequences. Letter to the Editor. *Significance* **4**, 46.

Longford, N.T. (2008). An alternative analysis of variance. *SORT*, Journal of the Catalan Institute of Statistics **32**, 77–91.

Longford, N.T. (2009). Analysis of all-zero binomial outcomes. *Journal of Applied Statistics* **36**, 1259–1265.

Longford, N.T. (2010). Bayesian decision making about small binomial rates with uncertainty about the prior. *The American Statistician* **64**, 164–169.

Longford, N.T. (2012a). Comparing normal random samples, with uncertainty about the priors and utilities. *Scandinavian Journal of Statistics* **39**, 729–742.

Longford, N.T. (2012b). 'Which model?' is the wrong question. *Statistica Neerlandica* **66**, 237–252.

Longford, N.T. (2012c). Handling the limit of detection by extrapolation. *Statistics in Medicine* **31**, 3133–3146.

Longford, N.T. (2013a). Searching for contaminants. *Journal of Applied Statistics* **40**, 2041–2055.

Longford, N.T. (2013b). Screening as an application of decision theory. *Statistics in Medicine* **32**, 849–863.

Longford, N.T. (2013c). *Statistical Decision Theory.* Springer-Verlag, Heidelberg, Germany.

Longford, N.T. (2014a). A decision-theoretical alternative to testing many hypotheses. *Biostatistics* **15**, 154–169.

Longford, N.T. (2014b). Sample size calculation for comparing two normal random samples using equilibrium priors. *Communications in Statistics — Simulation and Computation* **43**, 2149–2161.

Longford, N.T. (2014c). Screening test items for differential item functioning. *Journal of Educational and Behavioral Statistics* **39**, 3–21.

Longford, N.T. (2015a). On the inefficiency of the restricted maximum likelihood. *Statistica Neerlandica* **69**, 171–196.

Longford, N.T. (2015b). Classification in two-stage screening. *Statistics in Medicine* **34**, 3281–3297.

Longford, N.T. (2016a). Decision theory for selecting the winners, ranking and classification. *Journal of Educational and Behavioral Statistics* **41**, 420–442.

Longford, N.T. (2016b). Comparing two treatments by decision theory. *Pharmaceutical Statistics* **15**, 387–395.

Longford, N.T. (2017a). A decision-theoretical perspective on bioequivalence and bioavailability. *Journal of Bioequivalence and Bioavailability* **9**, 437–438.

Longford, N.T. (2017b). An index for proximity of two distributions. *Metron* **75**, 181–194.

Longford, N.T. (2017c). Estimation under model uncertainty. *Statistica Sinica* **27**, 859–877.

Longford, N.T. (2018). Decision theory for comparing institutions. *Statistics in Medicine* **37**, 437–456.

Longford, N.T. (2020). Performance assessment as an application of causal inference. *Journal of the Royal Statistical Society* Ser. A **183**, 1363–1385.

Longford, N.T., and Andrade Bejarano, M. (2015). Decision theory for the variance ratio in one-way ANOVA with random effects. *Revista Colombiana de Estadística* **38**, 181–207.

McLachlan, G., and Peel, D. (2000). *Finite Mixture Models.* Wiley, New York.

Meng, X.-L., and van Dyk, D. (1997). The EM algorithm – an old folk-song sung to a fast new tune. *Journal of the Royal Statistical Society* Ser. B **59**, 511–567.

Molanes-López, E.M., and Letón, E. (2011). Inference of the Youden index and associated threshold using empirical likelihood for quantiles. *Statistics in Medicine* **30**, 2467–2480.

O'Hagan, A. (1998). Eliciting expert beliefs in substantial practical applications. *Journal of the Royal Statistical Society* Ser. D **47**, 21–68.

O'Hagan, A. (2019). Expert knowledge elicitation: subjective but scientific. *The American Statistician* **73**, Suppl. 1, 69–81.

Ohlssen, D.I., Sharples, L.D., and Spiegelhalter, D.J. (2007). A hierarchical modelling framework for identifying unusual performance in health care providers. *Journal of the Royal Statistical Society* Ser. A **170**, 865–890.

Parmigiani, G., and Inoue, L.Y.T. (2009). *Decision Theory. Principles and Approaches*. Wiley, Chichester, UK.

Pearl, J., and MacKenzie, D. (2018). *The Book of Why. The New Science of Cause and Effect*. Basic Books, New York.

Peterson, M. (2017). *An Introduction to Decision Theory*, 2nd ed. Cambridge University Press, Cambridge, UK.

R Development Core Team. (2016). R: A language and environment for statistical computing. R Foundation for Statistical Computing, Vienna, Austria.

Rosenbaum, P.R. (2017). *Observation & Experiment. An Introduction to Causal Inference*. Harvard University Press, Cambridge, MA.

Rosenbaum, P.R., and Rubin, D.B. (1983). The central role of propensity score in observational studies for causal effects. *Biometrika* **70**, 41–55.

Rubin, D.B. (2006). *Matched Sampling for Causal Effects*. Wiley, New York.

Rubin, D.B. (2008). For objective causal inference, design trumps analysis. *Annals of Applied Statistics* **2**, 808–840.

Schervish, M.J. (1995). *Theory of Statistics*. Springer-Verlag, New York.

Schuirmann D.J. (1987). A comparison of the two one-sided tests procedure and the power approach for assessing the equivalence of average bioavailability. *Journal of Pharmacokinetics and Biopharmaceutics* **15**, 657–680.

Schwartz, G.E. (1978). Estimating the dimension of a model. *Annals of Statistics* **6**, 461–464.

Seber, G.A.F. (1977). *Linear Regression Analysis*. Wiley, New York.

Seidenfeld, T. (1992). R. A. Fisher's fiducial argument and Bayes' theorem. *Statistical Science* **7**, 358–368.

Shen, W., and Louis, T.A. (1998). Triple-goal estimates in two-stage hierarchical models. *Journal of the Royal Statistical Society* Ser. B **60**, 455–471.

Silber, J., Rosenbaum, P.R., Ross, R.N., Ludwig, J.M., Wang, W., Niknam, B.A., Mukherjee, N., Saynich, P.A., Even-Shoshan, O., Kelz, R.R., and Fleisher, L.A. (2014). Template matching for auditing hospital cost and quality. *Health Services Research* **49**, 1446–1474.

Spiegelhalter, D.J. (2005). Funnel plots for comparing institutional performance. *Statistics in Medicine* **24**, 1185–1202.

Spiegelhalter, D.J., Aylin, P., Best, N.G., Evans, S.J.W., and Murray, G.D. (2002). Commissioned analysis of surgical performance using routine data:

Lessons from the Bristol inquiry. *Journal of the Royal Statistical Society* Ser. A **165**, 191–231.

Spiegelhalter, D.J., Best, N.G., Carlin, B.P., and van der Linde, A. (2002). Bayesian measures of complexity and fit. *Journal of the Royal Statistical Society* Ser. B **64**, 583–639.

Spiegelhalter, D., Sherlaw-Johnson, C., Bardsley, M., Blunt, I., Wood, C., and Grigg, O. (2012). Statistical methods for healthcare regulation: rating, screening and surveillance. *Journal of the Royal Statistical Society* Ser. A **175**, 1–47.

Storey, J.D. (2002). A direct approach to false discovery rates. *Journal of the Royal Statistical Society* Ser. B **64**, 479–498.

Stuart, E.A. (2010). Matching methods for causal inference. A review and a look forward. *Statistical Science* **25**, 1–21.

Thomas N., Longford N.T., and Rolph J.E. (1994). Empirical Bayes methods for estimating hospital-specific mortality rates. *Statistics in Medicine* **13**, 889–903.

Tibshirani, R. (1996). Regression shrinkage and selection via the lasso. *Journal of the Royal Statistical Society* Ser. B **58**, 267–288.

Tuyl, F., Gerlach, R., and Mengersen, K. (2008). A comparison of Bayes-Laplace, Jeffreys and other priors: the case of zero events. *The American Statistician* **62**, 40–44.

Varian, H.R. (1975). A Bayesian approach to real estate assessment. In S.E. Fienberg and A. Zellner (eds.). *Studies in Bayesian Econometrics and Statistics in Honor of Leonard J. Savage*, pp. 195–208. Nord-Holland, Amsterdam.

Venables, W.N., and Ripley, B.D. (2000). S *Programming*. Springer-Verlag, New York.

White, D.J. (1969). *Decision Theory*. Aldine, Chicago, IL.

Youden, W.J. (1950). Index for rating diagnostic tests. *Cancer* **3**, 32–35.

Zabell, S.L. (1992). R.A. Fisher and the fiducial argument. *Statistical Science* **7**, 369–387.

Solutions to exercises

Chapter 1

Exercise 1.4. The choices made may differ because your friend may misjudge your perspective (when deciding on your behalf), and may have a different perspective (when deciding for self).

Exercise 1.7. Patients have different values and preferences. Some may prefer more involved treatments that entail greater discomfort and risk of side effects but have a better long-term prognosis; others may be more averse to such risks.

Exercise 1.12. Incentives to enlist are patriotism, interest in a career in the military, assessment of how just the cause pursued by the war is, the prospect of rewards and experiences, contribution to the country's prosperous and peaceful future. Disincentives are fear of injury or death, commitment to one's (developing) career, abhorrence of violence and the belief that a war cannot resolve the underlying conflict and the soldiers returning from the war will indirectly introduce violence and brutality into civilian life. The balance of the incentives and disincentives is perceived by prospective recruits in very different ways.

The loss with recruitment is injury, disability or death, participation in an unpleasant venture, discontinuity in the relationship with the family, friends and with one's civilian career. The loss with abstention is a stigma in the face of propaganda or prevailing attitudes (helping the country in its hour of need). The two entries of the loss matrix are highly personal, especially when the public opinion on waging the war is very divided.

Exercise 1.13. Suppose the best choice is L. Then L is the 'winner' in the first contest in which it is involved, and will be the winner in every subsequent contest because it is a choice that is better than any other. If a comparison is subject to error, then every comparison in which L participates represents a risk (jeopardy) of an incorrect choice. Therefore, L has a greater chance of winning if it is involved in fewer comparisons, and the greatest when it is involved in only the last comparison.

Exercise 1.14. Even if the estimate ($\hat{\theta}$) exceeds the threshold T (24 000 Euro), the target (θ) may fall short of the threshold. The sampling variance of $\hat{\theta}$ is essential for testing the relevant hypothesis. The perspective, from which the losses for the two kinds of inappropriate decisions could be deduced, are essential for choosing between $\theta < T$ and $\theta > T$.

Exercise 1.18. Even if a state has a small probability, the corresponding verdict may be appropriate if the consequences of selecting it inappropriately are innocuous — if selecting the other option inappropriately has catastrophic consequences.

Exercise 1.19. Define $\theta = \mu_2 - \mu_1$. Then the states (for θ) are $(-\infty, 0)$ and $0, +\infty)$ and the cutpoint is $T_1 = 0$. The nature of the variables, continuous or binary, is irrelevant; the problem revolves around their means and, in fact, only their difference matters.

Chapter 2

Exercise 2.1. An estimator is a random variable. It is defined as a function of the data, and may be described by an algorithm, math formula or a computer programme. It is defined, or selected for (future) application, at the design stage. The estimate is obtained by applying the estimator to the realised data, after data collection. The estimate is a value (a number).

The value of the estimation error is established only when both values of the estimate and the target are established. When the target is known to be in a given range but the estimate is distant from the range, then there are good grounds for suspecting that the estimation error is large. The distance should always be considered in relation to the standard deviation or another measure of the dispersion of the estimator.

Exercise 2.3. The shopping bills are not replications if the shopping lists on consecutive Saturdays are dependent. For instance, what you buy one Saturday (more expensive items in particular) you are less likely to buy the next Saturday if you still have ample supply. The question could be answered with greater confidence if we knew the subject's (household's) consumer habits and the variety of goods that are stocked by the shops visited on Saturdays.

Exercise 2.4. The reason why times taken for commuting are not replications is that on some days of the week there is more congestion than on others. Less plausibly a car driver's behaviour may be affected by the day of the week. These arguments are rather tenuous for a pedestrian or a cyclist.

Exercise 2.5. The key information required is about the rule for stopping the replications. If the number of replications is fixed, then the number of successes has binomial distribution. If they are stopped immediately after the first success, then the number of failures has geometric distribution. A variety of other stopping rules are compatible with the observed outcome.

Exercise 2.6. The estimator with error uniformly distributed on the interval $(-0.1, 0.4)$ is preferred because its realisation is always closer to the target than the estimator with error on the set $(-1, 1)$. The latter estimator would be preferred in the following esoteric setting. Suppose a negative estimation error has negligible consequences and the consequences of a positive error are far greater, equal to A, and do not depend on the magnitude of the error $\hat{\theta} - \theta$.

Then the uniformly distributed estimator is inferior because its expected error is 0.8A, whereas the binary estimator has expected error of 0.5A.

The estimators have respective biases 0.15 and 0.0, variances $1/48$ and 1.0 and MSEs $41/240$ and 1.0.

Exercise 2.8. Draw a histogram of a random sample from the binomial distribution in R by the expression

```
hist(rbinom(50000, 250, 0.35))
```

and improve on this (rudimentary) histogram by making all the labels nice, setting the width of the bin, and the like.

Superimpose the normal density by the expression

```
pts <- seq(50, 250); me <- 250*0.35
lines(pts, mult * dnorm(pts, me, sqrt(me*0.65)),
      lwd=0.6, col="red")
```

with the object `mult` set to a suitable positive number.

Exercise 2.10. Even a relatively small expenditure may have dire consequences when only limited funds are available. Borrowing money is often expensive, especially when it is on flexible terms and improvised, without a prior arrangement. A sudden windfall makes us feel great and we forget that we are in luck after numerous instances of losses, for example, when regularly buying a lottery ticket. With such a windfall, a persistent everyday concern is allayed for the foreseeable future.

Exercise 2.11. The population is all the students eligible to vote. The parameter of interest is the proportion of students who would vote for the proposition. The obvious estimator is the proportion of votes cast for the proposition. If everybody votes, then this proportion can be established with precision. This proportion, as a random variable, is the estimator, and its realisation, the actual proportion (or percentage), is the estimate.

A problem is how to deal with abstentions. If more than half of the students vote for (or against) the proposition, then the outcome is without contention. Otherwise the outcome depends on how the abstentions are treated. The two extreme ways are to regard all abstentions as votes for, and to regard them as against. Distributing the abstentions half and half to the proposition against it is poorly justified. The absentees may be distinctly different from the voters.

Exercise 2.12. As an analytical solution, the lognormal distribution derived from $\mathcal{N}(\mu, \sigma^2)$ has expectation $\exp(\mu + \frac{1}{2}\sigma^2)$ and variance $\exp(2\mu + \sigma^2)\{\exp(\sigma^2)-1\}$, obtained by evaluating the integrals $\int_{-\infty}^{+\infty} \exp(hx)\varphi(x; \mu, \sigma^2)$ for $h = 1, 2$. The bias of the mean of the exponentials is $\exp(\mu)\{\exp(\frac{1}{2}\sigma^2)-1\}$. The factor $\exp(\frac{1}{2}\sigma^2)$ is a steeply increasing function of σ^2. $\exp(\mu)$ is a multiplicative factor in the expectation and standard deviation of the log-normal distribution.

Exercise 2.13. The numerator in the Bayes theorem is equal to

$$C_1 \exp\left[-\frac{1}{2}\left\{\sum_{i=1}^{n} \frac{(x_i - \mu)^2}{\sigma^2} + \frac{(\mu - \theta)^2}{\tau^2}\right\}\right],$$

where C_1 is a constant that does not involve μ. Consolidate this expression as a transformation of a quadratic function of μ:

$$C_2 \exp\left\{-\frac{(\mu - \mu_{\text{post}})^2}{2 v_{\text{post}}}\right\},$$

where C_2 is a constant that does not involve μ and

$$v_{\text{post}} = \frac{1}{\frac{n}{\sigma^2} + \frac{1}{\tau^2}}$$

$$\mu_{\text{post}} = \frac{1}{v_{\text{post}}}\left(\sum_{i} \frac{x_i}{\sigma^2} + \frac{\theta}{\tau^2}\right).$$

Hence the posterior distribution is $\mathcal{N}(\mu_{\text{post}}, v_{\text{post}})$. Regarding the prior mean θ as an extra-data observation with variance τ^2, the posterior mean μ_{post} is a weighted mean of the observations, with the weights proportional to the reciprocals of the variances. The concentration is defined as the reciprocal of the variance. The posterior concentration is equal to the total of the concentrations of the observations and the prior.

Exercise 2.14. Let $s = x_1 + \cdots + x_n$. The numerator in the Bayes theorem is

$$\prod_{i=1}^{n} \left\{p^{x_i}(1-p)^{m-x_i}\right\} p^{\alpha-1}(1-p)^{\beta-1} = p^{s+\alpha-1}(1-\alpha)^{nm-s+\beta-1},$$

except for the multiplicative constant. Hence the posterior distribution of p is beta with parameters $s + \alpha$ and $nm - s + \beta$. The prior parameters α and β can be interpreted as extra-data numbers of successes and failures.

Exercise 2.16. The standard normal distribution $\mathcal{N}(0,1)$ has the density

$$D_1 \exp\left(-\frac{1}{2}x^2\right)$$

The transformation $f(x) = x^2$, which defines χ_1^2 leads to the density

$$D_2\, y^{-\frac{1}{2}} \exp\left(-\frac{y}{2}\right),$$

and the constant D_2 is found by matching this expression to the gamma function.

The density of the convolution of two χ_1^2 distributions is

$$E_1 \int_0^y x^{-\frac{1}{2}} \exp\left(-\frac{x}{2}\right) (y-x)^{-\frac{1}{2}} \exp\left(-\frac{y-x}{2}\right) \, dx = E_2 \exp\left(-\frac{y}{2}\right),$$

and the constant $E_2 = \frac{1}{2}$ is found by matching to the exponential density.

Exercise 2.17. In the first line of response, the manufacturer should produce 250 items, so that they could expect to have (80% of 250) 200 items that would pass the quality inspection. Since it is important to satisfy the order, some additional items should be produced, depending on the relative costs of storing unsold items and expectation of new orders. Also, the production is probably not instant, and the number of items to be produced may be adjusted during the production, especially if the quality control does not take a lot of time.

Exercise 2.19.
There is no difference between/among the studied groups:
— Incorrect, because the conclusion is a failure to reject the null hypothesis, yielding no support for the hypothesis. The conclusion is best described as a state of ignorance.
The expectations of the groups do not differ substantially:
— Incorrect, because the qualifier 'substantial' is not defined.
There is no evidence of any difference between/among the expectations of the groups:
— This statement is correct by the standard of identifying evidence with significance (i.e., evidence $\equiv$ significance).
The sizes of (some of) the groups are not sufficiently large:
— This statement is correct if we take for granted that the two groups have unequal expectations.
The differences between/among the within-group expectations, even if there are some, can for all reasonable purposes be ignored:
— Incorrect, unless the sample sizes have been set (by design) to be sufficiently large.

Exercise 2.20.
The probability that this result is incorrect is 0.05:
— Incorrect. The statement would become correct with the presumption that the null hypothesis is valid.
If the two expectations do not differ, then the probability that the conclusion is correct is 0.95.
— Correct, as a statement about a hypothetical probability.
If the two expectations do not differ, then the conclusion is correct:
— Incorrect. It is correct with the probability equal to the level of significance, but that is smaller than 1.0.

The hypothesis is contradicted by the text:
Incorrect. The statement of a hypothesis is correct and presents no contradiction even if the hypothesis is not valid.

Exercise 2.21.

```
Sol0225 <- function(nn, mm, vv, alp=0.05)
{
##   nn    The two sample sizes
##   mm    The two means
##   vv    The two variances
##   alp   The size of the test

##   Each vector of length 2
nn <- rep(nn, length=2);  mm <- rep(mm, length=2)
vv <- rep(vv, length=2);  ss <- sqrt(vv)

xx <- c()

##   The two sample means
for (i in seq(2))
xx <- c(xx, mean(rnorm(nn[i], mm[i], ss[i])))

##   The scaled difference
df <- diff(xx) / sqrt(sum(vv/nn))

##   The critical value for the two-sided test
cri <- qnorm(1 - alp / 2)

##   The result (significance)
abs(df) > cri
}

##   Set the clock
Time <- proc.time()

##   Application
Sol0225r <- table(replicate(10000,
          Sol0225(c(20, 24), c(-0.2, -0.2), c(0.25, 0.36))))

##   How long it takes
Time <- proc.time() - Time
```

Exercise 2.22.

```
Sol0226 <- function(pts, mm, vv, ttl="")
{
##   The tail probabilities and intersection of two densities

##   pts   The evaluation points
##   mm    The two means
##   vv    The two variances

mm <- rep(mm, length=2); vv <- rep(vv, length=2)
ss <- sqrt(vv)
```

```
## The tail probabilities
prs <- 1 - pnorm(pts, mm[1], ss[1]) + dnorm(pts, mm[2], ss[2])
names(prs) <- pts

## The two densities
dns <- cbind(dnorm(pts, mm[1], ss[1]), dnorm(pts, mm[2], ss[2]))
dimnames(dns) <- list(pts, seq(2))

## Plot of the tail probs
plot(pts, prs, type="l", xlab="x", ylab="p",
     main=paste(ttl, "- Tail probability"))

## The empty shell of the plot for the two densities
plot(range(pts), c(0, max(dns)), type="n",
     main="Densities", xlab="x", ylab="Density")

for (i in seq(2))
lines(pts, dns[, i])

dns
} ## End of function Sol0226

par(mfrow=c(2,2), mar=c(4,4,2.4,1), lab=c(3,3,1))
Sol0226r <- list()

## Application (two rounds)
Sol0226r[[1]] <- Sol0226(seq(0,6,0.1), c(1,5), c(3,2),
                         ttl="Round 1")
Sol0226r[[2]] <- Sol0226(seq(2.8,3.2,0.01), c(1,5), c(3,2),
                         ttl="Round 2")
```

The logarithm of the normal density is a quadratic function, and so two normal densities have two intersections (except when they have identical variances). The sum of the tail probabilities has a similar feature (see the plot for Round 1), although there is no analytical solution.

Chapter 3

Exercise 3.6. The rule originates from the formula for differentiating the product of two functions:

$$\frac{\partial\{u(x)\,v(x)\}}{\partial x} = u'(x)\,v(x) + u(x)\,v'(x)$$

for any differentiable functions u and v. Hence

$$\int u'(x)\,v(x)\,\mathrm{d}x = u(x)\,v(x) - \int u(x)\,v(x)'(x)\,\mathrm{d}x,$$

so long as both integrals are well defined.

Exercise 3.8. $\{x\,\Phi(x)\}' = \Phi(x) + x\,\varphi(x)$ and $\varphi'(x) = -x\,\varphi(x)$. Therefore $\{x\,\Phi(x) + \varphi(x)\}' = \Phi(x)$, and so $\Phi_1(x) = x\,\Phi(x) + \varphi(x)$.

Exercise 3.9. By differentiating $x\,\Phi_1(x)$ we obtain $\Phi_1(x) + x\,\Phi(x)$. Therefore $\{x\,\Phi_1(x) + \Phi(x)\}' = 2\Phi_1(x)$, and so $2\Phi_2(x) = x\,\Phi_1(x) + \Phi(x) = (1+x^2)\,\Phi(x) + x\varphi(x)$.

Exercise 3.10. The operations of expectation and linear combination can be interchanged:

$$\int \{a\,u(x) + b\,v(x)\}\,\mathrm{d}x = \int a\,u(x)\,\mathrm{d}x + \int b\,v(x)\,\mathrm{d}x,$$

for any constants a and b, so long as two of these three integrals are well defined. The expectation of a linear combination of losses is equal to the same linear combination of the expected losses.

Exercise 3.13. In general, the properties of no bias and efficiency are not retained by nonlinear transformations. This can be confirmed by simulations. The properties of the lognormal distribution provide ample examples. Suppose a real parameter $\theta \in (0, +\infty)$ is estimated without bias by $\hat{\theta}$. Then $\mathrm{E}(\hat{\theta}^2) = \theta^2 + \mathrm{var}(\hat{\theta})$, so $\hat{\theta}^2$ is not unbiased for θ^2. This example is particularly telling for $\theta = 0$.

Exercise 3.14. The inversion for the fiducial distribution yields the identity $\sigma^2 = k\hat{\sigma}^2/X$, where $X \sim \chi_k^2$ and k is the number of degrees of freedom. From the density of X, equation (2.3), the fiducial density of σ^2 is obtained by the transformation $y = k\hat{\sigma}^2/x$. The differential is $-k\hat{\sigma}^2/x^2$.

Exercise 3.17. The loss with verdict V_2 is R when $\theta < 0$ and additionally R' when $\theta < 0.25$. The expected loss with verdict V_2 is

$$
\begin{aligned}
Q_2 &= R\int_{-\infty}^{0} \varphi\left(x; \hat{\theta}, \tau\right)\mathrm{d}x + R'\int_{-\infty}^{-0.25} \varphi\left(x; \hat{\theta}, \tau\right)\mathrm{d}x \\
&= R\Phi(z_1) + R'\Phi(z_1'),
\end{aligned}
$$

where $z_1 = -\hat{\theta}/\tau$ and $z_1' = -(0.25 + \hat{\theta})/\tau$. Further, $Q_1 = 1 - \Phi(z_1)$. The balance equation is $(R+1)\Phi(z_1) + R'\Phi(z_1') = 1$. It is solved by the Newton-Raphson algorithm, using the expression for the differential of the imbalance function

$$-\frac{1}{\tau}(R+1)\varphi(z_1) - \frac{1}{\tau}R'\varphi)(z_1').$$

The following R function is for the Newton-Raphson algorithm, with the function and its differential as the arguments.

```
NewRa <- function(df, d2f, x0, tol=6, maxit=20, dgt=6)
{
## The Newton-Raphson algorithm
```

```
##  df, d2f  The objective function and its differential
##  x0        The initial solution
##  tol, maxit  The convergence criterion and max. iterations
##  dgt       The number of digits in rounding

##  The initial solution
xnew <- x0;  dfnew <- df(xnew)

##  The iterations control
iter <- 0;  dist <- 1

##  Iteration loop
while (dist < tol & iter < maxit)
{
##  The iteration counter
iter <- iter + 1

##  Store the current solution
xold <- xnew;  dfold <- dfnew

##  The next solution
xnew <- xnew - dfnew / d2f(xnew);  dfnew <- df(xnew)

##  The number of decimal places in precision
dist <- -log(abs(dfnew - dfold) + abs(xnew - xold), base=10)
}  ##  End of the iteration loop

if (iter == maxit)
    warning(paste("No convergence after", iter, "iterations".))

##  Check in the output that  dfnew  is equal to zero.
round(c(X=xnew, df=dfnew, ddf=d2f(xnew), Conv.crit=dist,
        Iterations=iter), dgt)
}  ##  End of function NewRa

##  An example/application
NewRaR <- NewRa(function(x) x^3-4*x+1, function(x) 2*x^2-4, 1)
```

Exercise 3.20. Denote $z = -\hat{\theta}/\tau$. The expected loss with verdict V_+ is

$$Q_+ \;=\; \int_{-\infty}^{0} \varphi\left(x; \hat{\theta}, \tau\right) \mathrm{d}x \;=\; \int_{-\infty}^{z} \varphi(y)\,\mathrm{d}y \;=\; \Phi(z).$$

The expected loss with verdict V_- is

$$Q_- \;=\; a \int_{0}^{+\infty} \exp(bx)\, \varphi\left(x; \hat{\theta}, \tau\right) \mathrm{d}x$$

$$\;=\; a \exp\left(b\hat{\theta}\right) \int_{z}^{+\infty} \frac{1}{\sqrt{2\pi}} \exp\left\{-\frac{1}{2}\left(y^2 - \tau y\right)\right\} \mathrm{d}y,$$

after substituting the expression for the normal density. By completing the square under the exponential we obtain an expression for the density of a different normal distribution:

$$
\begin{aligned}
Q_- &= a\exp\left(b\hat{\theta} + \frac{1}{2}\tau^2\right)\frac{1}{\sqrt{2\pi}}\int_z^{+\infty}\exp\left\{-\frac{1}{2}\left(y - \frac{1}{2}\tau\right)^2\right\}dy \\
&= a\exp\left(b\hat{\theta} + \frac{1}{2}\tau^2\right)\left\{1 - \Phi\left(z - \frac{1}{2}\tau\right)\right\}.
\end{aligned}
$$

The balance equation is

$$
a\exp\left(b\hat{\theta} + \frac{1}{2}\tau^2\right)\left\{1 - \Phi\left(z - \frac{1}{2}\tau\right)\right\} - \Phi(z) = 0,
$$

that is,

$$
\frac{\Phi(z)}{1 - \Phi\left(z - \frac{1}{2}\tau\right)} = a\exp\left(b\hat{\theta} + \frac{1}{2}\tau^2\right)
$$

It has a unique solution because the left-hand side is an increasing function of z with respective limits 0 and $+\infty$ as z diverges to $-\infty$ and $+\infty$. Check that the differential of the left-hand side is positive.

Exercise 3.22. Averaging of the plausible range is equivalent to collapsing it to a single value. That would fail to represent the uncertainty about the parameter concerned.

Exercise 3.26. Some principles: we want each member of the committee to contribute in equal measure and to focus on the same issue. Therefore, conferring is a good idea to discuss the task and to set the rules, but it is a bad idea to discuss the value of the parameter. Less senior and less vocal members may defer to others, reducing the variety of opinions. Anonymity has to be handled carefully to ensure that everybody is well informed and focused on the task and to keep influencing the statements of other members to minimum. Reconvening the committee is useful when new facts emerge that should have been taken into account in the original meeting.

Exercise 3.28. The values of the equilibrium function are the equilibria that separate the values of the estimates associated with the two verdicts. The borderline is the value of a parameter (such as the loss ratio R) which, for a given estimate, separates the regions in which one verdict or the other is issued, or which separates the regions of unequivocal verdict and impasse.

Chapter 4

Exercise 4.1. A scalar multiple of a loss function is also a loss function; therefore a convex combination of loss functions is equivalent to the total of a set of loss functions. The pointwise limit of a sequence of loss functions is, in general, not a loss function, because the sequence may converge to zero. If the sequence has a positive lower bound, then its limit is a loss function because the properties of loss functions (non-negativity and piecewise monotonicity) are retained by limits.

Exercise 4.3. Apply mathematical induction and equation (4.7),

$$x \cdot x^{m-1} 2^{m-1} \frac{\Gamma_2(k+2m-2)}{\Gamma_2(k)} f_{k+2m-2}(x) = 2^m \frac{\Gamma_2(k+2m)}{\Gamma_2(k)} f_{k+2m}.$$

Exercise 4.4. Integrate both sides of each equation over $(0, +\infty)$ and substitute unity for the integral of the density. Equation (4.7) yields $E(X) = k$ for $X \sim \chi_k^2$. Further, $\text{var}(X) = E(X^2) - \{E(X)\}^2 = 2k$ and $\text{var}(1/X) = 2/\{(k-2)^2(k-4)\}$.

Exercise 4.5. The density f_k of the χ_k^2 distribution is given by equation (2.3) in Section 2.5.1. Suppose variable Y has this distribution. Then

$$P\left\{\sqrt{\frac{Y}{k}} < x\right\} = P\left(Y < kx^2\right),$$

and the differential of the right-hand side is $2kx f(kx^2)$. Check separately the powers of 2, k and x in the factors of this expression.

To evaluate a moment of $\sqrt{Y/k}$, $\int_0^{+\infty} (\sqrt{y/k})^h f_k(y) \, dy$, $h = 1, 2$, match $y^h f_k(y)$ with the density of another χ^2 distribution according to equation (4.9), and consolidate the constant factors.

Inspect the output of application of this function Sol0405

```
Sol0405 <- function(k,nr, dgt=6)
{
##  k, nr   The degrees of freedom and number of replications
##  dgt     The precision (rounding)

##  A random sample of  Y
Y <- sqrt(rchisq(nr, k) / k)

##  The expectation
eU <- sqrt(2/k) * gamma((k+1)/2) / gamma(k/2)

##  The empirical and the analytical
res <- rbind(c(mean(Y), var(Y)), c(eU, 1 - eU^2))
dimnames(res) <- list(c("Empirical", "Analytical"),
                      c("Mean", "Variance"))

round(res, dgt)
} ##  End of function Sol0405

##  Application
Sol0405r <- Sol0405(19, 10^5)
```

Exercise 4.7. The estimator $\hat{\sigma}^2$ is unbiased, $E(\hat{\sigma}^2) = \sigma^2$, and its variance is

$\mathrm{var}(\hat{\sigma}^2) = 2\sigma^4/k$. The MSE of $c\hat{\sigma}^2$ is

$$\mathrm{MSE}\left(c\hat{\sigma}^2, \sigma^2\right) = \left\{(c-1)^2 + \frac{2c^2}{k}\right\}\sigma^4.$$

This quadratic function of c attains its minimum at $k/(k+2)$. Therefore, the most 'efficient' divisor for the corrected sum of squares is $k+2$, that is, $n+1$ (and not $n-1$) when only one degree of freedom is lost.

The expectation of $\sqrt{X/k}$, $\sqrt{2/k}\,\sigma\Gamma_2(k+1)/\Gamma_2(k)$, is obtained from equation (4.9). Thus, the bias is removed by the factor $d_1 = \sqrt{k/2}\,\Gamma_2(k)/\Gamma_2(k+1)$. The variance of $\sqrt{X/k}$ is $\mathrm{E}(X/k) - \sigma^2/d_1^2 = \sigma^2(1 - 1/d_1^2)$. The MSE of $d_2\sqrt{X/k}$ is

$$\mathrm{MSE}\left\{d_2\sqrt{\frac{X}{k}};\sigma\right\} = \sigma^2\left\{d_2^2\left(1 - \frac{1}{d_1^2}\right) + \left(\frac{d_2}{d_1} - 1\right)^2\right\},$$

and this quadratic function of d_2 attains its minimum for $d_2 = 1/d_1$.

Exercise 4.8. Apply the transformation for which the argument of f_k simplifies, and then apply equation (4.9):

$$\int_0^{+\infty} \frac{k\hat{\sigma}^2}{x} f_k\left(\frac{k\hat{\sigma}^2}{x}\right) \mathrm{d}x = \int_0^{+\infty} \frac{k\hat{\sigma}^2}{y} f_k(y)\,\mathrm{d}y = \frac{k\hat{\sigma}^2}{k-2}$$

The operations of expectation (integration) and nonlinear transformation (reciprocal) cannot be interchanged.

Exercise 4.9. For $\sigma_0^2 \to 0$ we obtain the expressions for the expectations.

Exercise 4.11. Let $\rho_0 = (n-1)\hat{\sigma}^2/\sigma_0^2$. The expected losses are

$$
\begin{aligned}
Q_s &= \int_{\sigma_0^2}^{+\infty} \frac{(n-1)\hat{\sigma}^2}{x^2} f_{n-1}\left\{\frac{(n-1)\hat{\sigma}^2}{x}\right\}\mathrm{d}x \\
&= \int_0^{\rho_0} f_{n-1}(y)\mathrm{d}y = F_{n-1}(\rho_0) \\
Q_l &= \int_0^{\sigma_0^2} \left\{R + S\left(\sigma_0^2 - x\right)\right\}\frac{(n-1)\hat{\sigma}^2}{x^2} f_{n-1}\left\{\frac{(n-1)\hat{\sigma}^2}{x}\right\}\mathrm{d}x \\
&= \int_{\rho_0}^{+\infty} \left\{R + S\sigma_0^2 - \frac{(n-1)S\hat{\sigma}^2}{y}\right\} f_{n-1}(y)\,\mathrm{d}y \\
&= \left(R + S\sigma_0^2\right)\left\{1 - F_{n-1}(\rho_0)\right\} - \frac{\rho_0}{n-1}\left\{1 - F_{n-3}(\rho_0)\right\}.
\end{aligned}
$$

The balance equation is $Q_l - Q_s = 0$. Apply the Newton-Raphson algorithm (see solution of Exercise 3.17).

Exercise 4.14. The linear loss for the variance is $\sigma^2 - \sigma_0^2$ when $\sigma^2 > \sigma_0^2$ and we claim the contrary. The quadratic loss for the standard deviation,

$a(\sigma - \sigma_0)^2 = a(\sigma^2 + \sigma_0^2 - 2\sigma\sigma_0)$ is a different function for any $a > 0$ because it contains a linear term in σ with a non-zero coefficient. However, the linear loss for the variance is matched by the sum of the quadratic loss $(\sigma - \sigma_0)^2$ and the linear loss $2\sigma_0(\sigma - \sigma_0)$ for the standard deviation.

Exercise 4.15. Plot the balance functions for a set of sample sizes n and compare their roots. For a more comprehensive solution, plot the equilibrium as a function of the sample size. If you choose a linear or quadratic loss function, apply the function for the Newton-Raphson algorithm from Exercise 3.17, if the equilibrium does not have a closed form.

Exercise 4.17. We prefer the instrument with smaller measurement error variance. We therefore compare the two variances. If the old instrument (O) has served the users well, then the new instrument (N) should be introduced only if it is very likely to be superior. This perspective is represented by a loss ratio $R = L_{nO}/L_{oN} > 1$. If R is set too large, then even some good proposals may be rejected. In a study comparing the two variances, the F statistic is used, see Example 7.

Exercise 4.20. The limiting prior does not exist, and therefore the posterior distribution is not defined either. The application of the Bayes theorem and taking limits for prior distributions cannot be interchanged.

Exercise 4.21. The prior density for the Poisson rate λ is

$$f_{\text{pri}}(x) = \frac{1}{\nu^\gamma \Gamma(\gamma)} x^{\gamma-1} \exp\left(-\frac{x}{\gamma}\right),$$

the Poisson probabilities are $p(k) = e^{-\lambda}\lambda^k/k!$. Therefore the posterior density is

$$C x^{k+\gamma-1} \exp\left\{-x\left(1 + \frac{1}{\gamma}\right)\right\}$$

for the constant C for which this is a density (of a gamma distribution). The interpretation: γ is the number of extra-data events and the variance is deflated $\nu/(1+\nu)$ times. This is perhaps a bit less elegant than the interpretation with $\delta = 1/\nu$, but the difference is not great.

Chapter 5

Exercise 5.1. Apart from some examples, such as replicate tosses of a coin or casts of a die and their derivatives, the model is known only when the dataset is generated by design, according to the model. In particular, the model is known in computer simulations, where it is set by the analyst.

Exercise 5.2. The more elementary selections you apply the greater the chance of an erroneous selection which cannot be compensated (undone) by any of the selections that follow. Conduct a simulation study with a simple model, such as ordinary regression, and count the replications in which you select the parsimonious valid model.

Exercise 5.7. The function $\Phi(z + \Delta) - \Phi(z)$ has zero limits at $\pm\infty$, and attains its maximum for $z = -\frac{1}{2}\Delta$, where its differential $\varphi(z + \Delta) - \varphi(z)$ vanishes. It is symmetric around $z = -\frac{1}{2}\Delta$. The same description applies to any t distribution, with the density of t in place of φ.

Exercise 5.9. The balance function is $(R + 1)\{\Phi(z_2) - \Phi(z_1)\} - 1$, where $z_1 = (T_1 - \hat{\theta})/\tau$ and $z_2 = (T_2 - \hat{\theta})/\tau$.

Exercise 5.10. The quadratic function $q(x) = ax^2 - 2bx + c$ for constants $a > 0$, b and c, attains its minimum at $x^* = b/a$. The first iteration of the Newton-Raphson algorithm is $x_0 - (2ax_0 - 2b)/(2a) = x^*$ for an arbitrary initial solution x_0. Geometric interpretation: the Newton-Raphson algorithm fits a quadratic function at the initial (or current) solution. For a quadratic objective function this fit is exact.

Exercise 5.13. For very large τ, the estimator $\hat{\theta}$ conveys very little information, so it is of little use in the choice of an option. That makes L, which covers (infinitely) more length that S, more attractive.

For loss functions that diverge to $+\infty$ with errors diverging to $\pm\infty$, S is more attractive because the loss is bounded for V_l, but not for V_s.

Exercise 5.18. If the new school has too few classrooms, expensive and unsatisfactory arrangements have to be made to accommodate the extra classes/students. Too many classrooms are associated with greater cost of construction, greater cost of maintenance (although some income can be earned by renting out the space that is not used by the school. Prediction of numbers of students in the future has to be considered in conjunction with the lifespan of the new building. Consider also the limitations, such as the building space available, and restrictions imposed by the zoning laws and regulations for (new) school buildings.

Exercise 5.19. First, consider how steep should the loss be as a function of the error. The powers $f(e) = |e|^q$ for $q > 0$ cater for a variety of 'steepness'. MAE and MSE correspond to $q = 1$ and $q = 2$, respectively. Adaptations for asymmetry include a factor dependent of the sign or the error and $f(e) = a_{\text{sign}(e)}|e|^q$, where the constant $a > 0$ depends on the sign of the error, and a quotient dependent on the sign, $f(e) = a_{\text{sign}(e)}|e|^{q\text{sign}(e)}$.

Exercise 5.21. The expression for MAE is

$$\tau\{R\Phi_1(z) + \Phi_1(-z)\} = -\tau z + (R + 1)\tau\{z\Phi(z) + \varphi(z)\},$$

obtained by substituting $\Phi_1(z) = z\Phi(z) + \varphi(z)$. Now substitute $z^* = \Phi^{-1}(r)$ for z; $r = 1/(R + 1)$.

$$\text{MAE}^* = -c^* + c^*\frac{1}{r}\Phi\{\Phi^{-1}(r)\} + (R + 1)\varphi\{\Phi^{-1}(r)\},$$

and the first two terms cancel out.

Exercise 5.22. Solve the problem for z. The solution for $R = 1$ is $z^* = 0$. For

$R \neq 1$, use the expression $1 + (R-1)\Phi(z)$ for the differential of the objective function.

Exercise 5.23. The single overarching rule is: more serious error — greater loss, although the ordering for the errors is only partial. Further, a gross error should be proportionate to the minor errors it comprises; for example, L_{13} should be related to $L_{12} + L_{23}$.

Exercise 5.24. Part of this exercise for the expected losses can be organised by compiling an (R) function for drawing a pair of functions. Its arguments can be the two functions, or their values on a fine grid of points $\hat{\theta}$.

Chapter 6

Exercise 6.6. The key condition is a list of events that would in the future be regarded as remarkable. (No bookie will accept a bet on an event that occurred in the past.) Compile such a list for the next week. The longer the list the more likely you are to guess something, but the less remarkable it is that you guess something.

Exercise 6.8. Denote $z = \sqrt{n}\,\hat{\mu}/\sigma$. The expected losses are $Q_+ = 1 - \Phi(z)$ and $Q_- = R\Phi(z)$, so the balance function is $(R+1)\Phi(z) - 1$. Solving for $\hat{\mu} = z\sigma/\sqrt{n}$ yields equation (6.1).

Exercise 6.10. The bounds of the uniform distribution can be set wider or narrower, and the support of the distribution need not be symmetric. The class of beta distributions, linearly transformed, are another alternative.

Exercise 6.14. The derivations involve σ^2 and n only through $\text{var}(\hat{\Delta}) = 2\sigma^2/n$. When the two groups have variances $\sigma_1^2 = \sigma^2$ and $\sigma_2^2 = \rho\sigma^2$, then $\text{var}(\hat{\Delta}) = (1+\rho)\sigma^2/n$. Follow through the steps in derivations to obtain the condition

$$n \geq (1+\rho)\frac{\sigma^2}{\Delta^{\dagger 2}}\left\{\Phi^{-1}\left(1 - \frac{\alpha}{2}\right) + \Phi^{-1}(\beta)\right\}^2.$$

When $n_1 = n$ and $n_2 = \pi n$, we have $\text{var}(\hat{\Delta}) = (1/n + \lambda)\sigma^2/n$, where $\lambda = \rho/\pi$. So, replace ρ with λ in the displayed condition. For a one-sided test, replace $\frac{1}{2}\alpha$ with α.

Exercise 6.15. Consider a plausible range of values of σ^2. It amounts to working with the upper bound of the plausible range. Instead of the normal quantiles, consider the corresponding quantiles of a t distribution, which will be used in estimation. Iterate setting n, resetting the number of degrees of freedom k of the t distribution until convergence. Two or three cycles suffice because only integer values of k have to be considered.

Exercise 6.17. Study the variance $v = \sigma^2(1/n_1 + 1/n_2)$ as a bivariate function of n_1 and n_2. If $n_2 \ll n_1$, then $v \doteq \sigma^2/n_1$. Substantial increase of n_2 will not make much difference to v. A similar increase on n_1 (either multiplicatively or additively), will reduce v substantially. By way of an example, compare $v = 1/10 + 1/100$ with $1/10 + 1/200$ and $1/20 + 1/100$.

Why within-group sample sizes are planned to be unequal: the intervention in one group is more expensive, less desirable, or there is some other disincentive to apply it.

Exercise 6.19. Let $r = m/q$. Then

$$\frac{q(q+m)}{\left(\frac{1}{2}m+q\right)^2} = \frac{1+r}{1+r+\frac{1}{4}r^2}.$$

This is a decreasing function, equal to unity for $r = 0$ and 0.89 for $r = 1$. It exceeds 0.99 for $r < 0.22$.

Exercise 6.20. For scenario B, the lines that correspond to the borderline for impasse touch the plausible rectangle at (q_-, δ_-) and (q_-, δ_+), see Figure 6.6. Then proceed as in equation (6.4), solving the adaptation of equation (6.3).

The bounds (e.g., B_- and B_+) would be solutions of nonlinear equations. The 'linear' solutions would serve as good initial solutions and convergence would be achieved after a few iterations.

   If a loss structure different from piecewise linear is used, the only difference is that z_R has to be evaluated iteratively, replacing equation (6.2).

Exercise 6.21. The advantage of this scheme is that it involves no conditions, usually a nuisance in programming.

Exercise 6.22. Differentiate the density of the t_k distribution:

$$
\begin{aligned}
\psi_k'(x) &= \frac{\Gamma_2(k+1)}{\sqrt{k\pi}\,\Gamma_2(k)} \frac{-(k+1)}{2} \frac{2x}{k} \left(1+\frac{x^2}{k}\right)^{-\frac{1}{2}(k+3)} \\
&= -x\psi_{k+2}\left(x\sqrt{\frac{k+2}{k}}\right) \frac{k+1}{k} \sqrt{\frac{k+2}{k}} \frac{\Gamma_2(k+1)\Gamma_2(k+2)}{\Gamma_2(k)\Gamma_2(k+3)} \\
&= -x\sqrt{\frac{k+2}{k}} \, \psi_{k+2}\left\{x\sqrt{\frac{k+2}{k}}\right\}.
\end{aligned}
$$

Differentiate the function $x\Psi_k(x) + c\psi_{k-2}(x\sqrt{(k-2)/k})$, and then set the constant c so that the result is $\Psi_k(x)$, the distribution function of t_k. The differential is

$$\Psi_k(x) + x\psi_k(x) - cx\sqrt{\frac{k-2}{k}} \sqrt{\frac{k}{k-2}} \sqrt{\frac{k-2}{k}} \, \psi_k(x),$$

and this reduces to $\Psi_k(x)$ for $c = \sqrt{k/(k-2)}$. So,

$$\Psi_1^{(k)}(x) = x\Psi_k(x) + \sqrt{\frac{k}{k-2}} \, \psi_{k-2}\left(x\sqrt{\frac{k-2}{k}}\right).$$

To derive $\Psi_2^{(k)}$, start by differentiating function $\frac{1}{2} x^2 \Psi_k(x)$ and compensate for the term $\frac{1}{2} x^2 \psi_k(x)$ by adjusting $\frac{1}{2} x^2 \Psi_k(x)$ by $\frac{1}{2} x \sqrt{k/(k-2)} \psi_{k-2}(x\sqrt{(k-2)/k})$. By another adjustment we obtain

$$\Psi_2^{(k)}(x) = \frac{x^2}{2} \Psi_k(x) + \frac{k}{2(k-2)} \Psi_{k-2}\left(x\sqrt{\frac{k-2}{k}}\right)$$
$$+ \frac{1}{2}\sqrt{\frac{k}{k-2}} \psi_{k-2}\left(x\sqrt{\frac{k-2}{k}}\right).$$

If you do not follow these steps, check the results by differentiation.

Exercise 6.24. The calculations depend on R only through $z_R = \Phi^{-1}\{R/(1+R)\}$. The ratio $R/(1+R)$ with $r = 1/R$ substituted for R becomes $1/(1+r)$, and $z_r = -z_R$. This change of sign cancels with the changes of the signs of δ, δ_- and δ_+ and $\hat{\Delta}$. This property of invariance is retained when $n_1 \neq n_2$ because all that matters are the factors m and q. However, it breaks down when either the prior distribution or the distribution of the estimator is not symmetric.

Chapter 7

Exercise 7.3. Piecewise constant loss is appropriate when the magnitude of the error is irrelevant (when all that matters and is disclosed is the binary assessment: healthy/diseased). The loss ratio may be set differently in populations that have different prevalence, or where the consequences of the two kinds of error are perceived differently. More frequent screening makes the consequences of a false negative less serious. Periodic revision of the loss ratio is well founded if the context changes—if treatment becomes cheaper, easier and more successful, the prevalence of the disease changes, etc.

Exercise 7.4. With greater R we are more averse to false negatives, and the response to such aversion is to reduce the threshold T – to err more on the side of false positives.

Apply the implicit function theorem to confirm that the minimum expected loss increases with R.

Exercise 7.5. Compare the expressions for $Q'(T)$ and $Q''(T)$. The latter is obtained by multiplying the terms of Q' by $-z_D/\sigma_D$ or $-z_H/\sigma_H$, so problems may arise when these factors are similar.

Implement the solution, minimising $Q(T)$, and explore it for extreme settings of R and p_D. Try to 'break' the algorithm.

Exercise 7.7. Use the identity $\partial f/\partial x = f(x) \partial[\log\{f(x)\}]/\partial x$, and exploit the simple expression of $\log(f)$. For example, for a beta density $f(x; \alpha, \beta)$ we have

$$\frac{\partial f}{\partial x} = f(x; \alpha, \beta) \left(\frac{\alpha - 1}{x} - \frac{\beta - 1}{1 - x} \right),$$

and $f(x; \alpha, \beta)/x$, $f(x; \alpha, \beta)/(1-x)$ can be related to other beta densities (for $\alpha > 1$ and $\beta > 1$).

Exercise 7.8. The expectation of the t_k distribution is zero for $k > 1$, owing to symmetry. Also, integration of the right-hand side of the equation in Exercise 7.7 yields this result. For a t_k-distributed variable X, $\text{var}(X) = \text{E}(X^2)$. Express the integrand $x^2 \psi_k(y)$ in terms of $y\psi_{k-2}(y\sqrt{(k-2)/k})$ and integrate the result by parts. As an alternative, express $x^2\psi_k^2$ as a linear combination of the densities ψ_k and ψ_{k-2}.

Exercise 7.10. Let I be the (binary) indicator of the component of a mixture, and denote by p its probability for the first component. The key assumption is that I is independent of the mixture components X and Y. Then $\text{E}\{XI + Y(1-I)\} = p\text{E}(X) + (1-p)\text{E}(Y)$. Since $I(1-I) = 0$,

$$\text{E}\left[\{XI + Y(1-I)\}^2\right] = p\text{E}(X^2) + (1-p)\text{E}(Y^2)$$

and adjustment by $\{p\text{E}(X) + (1-p)\text{E}(Y)\}^2$ yields

$$\text{var}\{XI + Y(1-I)\} = p\,\text{var}(X) + (1-p)\,\text{var}(Y) + p(1-p)\{\text{E}(X) - \text{E}(Y)\}^2.$$

Exercise 7.11. The verdict is equivocal for all subjects with outcomes Y between $T^*(800) = 0.521$ and $T^*(200) = 0.662$. They form

$$99.8 \times \left\{\Phi(0.662; 0.25, \sqrt{0.05}) - \Phi(0.521; 0.25, \sqrt{0.05})\right\}$$
$$+ 0.2 \times \left\{\Phi(0.662; 0.95, \sqrt{0.12}) - \Phi(0.521; 0.95, \sqrt{0.12})\right\} = 8.01$$

percent of the screened population. The vast majority of them are negative (7.99%).

Exercise 7.12. With the proposed notation for the odds ratio, $p_{\text{D}}^{\dagger} = r_{\text{D}}/(d + \text{D})$, and so $r_{\text{D}}^{\dagger} = r_{\text{D}}/d$. Hence, d is the factor by which the odds ratio is reduced.

Exercise 7.14. Any (continuous) distribution can be approximated by a mixture of normals. Therefore the interpretation of mixture components as some subpopulation of importance is far fetched, perhaps with the exception when the components are well separated.

Chapter 8

Exercise 8.2. Let $P_1, \ldots, P_K$ be the events of rejecting the null hypotheses $H_1, \ldots, H_K$. Then the probability $\text{P}(P_1, P_2, \ldots, P_K)$ depends not solely on the marginal probabilities $P_1, \ldots, P_K$ but also on the joint probabilities. In contrast, the overall loss involved with the corresponding K verdicts is $Q = Q_1 + \cdots + Q_K$, irrespective of the dependence of the statistics involved.

Exercise 8.5. Unit-wise minimisation cannot be applied.

Exercise 8.6. The expected losses are

$$
\begin{aligned}
Q_{k,\mathrm{x}} &= \int_{\theta-\Delta\theta}^{\theta+\Delta\theta} \varphi\left(x; \hat{\theta}_k, \tau_k\right) \mathrm{d}x \\
&= \int_{z_{k-}}^{z_{k+}} \varphi(y)\mathrm{d}y = \Phi(z_{k+}) - \Phi(z_{k+}) \\
Q_{k,\mathrm{o}} &= R\int_{-\infty}^{\theta-\Delta\theta} \varphi\left(x; \hat{\theta}_k, \tau_k\right) \mathrm{d}x + \int_{\theta+\Delta\theta}^{+\infty} \varphi\left(x; \hat{\theta}_k, \tau_k\right) \mathrm{d}x \\
&= R\{1 - \Phi(z_{k+}) + \Phi(z_{k+})\}.
\end{aligned}
$$

The verdict $V_{k,\mathrm{x}}$ is issued when $Q_{k,\mathrm{x}} < Q_{k,\mathrm{o}}$.

Exercise 8.7. By increasing $\Delta\theta$ we widen the ordinary range, and so make V_o more attractive; $\Phi(u_+) - \Phi(u_-)$ is increased. By increasing σ we 'flatten' the density $\varphi(x; \mu, \sigma)$. If $\mu \in (\theta - \Delta\theta, \theta + \Delta\theta)$ and this range is narrow in relation to σ, then increasing σ reduces $\Phi(u_+) - \Phi(u_-)$.

Exercise 8.11. The expected loss with verdict V_x is $Q_\mathrm{x} = Q_{\mathrm{x}+} + Q_{\mathrm{x}-}$, where (applying a linear transformation followed by integration by parts)

$$
\begin{aligned}
Q_{\mathrm{x}+} &= R\int_{\theta}^{\theta+\Delta\theta} (\theta + \Delta\theta - x)\, \varphi\left(x; \hat{\theta}_k, \tau_k\right) \mathrm{d}x \\
&= R\tau_k \int_{z_k}^{z_{k+}} (z_{k+} - y)\, \varphi(y)\, \mathrm{d}y \\
&= R\tau_k \left[(z_{k+} - y)\, \Phi(y)\right]_{z_k}^{z_{k+}} + R\tau_k \int_{z_k}^{z_{k+}} \Phi(y)\, \mathrm{d}y \\
&= -R\Delta\theta\, \Phi(z_k) + R\tau_k \left\{\Phi_1(z_{k+}) - \Phi_1(z_k)\right\},
\end{aligned}
$$

where $z_{k+} = (\theta + \Delta\theta - \hat{\theta}_k)/\tau_k$, $z_{k-} = (\theta - \Delta\theta - \hat{\theta}_k)/\tau_k$, and $z_k = (\theta - \hat{\theta}_k)/\tau_k$. By similar steps, we obtain

$$
\begin{aligned}
Q_{\mathrm{x}-} &= R\int_{\theta-\Delta\theta}^{\theta} (x - \theta + \Delta\theta)\, \varphi\left(x; \hat{\theta}_k, \tau_k\right) \mathrm{d}x \\
&= R\Delta\theta\, \Phi(z_k) - R\tau_k \left\{\Phi_1(z_k) - \Phi_1(z_{k-})\right\},
\end{aligned}
$$

so that $Q_\mathrm{x} = R\tau_k \{\Phi_1(z_{k+}) - 2\Phi_1(z_k) + \Phi_1(z_{k-})\}$. The expected loss with V_o

also has two components,

$$Q_{o-} = \int_{-\infty}^{\theta - \Delta\theta} (\theta - \Delta\theta - x)\, \varphi\left(x; \hat{\theta}_k, \tau_k\right) dx$$

$$= \tau_k \int_{-\infty}^{z_{k-}} \Phi(y)\, dy = \tau_k\, \Phi_1(z_{k-}),$$

$$Q_{o+} = \int_{\theta + \Delta\theta}^{+\infty} (x - \theta - \Delta\theta)\, \varphi\left(x; \hat{\theta}_k, \tau_k\right) dx$$

$$= \tau_k \int_{-\infty}^{-z_{k+}} (-y - z_{k+})\, \Phi(y)\, dy = \tau_k\, \Phi_1(-z_{k+}),$$

and so $Q_o = \tau_k\{\Phi_1(z_{k-}) - z_{k+} + \Phi_1(z_{k+})\}$. Next, we solve the balance equation $Q_x - Q_o = 0$ by the Newton-Raphson algorithm, using the expressions

$$\frac{\partial Q_x}{\partial \hat{\theta}_k} = R\{2\Phi(z_k) - \Phi(z_{k+}) - \Phi(z_{k-})\}$$

$$\frac{\partial Q_o}{\partial \hat{\theta}_k} = 1 - \Phi(z_{k+}) - \Phi(z_{k-}).$$

There is no explicit solution.

Exercise 8.13. Cost of estimation at stage 2 makes selection at stage 1 less attractive. Units selected at stage 1 by a margin narrower than the cost of estimation at stage 2 should therefore be discarded, treated as if they were not selected.

Second-stage selection could be improved by combining the two estimators. The first candidate for this is the combination with the reciprocals of the sampling variances as the weights. This entails some bias because the first-stage estimates used are a selection.

Exercise 8.14. Selection of exactly H units may look good because it does not rule out the possibility of no loss (perfect choice). However, by choosing a different (usually greater) number of units the expected loss may be smaller. Even if there is a cost (penalty) for selecting more than H units, the smallest expected loss may be attained for a selection of more (or fewer) than H units.

Exercise 8.16. Raise or lower the standard for selection depending on the relevant factors. The standard is lowered as the director starts running out of applicants, especially if readvertising is not an attractive proposition. The problem is amenable to simulation on the computer.

Exercise 8.18. Among the many cardinal differences we have that the potential outcomes framework is concerned with a particular (fixed) set of subjects, whereas modelling refers to a superpopulation. The potential outcomes framework does not have any device for extrapolating from the study sample to the relevant population. Modelling pretends to have it, but falsely assumes that the sample of subjects, recruited for the study, is a random sample from the relevant (or some other meaningful) population.

Chapter 9

Exercise 9.3. Under the hypothesis that $p = p_0 = 0.96$, the standard error of the estimator $\hat{p}$ of the rate of compliance is $s_n = \sqrt{p_0(1 - p_0)/n}$. Substitute $n = 600$ and $n = 4000$ for the two institutions. By one method, the performance of the institution is not satisfactory when $\hat{p} + 1.645s < p_0$. The probability of this event is

$$r_n = \mathrm{P}(\hat{p} < p_0 - 1.645s) = \Phi\left(\frac{p_0 - p - 1.645s_n}{\sqrt{p(1 - p)}}\sqrt{n}\right),$$

with $p = 0.938$ and $n = 600$ or 4000. The problem: the probability depends strongly on the caseload n; for the smaller institution, $r_n = 0.815$, whereas for the larger institution $r_n > 0.999$.

Exercise 9.5. Exercise 9.3 suggests that the caseload (n) is an important factor. If the caseloads are enormous, then about half of the institutions will be flagged; if the caseloads are very small, none will be flagged. Make up on the computer a set of 25 rates as a random sample from $\mathcal{N}(0.65, 0.05^2)$ and simulate the estimates $\hat{p}$, first for institutions with equal caseload.

Exercise 9.8. Any maximum M can be adopted, and the probabilities evaluated conditionally on this value M. Of all the candidate values of M, the one for which $P_k(M)$ is the largest should be adopted.

Exercise 9.10. The set $\mathcal{O}$ contains all the institutions that are as good as the best, and contains no others. This is the least forgiving event; we have to be right about every institution. Its probability is

$$\prod_{k \in \mathcal{O}} P_k(M) \prod_{k \notin \mathcal{O}} \{1 - P_k(M)\}$$

($P_k(M)$ is defined in Section 9.3).

The set $\mathcal{O}$ contains all the institutions that are as good as the best. Here we are not concerned about the complement of $\mathcal{O}$. The probability of this event is

$$\prod_{k \in \mathcal{O}} P_k(M).$$

Every institution in set $\mathcal{O}$ is as good as the best. That is, no institution in the complement of $\mathcal{O}$ is as good as the best. The probability of this event is

$$\prod_{k \notin \mathcal{O}} \{1 - P_k(M)\}.$$

Set $\mathcal{O}$ contains the best performer; the probability of this event is difficult to evaluate but it can be approximated by the probability that the complement contains no performer within a small margin δ.

$$\prod_{k \notin O} \mathrm{P}\left(\theta_k < M - \delta \mid \hat{\theta}_k\right) ;$$

give careful thought to how the value of M should be set.

Exercise 9.12. By adjustment for some covariates, the variance ν^2 is hopefully reduced, making shrinkage, when it is appropriate, more effective. In the conditional expectation, θ is replaced by the regression $\mathbf{x}_k \beta$; that is,

$$\left(\theta_k \mid \hat{\theta}_k\right) \sim \mathcal{N}\left\{\mathbf{x}_k \beta + \frac{\nu^2}{\nu^2 + \tau_k^2}\left(\hat{\theta}_k - \mathbf{x}_k \beta\right), \frac{\nu^2 \tau_k^2}{\nu^2 + \tau_k^2}\right\}.$$

Exercise 9.13. The national rate is a weighted mean of the institutions' rates, where the weights reflect the institutions' caseloads (sizes) N_k. In the average rate, each institution is equally important, irrespective of its size. Consider the superpopulation versions of the sizes N_k and rates θ_k, denoted by N and Θ. Then the issue is: when is $\mathrm{E}(N\Theta)/\mathrm{E}(N)$ equal to $\mathrm{E}(\Theta)$. This condition is equivalent to $\mathrm{cov}(N, \Theta) = 0$, that is, when the sizes and performances are not correlated.

Exercise 9.14. We have $\mathrm{E}(\hat{\theta}_k) = \theta_k$ and $\mathrm{var}(\hat{\theta}_k) = \tau_k^2$. Further, $\mathrm{var}(\theta) = 0$ and $\mathrm{E}(\theta - \hat{\theta}) = \theta - \hat{\theta}$. For the convex combination $\tilde{\theta}_k = b\hat{\theta}_k + (1 - b)\theta$, we have

$$\mathrm{E}\left(\tilde{\theta}\right) - \theta_k = (1 - b)\left(\theta - \theta_k\right)$$

$$\mathrm{var}\left(\tilde{\theta}\right) - \theta_k = b^2 \tau_k^2,$$

and so $\mathrm{MSE}(\tilde{\theta}_k ; \theta_k) = (1-b)^2(\theta-\theta_k)^2 + b^2\tau_k^2$. The dependence on θ_k is removed by averaging over its marginal distribution, $\mathrm{E}\{(\theta - \theta_k)^2\} = \nu^2$. Miminisation of the averaged MSE, $(1-b)^2\nu^2 + b^2\tau_k^2$, yields the coefficient $b_k^* = \nu^2/(\nu^2 + \tau_k^2)$. The shrinkage estimator is biased if θ_k is regarded as fixed; see Exercise 9.11; $\mathrm{E}(\tilde{\theta}_k \mid \theta_k) = \theta_k$ only when $\theta_k = \theta$. On average, the bias is absent: $\mathrm{E}\{\mathrm{E}(\tilde{\theta}_k \mid \theta)\} = 0$, where the outer expectation is over the superpopulation of institutions. If we are assessing institution k, then θ_k is fixed.

Exercise 9.16. When there are many institutions with large caseloads, especially when ν^2 is not large. Explore other settings by simulation.

Chapter 10

Exercise 10.1. An important example. Patients on the same ward receive everyday care that compensates for the (perceived) differences in the treatment applied by the clinical trial. The outcome depends not only on the treatment received but also on the treatment received by others on the ward.

Exercise 10.3. In R, evaluate   `2*pbinom(6, 20, 0.5)`.

The proposed designs restrict the variety of treatment assignments but avoid gross imbalance. Ruling out a small fraction of the randomisations creates no problems. The key is to retain the control over the probabilities of assignment. Accept or reject any particular randomisation *without* inspecting the outcomes.

Exercise 10.5. Re. balance with respect to background variables. We have one replication drawn at random from a large set in which there is balance on average—after averaging over replications. In a linear model, each subject (element) is associated with a random draw from a centred normal distribution. In a large study (sample), the average of these deviations is close to zero (has a distribution highly concentrated around zero).

Exercise 10.6. No evidence in one study (failure to reject the hypothesis) is not in any contradiction with evidence against the hypothesis in another study, because the outcome of the former study is 'do not know', not evidence supporting the null hypothesis.

Exercise 10.7. The exercise is simplified by realising that the sample size required depends on $\{\Phi^{-1}(1 - \frac{1}{2}\alpha) + \Phi^{-1}(\beta)\}^2$.

The sampling variance of the estimator of the treatment effect is $\sigma^2 = \tau^2(1 + 1/\rho)/n$, where the within-group sample sizes are $n_D = n$ and $n_E = n\rho$. In equation (10.1), replace the factor $2\tau^2$ with $(1 + 1/\rho)\tau^2$.

Exercise 10.8. There is a greater urgency to introduce a new treatment when there is no established treatment for the same condition, and this condition is life-threatening. This should be reflected by a smaller loss ratio $R = L_{aR}/L_{rA}$—relatively more harm is done by false rejection of the proposed treatment, although R is still much greater than 1.0.

Exercise 10.9.

$$Q_r = 1 - \Phi(z^*)$$

with constant loss ($L_{rA} = 1$). Solve the balance equation $Q_a = Q_r$ by an iterative method. You can start the iterations by solving the problem with R_0 reset to zero.

Exercise 10.11.

$$Q_r = 1 - \Phi(z^*) + S'\tau\{-z^* + \Phi_1(z^*)\}$$

with linear loss, $L_{aR} = 1 + S'(\Delta - \Delta^*)$. The constraints should be $R_0 > 1$ and $S > S'$. A choice of some merit is with $S/S' = R_0$, maintaining the same relative consequences of the two kinds of error.

Exercise 10.12.

$$Q_a = S \int_{-\infty}^{\Delta^*} (\Delta^* - x)^2 \varphi\left(x; \hat{\Delta}, \tau\right) dx$$

$$= S\tau^2 \int_{-\infty}^{z^*} (z^* - y)^2 \varphi(y) dy$$

$$= 2S\tau^2 \Phi_2(z^*) ,$$

where $z^* = (\Delta^* - \hat{\Delta})/\tau$.

Exercise 10.15. The key step in the (explicit) solution is $\Phi(\delta) - \Phi(-\delta) = 2\Phi(\delta) - 1$ for $\delta = \Delta^*/\tau$. Without symmetry this step is not possible. A way of resolving this is by transformation to symmetry, working with $\Phi\{f(\gamma)\} - \Phi\{f(-\gamma)\}$; finding a suitable function f is not straightforward; nor is specification of the loss function on the transformed scale.

Exercise 10.16. The loss functions are $R_0 + S_0(-\Delta^* - \theta)$, $R_1 + S_1(\theta + \Delta^*)$, $R_2 + S_2(\Delta^* - \theta)$ and $R_3 + S_3(\theta - \Delta^*)$ in the respective intervals $(-\infty, -\Delta^*)$, $(-\Delta^*, 0)$, $(0, \Delta^*)$ and $(\Delta^*, +\infty)$. The expected losses are

$$Q_{a-} = \int_{-\infty}^{-\Delta^*} \{R_0 + S_0(-\Delta^* - x)\} \varphi\left(x, \hat{\theta}, \tau\right) dx$$

$$= R_0 \Phi\left(-z_-^*\right) + S_0 \tau \Phi_1\left(-z_-^*\right)$$

$$Q_{a+} = \int_{\Delta^*}^{+\infty} \{R_3 + S_3(x - \Delta^*)\} \varphi\left(x, \hat{\theta}, \tau\right) dx$$

$$= R_3 \Phi\left(-z_+^*\right) + S_3 \tau \Phi_1\left(-z_+^*\right)$$

$$Q_{r-} = \int_{-\Delta^*}^{0} \{R_1 + S_1(x + \Delta^*)\} \varphi\left(x; \hat{\Delta}, \tau\right) dx$$

$$= R_1\{\Phi(z_0) - \Phi(z_-^*)\} + S_1 \tau (z_0 - z_-^*) \Phi(z_0) - \Phi_1(z_0) + \Phi_1(z_-^*)$$

$$Q_{r+} = \int_{0}^{\Delta^*} \{R_2 + S_2(\Delta^* - x)\} \varphi\left(x; \hat{\Delta}, \tau\right) dx$$

$$= R_2\{\Phi(z_+^*) - \Phi(z_0)\} - S_2 \tau (z_+^* - z_0) \Phi(z_0) + \Phi_1(z_+^*) - \Phi_1(z_0) ,$$

where $z_+^* = (\Delta^* - \hat{\Delta})/\tau$, $z_-^* = (-\Delta^* - \hat{\Delta})/\tau$ and $z_0 = -\hat{\Delta}/\tau$.

Exercise 10.17. Without a washout period, there is an effect of the treatment received immediately prior to the first period. However, there are settings in which a washout period is impossible or unethical—a patient has to receive *some* treatment. Further, one of the treatments may be the same as the treatment provided in the washout period.

Exercise 10.19. Minimise the MSE

$$(1 - b)^2 \tau_1^2 + b^2 \tau_2^2 + b^2 \gamma_{max}^2 ,$$

where τ_1^2 and τ_2^2 are the within-treatment variances. The solution is $b^* = \tau_1^2/(\tau_1^2 + \tau_2^2 + \gamma_{\max}^2)$.

Exercise 10.20. The largest MSE is

$$M(\tau^2) = \frac{\tau^2}{2}\left(1 + \frac{\gamma_{\max}^2}{2\tau^2 + \gamma_{\max}^2}\right),$$

see the passage between equations (10.5) and (10.6). Solve the equation $M(\tau^2) = H$, where H is the largest permissible MSE. This leads to the quadratic equation

$$\tau^4 + \tau^2\left(\gamma_{\max}^2 - 2H\right) - H\gamma_{\max}^2 = 0,$$

which has a positive and a negative solution. (The discriminant does not afford any simplification.) Find the sample size that corresponds to the positive solution:

$$\tau^{*2} = \tfrac{1}{2}\left(-\gamma_{\max}^2 + 2H + \sqrt{\gamma_{\max}^4 + 4H^2}\right) = \tfrac{1}{2}\gamma_{\max}^2\left(-1 + \rho + \sqrt{1 + \rho^2}\right),$$

where $\rho = 2H/\gamma_{\max}^2$. Then $n^* = \tau^{*2}/(2\sigma^2)$. When ρ is much smaller 1.0, $\tau^{*2} \doteq \tfrac{1}{4}\rho(2 + \rho)\gamma_{\max}^2 = H(1 + \tfrac{1}{2}\rho)$.

Chapter 11

Exercise 11.1. If $\mathbf{X}^\top\mathbf{X}$ is nonsingular, then

$$\mathbf{P}^2 = \mathbf{X}\left(\mathbf{X}^\top\mathbf{X}\right)^{-1}\mathbf{X}^\top\mathbf{X}\left(\mathbf{X}^\top\mathbf{X}\right)^{-1}\mathbf{X}^\top = \mathbf{P}.$$

So, $\mathbf{P}$ is idempotent. Its eigenvalues are zero and one, so $\mathrm{tr}(\mathbf{P})$ is equal to the multiplicity of the unit eigenvalue. Since $\mathbf{PX} = \mathbf{X}$, any basis of the subspace generated by the columns of $\mathbf{X}$ forms a set of eigenvectors with unit eigenvalue. Hence there are at least $K + 1$ unit eigenvalues. Let $\mathbf{v}$ be a vector orthogonal to this subspace, so that $\mathbf{X}^\top\mathbf{v} = \mathbf{0}$. Then $\mathbf{Pv} = \mathbf{0}$, so the multiplicity of the zero eigenvalue of $\mathbf{P}$ is $n - K - 1$.

If $\mathbf{Pv} = \lambda\mathbf{v}$, then $(\mathbf{I}-\mathbf{P})\mathbf{v} = (1-\lambda)\mathbf{v}$. So, a set of eigenvectors of $\mathbf{P}$ is also a set of eigenvectors of $\mathbf{I} - \mathbf{P}$. Therefore $\mathbf{I} - \mathbf{P}$ has eigenvalue zero with multiplicity $K + 1$ and unit eigenvalue with multiplicity $n - K - 1$.

Exercise 11.2. The OLS minimises $S = \mathbf{e}^\top\mathbf{e}$, where $\mathbf{e} = \mathbf{y} - \mathbf{X}\beta$. Express S in the form

$$S = (\mathbf{u} - \beta)^\top\mathbf{X}^\top\mathbf{X}\,(\mathbf{u} - \beta) + U$$

for suitable vector $\mathbf{u}$ and scalar U that does not involve β. To absorb the term linear in β, the vector $\mathbf{u}$ has to be $(\mathbf{X}^\top\mathbf{X})^{-1}\mathbf{X}^\top\mathbf{y}$, and then

$$U = \mathbf{y}^\top\mathbf{y} - \mathbf{y}\mathbf{X}^\top\left(\mathbf{X}^\top\mathbf{X}\right)^{-1}\mathbf{X}^\top\mathbf{y} = \mathbf{y}^\top(\mathbf{I} - \mathbf{P})\,\mathbf{y},$$

where $\mathbf{P} = \mathbf{X}^\top (\mathbf{X}^\top \mathbf{X})^{-1} \mathbf{X}$ is the projection matrix. Therefore $\hat{\beta} = \mathbf{u}$. This is also the maximum likelihood estimator because the log-likelihood, $\log L = -\frac{1}{2} n \log(\sigma^2) - \frac{1}{2} S/\sigma^2$, differs from $-\frac{1}{2} S/\sigma^2$ only by a term that does not involve β. A simpler derivation is by matrix differentiation: $\partial S/\partial \beta = 2\mathbf{e}^\top (-\mathbf{X})$, and this linear function of β has the sole root $\hat{\beta} = (\mathbf{X}^\top \mathbf{X})^{-1} \mathbf{X}^\top \mathbf{y}$.

Estimator $\hat{\beta}$ is normally distributed with $\mathrm{E}(\hat{\beta}) = (\mathbf{X}^\top \mathbf{X})^{-1} \mathbf{X}^\top \mathrm{E}(\mathbf{y}) = \beta$ and $\mathrm{var}(\hat{\beta}) = (\mathbf{X}^\top \mathbf{X})^{-1} \mathbf{X}^\top \mathrm{var}(\mathbf{y}) \mathbf{X} (\mathbf{X}^\top \mathbf{X})^{-1} = \sigma^2 (\mathbf{X}^\top \mathbf{X})^{-1}$.

Exercise 11.3. The variance σ^2 is estimated from the attained minimum of S, $\mathbf{y}^\top (\mathbf{I} - \mathbf{P}) \mathbf{y}$, scaled so as to be unbiased for σ^2.

$$\mathrm{E}\left(\mathbf{y}^\top \mathbf{P} \mathbf{y}\right) = \sigma^2 \mathrm{tr}(\mathbf{P}).$$

The trace of a (square) matrix is equal to the sum of its eigenvalues. Therefore $\mathbf{y}^\top (\mathbf{I} - \mathbf{P}) \mathbf{y}/(n - K - 1)$ is an unbiased estimator of σ^2.

The maximum likelihood estimator of σ^2 is the solution of the equation $\partial \log L/\partial \sigma^2 = 0$, that is, $-n/\sigma^2 + S/\sigma^4 = 0$. So, when β is also estimated, the MLE is $\sigma^2_{\mathrm{ML}} = \frac{1}{n} \mathbf{y}^\top (\mathbf{I} - \mathbf{P}) \mathbf{y}$.

The distribution of $\mathbf{y}^\top (\mathbf{I} - \mathbf{P}) \mathbf{y}$ is χ^2 with $n - K - 1$ degrees of freedom. This follows from the eigenvalue decomposition of $\mathbf{I} - \mathbf{P}$, $\sum_{k=1}^{n-K-1} (\mathbf{v}_k^\top \mathbf{y})^2$, in which $\mathbf{v}_k^\top \mathbf{v}_k = 1$ and $\mathbf{v}_k^\top \mathbf{v}_h = 0$ when $k \neq h$. Therefore, the terms $\mathbf{v}_k^\top \mathbf{y}$ are mutually independent and each has expectation zero and variance σ^2.

Exercise 11.4. If $\mathbf{y} = \mathbf{X}\beta + \varepsilon$, then $\mathbf{y} = \mathbf{X}\mathbf{A}\mathbf{A}^{-1}\beta + \varepsilon$ for any $(K+1) \times (K+1)$ nonsingular matrix $\mathbf{A}$. In this exercise, the matrix $\mathbf{A}$ that transforms $\mathbf{X}$ to orthogonality is $\mathbf{A} = \begin{pmatrix} 1 & -\bar{x} \\ 0 & 1 \end{pmatrix}$, and $\mathbf{A}^{-1} = \begin{pmatrix} 1 & \bar{x} \\ 0 & 1 \end{pmatrix}$. The vector β is changed to $\mathbf{A}^{-1}\beta$.

Exercise 11.5. Select for $\mathbf{A}$ a matrix that orthogonalises the columns of $\mathbf{X}$. Then β is transformed to $\mathbf{A}^{-1}\beta$ and $\hat{\beta}$ to $\mathbf{A}^{-1}\hat{\beta}$.

Exercise 11.6. The borderline is given by $\sigma\sqrt{D}$, see equation (11.6). If $\mathbf{z}$ is orthogonal to $\mathbf{X}$, that is, $\mathbf{z}^\top \mathbf{X} = 0$, then $D = 1/\mathbf{z}^\top \mathbf{z}$. In this exercise, $\mathbf{z}^\top \mathbf{z}$ is the corrected sum of squares of X, that is, $\sum_{i=1}^n (x_i - \bar{x})^2$.

Exercise 11.11. We have $D = 1/\{\mathbf{z}^\top (\mathbf{I} - \mathbf{P})\mathbf{z}\}$. All the eigenvalues of $\mathbf{I} - \mathbf{P}$ are nonnegative, so the denominator is also nonnegative. It is equal to zero only when $\mathbf{z}$ is in the subspace spanned by the columns of $\mathbf{X}$. But that is ruled out by the assumption that $\mathbf{Z} = (\mathbf{X}\ \mathbf{z})$ is nonsingular.

Exercise 11.12. Consider the partitioned matrix

$$\begin{pmatrix} \mathbf{V} & -\mathbf{b} \\ \mathbf{b}^\top & 1 \end{pmatrix}$$

and apply linear transformations that change it to a block-diagonal matrix. One way we obtain $\mathbf{V} + \mathbf{b}^\top \mathbf{b}$ in a diagonal block, and the other we obtain an expression for its inverse.

Exercise 11.14. First, $\text{var}(c_x^\top \hat{\beta}) = c_x^\top (X^\top X)^{-1} c_x$. No generality is lost by assuming orthogonality, that $z^\top X = 0$, and that $\sigma^2 = 1$. Then

$$
\begin{aligned}
\text{cov}\left(c^\top \hat{\beta}, c_x^\top \hat{\beta}_x\right) &= c^\top \left(Z^\top Z\right)^{-1} Z^\top \text{var}(y) \, X \left(X^\top X\right)^{-1} \\
&= c^\top \begin{pmatrix} \left(X^\top X\right)^{-1} & 0 \\ 0 & 1/(z^\top z) \end{pmatrix} \begin{pmatrix} X^\top X \\ 0 \end{pmatrix} \left(X^\top X\right)^{-1} c_x \\
&= c^\top \begin{pmatrix} I \\ 0 \end{pmatrix} c_x = \text{var}\left(c_x^\top y\right).
\end{aligned}
$$

Exercise 11.15. For $r = \sqrt{r_1 r_2}$,

$$
r_1 + r_2 + 2 r_1 r_2 - r(r_1 + r_2 + 2) = \left(\sqrt{r_1} - \sqrt{r_2}\right)^2 \left(1 - \sqrt{r_1 r_2}\right)
$$

This expression vanishes when $r_1 r_2 = 1$. Therefore overstating the slope β_z κ times inflates the MSE as much as understating it κ times for any $\kappa > 1$.

Exercise 11.19. In the text, we prove that any estimator in the proposed basis can be generated as a composition of the estimators in the original basis. Conversely, any estimator in the original basis can be expressed as a composition of estimators in the proposed basis. For example, the estimator based on model M_2 is composed as $\xi_1 + \xi_2 - \xi_0$, where ξ_k, $k = 0, \ldots, K$, are the estimators in the proposed basis and the covariates are in the same order in both bases. Denote $v_k = \text{var}(\xi_k)$, $u_0 = 1/v_0$ and $u_k = 1/(v_k - v_0)$. The variance matrix V for the proposed basis has the diagonal $(v_0, v_1, \ldots, v_K)$ and every off-diagonal element of V is equal to v_0. For example, for $K = 3$ we have

$$
V = \begin{pmatrix} v_0 & v_0 & v_0 & v_0 \\ v_0 & v_1 & v_0 & v_0 \\ v_0 & v_0 & v_2 & v_0 \\ v_0 & v_0 & v_0 & v_3 \end{pmatrix}.
$$

For arbitrary K, the inverse of such a matrix is

$$
V^{-1} = \begin{pmatrix} \sum_{k=0}^K u_k & -u_1 & -u_2 & \cdots & -u_K \\ -u_1 & u_1 & 0 & \cdots & 0 \\ -u_2 & 0 & u_2 & \cdots & 0 \\ \vdots & \vdots & \vdots & \ddots & \vdots \\ -u_K & 0 & \cdots & 0 & u_K \end{pmatrix}.
$$

That is, V^{-1} has the diagonal $(\sum u_k, u_1, u_2, \ldots u_K)$, and all its off-diagonal elements are zero, except for the first row and first column, which, omitting the element $(1,1)$, are $-(u_1, -u_2, \ldots, -u_K)$. Therefore, $1^\top V^{-1} 1 = 1/v_0$, $1^\top V^{-1} b = b_0/v_0$, and the derivation of the optimal composition proceeds by the same steps, and the same intermediate results, as in Section (11.3) with the original basis of estimators. The results for the original and proposed bases are the same because the bases generate the same space of estimators.

Exercise 11.21. A matrix $\mathbf{D}$ is positive definite if $\mathbf{v}^\top \mathbf{D} \mathbf{v} > 0$ for all vectors $\mathbf{v}$. $\mathbf{u}^\top \mathbf{A} \mathbf{u} > 0$ for all vectors $\mathbf{u}$ because $\mathbf{u}^\top \mathbf{A} \mathbf{u} = \mathbf{u}_0^\top \mathbf{D} \mathbf{u}_0 > 0$ for vector $\mathbf{u}_0$ formed from $\mathbf{u}$ by supplementing it with zero elements to be of length compatible with $\mathbf{D}$. The proof for $\mathbf{C}$ is the same. The matrix $\mathbf{C} - \mathbf{B}^\top \mathbf{A}^{-1} \mathbf{B}$ is obtained from $\mathbf{D}$ as $\mathbf{C} - \mathbf{B}^\top \mathbf{A}^{-1} \mathbf{B} = \mathbf{F}^\top \mathbf{D} \mathbf{F}$ for a suitable matrix $\mathbf{F}$. For any vector $\mathbf{t}$, $\mathbf{t}^\top \mathbf{F}^\top \mathbf{D} \mathbf{F} \mathbf{t} > 0$, therefore $\mathbf{C} - \mathbf{B}^\top \mathbf{A}^{-1} \mathbf{B}$ is positive definite.

Exercise 11.22. Postmultiply $\mathbf{D}$ by

$$\begin{pmatrix} \mathbf{I} & -\mathbf{A}^{-1}\mathbf{B} \\ \mathbf{0} & \mathbf{I} \end{pmatrix} \begin{pmatrix} \mathbf{I} & \mathbf{0} \\ -\mathbf{E}^{-1}\mathbf{B}^\top & \mathbf{I} \end{pmatrix} \begin{pmatrix} \mathbf{A}^{-1} & \mathbf{0} \\ \mathbf{0} & -\mathbf{E}^{-1} \end{pmatrix}$$

to obtain

$$\begin{pmatrix} \mathbf{A}^{-1} - \mathbf{A}^{-1}\mathbf{B}\mathbf{E}^{-1}\mathbf{B}^\top \mathbf{A}^{-1} & -\mathbf{A}^{-1}\mathbf{B}\mathbf{E}^{-1} \\ -\mathbf{E}^{-1}\mathbf{B}\mathbf{A}^{-1} & \mathbf{E}^{-1} \end{pmatrix}.$$

Exercise 11.23. Write a mixture as $X = \sum_{k=1}^{K} I_k X_k$ where I_k is the indicator of component k. These indicators are mutually exclusive, $\sum_{k=1}^{K} I_k = 1$, and they are binary, supported on the set $(0,1)$. Denote $p_k = \mathrm{E}(I_k)$. By conditioning on the set of indicators I_k, we have

$$\begin{aligned} \mathrm{E}(X) &= \sum_{k=1}^{K} \mathrm{E}\{I_k \mathrm{E}(X_k \mid I_k = 1)\,\mathrm{E}(I_k)\} \\ &= \sum_{k=1}^{K} p_k\, \mathrm{E}(X_k) . \end{aligned}$$

For evaluating the variance $\mathrm{var}(X) = \mathrm{E}(X^2) + \{\mathrm{E}(X)\}^2$, we apply this expression to X^2:

$$\begin{aligned} \mathrm{var}(X) &= \sum_{k=1}^{K} p_k\, \mathrm{E}\left(X_k^2\right) - \left\{ \sum_{k=1}^{K} p_k\, \mathrm{E}\left(X_k\right) \right\}^2 \\ &= \sum_{k=1}^{K} p_k\, \mathrm{var}\left(X_k\right) + \sum_{k=1}^{K} p_k(1-p_k)\,\{\mathrm{E}\left(X_k\right)\}^2 - \sum_{k \neq h} \mathrm{E}(X_k)\,\mathrm{E}(X_h) \\ &= \sum_{k=1}^{K} p_k\, \mathrm{var}\left(X_k\right) + \mathrm{E}(\mathbf{X})^\top \mathbf{V}_\mathrm{I}\, \mathrm{E}(\mathbf{X}), \end{aligned}$$

where $\mathbf{X}$ is the vector of the components, $(X_1, \ldots, X_K)^\top$ and $\mathbf{V}_\mathrm{I}$ is the variance matrix of the indicators $(I_1, \ldots, I_K)^\top$; $\mathrm{var}(I_k) = p_k(1 - p_k)$ and $\mathrm{cov}(I_k, I_h) = -p_k\,p_h$ for $k \neq h$.

Exercise 11.24. Compile a function which has the following arguments: the dataset, the number of mixture components (K), maximum number of iterations and the convergence criterion—the number of decimal places), and, as an option, the starting solution. Set first the initial solution for the means, variances and probabilities, unless they are given as optional arguments. Initialise the iteration counter and the indicator of 'continue to iterate'. In a loop (while in R), keep iterating until you reach the maximum number of iterations or the convergence criterion is reached. Within the loop, implement the E-step, in which the probabilities of belonging to each component are (re-)estimated, followed by estimation of the means, variances and marginal probabilities of the components.

Let $f_k(x) = \phi(x; \hat{\mu}_k, \hat{\sigma}_k^2)$, $k = 1, \ldots, K$, be the current density of component k. Then the E-step probabilities are updated by the formula

$$r_k(x) = \frac{\hat{p}_k f_k(x)}{\hat{p}_1 f_1(x) + \cdots + \hat{p}_K f_K(x)}.$$

In the M-step, the components' (marginal) probabilities are estimated by the averages of $r_k(x_i)$, $i = 1, \ldots, n$, the means of the components are estimated by the weighted means, with the weights set to the conditional probabilities $r_k(x_i)$, and the variances are also estimated by their weighted sample versions:

$$\hat{\sigma}_k^2 = \frac{1}{n} \sum_{i=1}^{n} r_k(x_i)(x_i - \bar{x}_k)^2.$$

For the convergence criterion, evaluate the Euclidean distance of the two successive solutions (probabilities, means and variances). Its logarithm with base 10, suitably scaled, can be interpreted as the number of decimal places to which the solution has converged.

Index

χ^2 distribution, 68, 220

additivity, 35, 108, 153, 248
adjustment by regression, 191
algebra, 228
    linear, 240
alternative, 29
ambiguity, 53
analysis
    objective, 2
    subjective, 3
analysis (of a dataset), 5
analysis of variance (ANOVA), xiii
anchor, 55, 109
asymmetric combination, 106
asymptotics, 21, 41, 223
aversion, 4, 29, 49, 73, 94, 106, 143,
    155, 158, 203

background, 120, 190, 197
    variable, 169
balance, 158
    equation, 49, 79, 204
    function, 49, 95, 214
        scaled, 117
balance (in matching), 171
    overall, 171
basis, 48, 228, 240
    distribution, 243
    of the prior, 83
Bayes
    empirical, 181
    rule, 42
    theorem, 20, 25
Bayesian paradigm, xii
beta
    distribution, 19, 82

    function, 82
    prior, 82
bias, 17, 29, 41, 91, 220
    adjustment (correction for), 18
    bounded, 224
binomial distribution, xii, 19, 81, 185
bioequivalence, 205, 214
borderline, 54, 70, 94, 99, 102, 114,
    159
    function, 57, 80
borrowing strength, 182
box plot, 29

caliper matching, 172
carryover effect, 208, 215
casemix, 191
chi-squared
    distribution, xii, 19, 67
        inverse, 26, 72
    scaled, 67
claim, 178
    appropriate, 178
class of equivalence, 48
classification, 139, 175
    equivocal, 152
client, xi, 2, 59, 247
clinical trial, 137, 164, 169
coarseness, 112
coherence, 21
composite estimator, 210, 215
composition, 228
concentration, 156, 182, 260
    relative, 86
confidence interval, 180, 187
confounding, 192, 195, 197
conjugate
    distribution, 82

Milton Keynes UK
Ingram Content Group UK Ltd.
UKHW021619071024
449327UK00020BA/1105